Animal–Sediment Relations
The Biogenic Alteration of Sediments

TOPICS IN GEOBIOLOGY
Series Editor: F. G. Stehli, University of Florida

Volume 1   SKELETAL GROWTH OF AQUATIC ORGANISMS
Biological Records of Environmental Change
Edited by Donald C. Rhoads and Richard A. Lutz

Volume 2   ANIMAL–SEDIMENT RELATIONS
The Biogenic Alteration of Sediments
Edited by Peter L. McCall and Michael J. S. Tevesz

Animal–Sediment Relations

The Biogenic Alteration of Sediments

Edited by
Peter L. McCall
Case Western Reserve University
Cleveland, Ohio

and
Michael J. S. Tevesz
Cleveland State University
Cleveland, Ohio

PLENUM PRESS · NEW YORK AND LONDON

Library of Congress Cataloging in Publication Data

Main entry under title:

Animal-sediment relations.

(Topics in geobiology; v. 2)
Includes bibliographical references and index.
1. Benthos. 2. Sediments (Geology). 3. Marine sediments. I. McCall, Peter L., 1948-   . II. Tevesz, Michael J. S. III. Series.
QH90.8.B46A54 1982   574.5'2636   82-16523
ISBN 0-306-41078-8

© 1982 Plenum Press, New York
A Division of Plenum Publishing Corporation
233 Spring Street, New York, N.Y. 10013

All rights reserved

No part of this book may be reproduced, stored in a retrieval system, or transmitted in any form or by any means, electronic, mechanical, photocopying, microfilming, recording, or otherwise, without written permission from the Publisher

Printed in the United States of America

Contributors

Robert C. Aller Department of the Geophysical Sciences, University of Chicago, Chicago, Illinois 60637

Larry F. Boyer Department of Geology and Geophysics, Yale University, New Haven, Connecticut 06511

Charles W. Byers Department of Geology and Geophysics, University of Wisconsin, Madison, Wisconsin 53706

J. Berton Fisher Department of Geological Sciences, Case Western Reserve University, Cleveland, Ohio 44106. *Present address*: Research Center, Amoco Production Company, Tulsa, Oklahoma 74102

Gerald Matisoff Department of Geological Sciences, Case Western Reserve University, Cleveland, Ohio 44106

Peter L. McCall Department of Geological Sciences, Case Western Reserve University, Cleveland, Ohio 44106

Donald C. Rhoads Department of Geology and Geophysics, Yale University, New Haven, Connecticut 06511

Michael J. S. Tevesz Department of Geological Sciences, Cleveland State University, Cleveland, Ohio 44115

Preface

In 1881, Charles Darwin published *The Formation of Vegetable Mould through the Action of Worms*. In his book he described the feeding activities of terrestrial oligochaetes and their effect on the physical and chemical properties of the soil and soil fertility. In his autobiography he confided:

> I have now (May 1, 1881) sent to the printers the MS. of a little book on The Formation of Vegetable Mould through the Action of Worms. This is a subject of but small importance; and I know not whether it will interest any readers, but it has interested me. It is the completion of a short paper read before the Geological Society more than forty years ago, and has revived old geological thoughts.

Darwin was wrong. In the two years following the publication of the book, 8500 copies were sold. The book was more immediately popular than *The Origin of Species*. (Of course, it also followed the publication of that by now famous book on evolution.) Later workers were to confirm Darwin's observations on the effects of oligochaetes on soil structure and fertility and further document soil alterations produced by other animals (Taylor, 1935; Satchell, 1958; Edwards and Lofty, 1972).

Still other studies have shown that the biogenic alteration of sediments is a pervasive process in aquatic environments as well. Dapples (1942) authored one of the first large-scale reviews of aquatic animal–sediment interactions and emphasized that benthic macroinvertebrates profoundly affect sediment properties and processes in the marine realm:

> In many, but not all, environments in which marine deposits accumulate benthonic life is abundant. Certain of these organisms burrow for shelter or food, or they may ingest the sediments in search of any contained organic material. Either of these two behaviors results in some alteration of the sediments already deposited. A list of such changes includes obliteration of stratification, destruction of gradation of grain size resulting from normal settling

through an aqueous medium, trituration and solution of rock masses and fragments, formation of tubes and burrows, addition of faecal matter, initiation of cementation, bleaching of the sediments, and reduction of the amount of contained organic material.

On the basis of present meagre quantitative information, there is reason to believe that the larger organisms which contribute to diagenetic changes are holothurians and worms, whereas, rock boring organisms and echinoderms are secondary in importance. Obviously, the degree to which any of the alterations already listed may take place is dependent upon the quantity of these organisms dwelling in any particular area and, hence, is in turn dependent upon the localities which possess the optimum living conditions.

Rhoads (1974) recently summarized the effects of marine bottom-dwelling organisms on the properties of mud bottoms in shallow marine environments, and Rowe (1974) produced a sketchier outline for the less well-known deep-sea environments. Their conclusion, based on quantitative results, was that benthic flora and fauna, especially macroinvertebrates (primarily molluscs, annelids, and arthropods) are able to produce large changes in the physical and chemical properties of marine sediments. Petr (1977) produced a short review of the early literature on the effects of invertebrates on freshwater sediments.

As a perusal of the reference lists of any of the following chapters demonstrates, there has been a great deal of work on animal–sediment relations in the last several years. And, as the text of these chapters demonstrates, there have appeared a number of new ideas concerning the ways in which organisms affect sediments. We have reviewed animal–sediment relations in comparable aquatic environments where organisms are liable to have large and observable effects on bottom properties—Recent marine (Part I) and freshwater (Part II) soft bottoms (unconsolidated fine clastics)—and have tried to add a time dimension by examining what kinds of information have been extracted from the rock record of these two environments (Part III). The volume and sophistication of our present knowledge of animal–sediment relations descend in the same order. There has been a greater number of workers working for a longer period of time on marine than on nonmarine animal–sediment relations. There is a smaller amount of less conclusive work on ancient marine animal–sediment relations. We know the least about ancient freshwater sediments.

In Chapter 1, Rhoads and Boyer eschew a review of the kinds and rates of sediment mixing in marine sediments, since Aller (1978) and Lee and Schwarz (1980) have already reviewed the area. They have instead concentrated on biogenic changes in physical and chemical properties that occur in the course of the ecological succession of organisms following some disturbance of the bottom. They pay detailed attention to the many activities of organisms that affect sediment transport and particularly emphasize the interrelated influence of tubes, pelletal layers, and microbial mucopolysaccharide binders on entrainment properties. In Chapter 2, Aller examines the numerous ways in which macrobenthos alter the horizontal and vertical distribution of particles and solutes in marine sediments and alter the rate of diagenesis of organic matter. Macrobenthos not only alter the flux of materials between sediments and the overlying water column, but also influence the temporal and spatial distributions of other benthic organisms.

In Chapter 3, McCall and Tevesz review the present state of knowledge concerning the effects of the major groups of freshwater lake macrobenthos on the physical properties and transport of clastic lake sediments. They conclude with speculation that the animal–sediment interactions they describe may be of great import to the study of the evolution of benthic communities as well as to the study of sedimentologic processes. In Chapter 4 Fisher reviews the present state of knowledge concerning the effects of the same major groups of freshwater benthos on chemical properties of the sediments and chemical mass transfer between sediments and overlying water.

In Chapter 5, Byers reviews recent work on the relation of marine trace fossils to sedimentology that has appeared since the volume of Frey (1975). He ends with some ideas about what these biogenic sedimentary structures can tell us about the history of life. In Chapter 6, Tevesz and McCall review the meager knowledge concerning freshwater trace fossils. The few excellent studies available serve as examples of the potential for reconstructing the physical–chemical properties of ancient aquatic substrata.

In Chapter 7, Matisoff reviews the various kinds of quantitative models that have been constructed to describe the effects of organisms on particle mixing and chemical mass transfer between sediment pore fluids and overlying waters. He compares results obtained from different kinds of models and suggests their utility in deciphering the sedimentary record of biological mixing.

Some readers may be unfamiliar with several terms that appear throughout this volume. The book concerns the effects of *benthos* (bottom-dwelling organisms) on sediment properties. Most authors have written about *macrobenthos* (the adult stages of which remain in a 1-mm mesh sieve during the process of separating organisms from sediment). Some attention is also paid to *meiobenthos* (organisms from 1 mm to about 100 μm in diameter)—including nematodes, harpacticoid copepods, ostracods, and various aschelminthes—and *microbenthos* (organisms smaller than 100 μm in diameter)—which include protists, fungi, and bacteria. Macrobenthos are further classified according to life habits (Fig. 1). *Epi-*

		Filter Feeder	Deposit Feeder	Herbivore	Carnivore
EPIFAUNAL	Mobile	Crustaceans		Gastropods* Echinoids	Gastropods Crustaceans Asteroids
	Sedentary				Coelenterates
	Attached	Sponges Bivalves Polychaetes Bryozoans			Coelenterates* (Corals)
INFAUNAL	Mobile	Polychaetes	Bivalves Polychaetes Oligochaetes		Gastropods Crustaceans Polychaetes
	Sedentary	Bivalves Chironomids	Polychaetes Crustaceans		Coelenterates
	Attached	Bivalves			

Figure 1. Life habits of benthic macroinvertebrates, with examples. There is no one-to-one correspondence between taxa and life habits, and this list of taxa is not exhaustive. Asterisk (*) indicates the predominant life habitat of the taxon. After Eicher and McAlester (1980).

fauna spend the majority of their time on the substratum surface; *infauna* live primarily beneath the substratum surface. *Vagile* benthos move actively on or in the substratum, *sedentary* benthos do not, and *attached* benthos are incapable of movement along the bottom. *Filter feeders* gather their food from the water overlying the bottom or, in rare cases, from sediment pore water. *Deposit feeders* gather their food by feeding directly on unconsolidated sediment deposits. *Herbivores* feed selectively on live plant material and *carnivores* prey on other animals.

We have not made many freshwater–marine comparisons of animal–sediment relations to identify or explain differences and commonalities. This attempt is premature given our present state of knowledge. We can only conclude now that functionally similar organisms have a similar effect on similar bottoms in both marine and freshwater environments. We think that future comparisons will be more cogent. However, we hope at least that the co-occurrence in one book of work on animal–sediment relations in freshwater and marine environments and in ancient rocks will promote a cross-fertilization of methods and ideas among workers that will result in geologically and ecologically interesting and useful conclusions.

Peter L. McCall
Michael J. S. Tevesz

References

Aller, R. C., 1978, Experimental studies of changes produced by deposit feeders on pore water, sediment, and overlying water chemistry, *Am. J. Sci.* **217**:1185–1234.

Dapples, E. C., 1942, The effect of macro-organisms upon near shore marine sediments, *J. Sediment. Petrol.* **12**:118–126.

Edwards, C. A., and Lofty, J. R., 1972, *Biology of Earthworms*, Chapman and Hall, London.

Eicher, D. L., and McAlester, A. L., 1980, *History of the Earth*, Prentice-Hall, Englewood Cliffs, New Jersey.

Frey, R. W. (ed.), 1975, *The Study of Trace Fossils*, Springer-Verlag, New York.

Lee, H., and Schwartz, R. C., 1980, Biological processes affecting the distribution of pollutants in marine sediments. Part II. Biodeposition and bioturbation, in: *Contaminants and Sediments* (R. A. Baker, ed.), Volume 2, pp. 555–605, Science Publishers, Ann Arbor, Michigan.

Petr, T., 1977, Bioturbation and exchange of chemicals in the mud–water interface, in: *Interactions Between Sediments and Fresh Water* (H. L. Golterman, ed.), pp. 216–226, Junk, The Hague.

Rhoads, D. C., 1974, Organism–sediment relations on the muddy sea floor, *Oceanogr. Mar. Biol. Annu. Rev.* **12**:263–300.

Rowe, G. T., 1974, The effects of the benthic fauna on the physical properties of deep-sea sediments, in: *Deep Sea Sediments* (A. L. Inderbitzen, ed.), pp. 381–400, Plenum Press, New York.

Satchell, J. E., 1958, Earthworm biology and soil fertility, *Soil Fert.* **21**:209–219.

Taylor, W. P., 1935, Some animal relations to soils, *Ecology* **16**:127–136.

Contents

I. Recent Marine Environments

Chapter 1 · The Effects of Marine Benthos on Physical Properties of Sediments: A Successional Perspective
Donald C. Rhoads and Larry F. Boyer

1.	Introduction	3
2.	Ecological Succession and Organism–Sediment Relations	5
3.	Ecological Succession and Geotechnical Properties: A Laboratory Experiment	10
4.	Sediment Transport and Biogenic Activity	19
5.	A Qualitative Predictive Model	36
6.	Recommendations for Future Work	40
	References	43

Chapter 2 · The Effects of Macrobenthos on Chemical Properties of Marine Sediment and Overlying Water
Robert C. Aller

1.	Introduction	53
2.	Diagenetic Reactions	54
3.	Reactive Particle Redistribution	56
4.	Solute Transport	71
5.	Macrofaunal Influence on Sediment–Water Exchange Rates	83
6.	Reaction Rates	87
7.	Chemistry of the Burrow Habitat	89

8.	Spatial and Temporal Patterns in Sediment Chemistry	92
9.	Summary	93
	Appendix: Solutions to Model Equations	94
	References	96

II. Recent Freshwater Environments

Chapter 3 · The Effects of Benthos on Physical Properties of Freshwater Sediments
Peter L. McCall and Michael J. S. Tevesz

1.	Introduction	105
2.	Freshwater Sediments and Macrobenthos	106
3.	Macrobenthos Life-Styles	113
4.	Mixing of Sediments	124
5.	Biogenic Modification of Sediment Transport	150
6.	Ecological Interactions and Sediment Properties	161
7.	Future Work	166
	References	168

Chapter 4 · Effects of Macrobenthos on the Chemical Diagenesis of Freshwater Sediments
J. Berton Fisher

1.	Introduction	177
2.	Freshwater Sediment System	178
3.	Mechanisms by Which Benthos Affect Chemical Diagenesis	186
4.	Observed Effects of Freshwater Macrobenthos on Chemical Diagenesis	200
5.	Conclusions	209
	References	211

III. Ancient Environments

Chapter 5 · Geological Significance of Marine Biogenic Sedimentary Structures
Charles W. Byers

1.	Introduction	221
2.	Traces and Sedimentology	222

3. Paleobathymetry 233
4. Precambrian Traces 243
   References 252

Chapter 6 · Geological Significance of Aquatic Nonmarine Trace Fossils
Michael J. S. Tevesz and Peter L. McCall

1. Introduction 257
2. Identification of Traces 259
3. Trophic Level Reconstruction 270
4. Paleoenvironmental Reconstruction 271
5. Discussion 277
6. Concluding Remarks 280
   References 281

IV. Models

Chapter 7 · Mathematical Models of Bioturbation
Gerald Matisoff

1. Introduction 289
2. Particle Transport Models 293
3. Fluid Transport Models 313
4. Conclusions 325
   References 327

Index 331

I

Recent Marine Environments

Chapter 1

The Effects of Marine Benthos on Physical Properties of Sediments
A Successional Perspective

DONALD C. RHOADS and LARRY F. BOYER

1. Introduction	3
2. Ecological Succession and Organism–Sediment Relations	5
2.1. Pioneering Stages	7
2.2. Equilibrium Stages	9
3. Ecological Succession and Geotechnical Properties: A Laboratory Experiment	10
3.1. Methods	13
3.2. Results	15
3.3. Conclusions	17
4. Sediment Transport and Biogenic Activity	19
4.1. Bed Roughness	19
4.2. Pelletal Textures	24
4.3. Geotechnical Mass Properties	29
5. A Qualitative Predictive Model	36
6. Recommendations for Future Work	40
References	43

1. Introduction

The effects of benthic organisms on the physical properties of granular substrata are well documented. The range of effects has been presented in H. B. Moore (1931, 1939), Schwartz (1932), Dapples (1942), D. G. Moore and Scruton (1957), McMaster (1967), Rhoads (1974), Rowe (1974), Powell (1974), Richards and Park (1976), Myers (1977a,b), Self and Jumars (1978), Lee and Swartz (1980), and Carney (1981). These papers relate the effects of benthic species to changes in grain size, sorting, fabric, water content,

DONALD C. RHOADS and LARRY F. BOYER • Department of Geology and Geophysics, Yale University, New Haven, Connecticut 06511.

compaction, shear strength, and bottom stability. Those autecologic parameters that appear to be most highly correlated with physical modifications of sediments include: method of feeding, feeding selectivity, feeding level relative to the sediment–water interface, degree of mobility, organism size and population density, burrowing depth, and, if the organism is a tube dweller, the density, spacing, and length of tubes.

The influence of most invertebrate benthos on the biological benthic boundary layer* is limited to a relatively small vertical zone, i.e., from a few centimeters above the sediment surface to a few decimeters below the bottom. Important exceptions exist to this generalization, most notably shrimp that burrow to depths of 3–4 m (Pemberton et al., 1976; Myers, 1979) and certain fish that can burrow to depths of 60–70 cm (Cool, 1971; Atkinson and Pullin, 1976).

In view of the generally surficial effects of benthos on the physical properties of sediments, why study this phenomenon? Our interest in the upper few centimeters of sediment is based on the fact that the interface is both biologically active and chemically reactive. The physical properties of sediments, as controlled by biological processes, may have a major effect on sedimentation, sediment transport, nutrient regeneration, and the fates and historical records of radiochemical species, pollutants, and fossils. Biology, chemistry, and sediments are coupled through the process of bioturbation, involving the transport of particles, as well as pumping water into, and out of, the bottom. Aller (1977, 1978) has reviewed the literature on bioturbation, and makes the observation that water pumping is about 100 times greater (on a weight-to-weight basis) than particle bioturbation. We will see that both fluid and particle bioturbation can affect the physical properties of sediments.

The dispersal or concentration of potentially harmful environmental contaminants may be controlled by the interactions of both particulate and dissolved pollutants with the seafloor. The way a particular benthic organism or suites of benthic organisms bioturbate, or otherwise process the seafloor, can determine if a pollutant is entrapped or mixed into the bottom or, alternatively, transported and dispersed (e.g., Phelps, 1966; Haven and Morales-Alamo, 1972; Swartz and Lee, 1980; Lee and Swartz, 1980; Berger and Heath, 1968; Hanor and Marshall, 1971; Aller and Cochran, 1976; Jumars et al., 1981).

A small but increasing body of data indicates that the response of sediments to fluid shear stress is significantly affected by their biological

---

* Here defined as the height above and below the sediment–water interface that is occupied, or otherwise influenced, by benthic organisms. The physical effects of organisms extend spatially beyond their immediate "biospace." For example, the current velocity profile, as measured several centimeters away from the bed, may be related to topographic boundary conditions controlled by biological features.

components. Plants and invertebrates are known to trap and bind sediments or promote sedimentation (Ginsburg and Lowenstam, 1958; Fager, 1964; Neuman et al., 1970; Marshall and Lucas, 1970; Scoffin, 1970; M. Lynch and Harrison, 1970; Holland et al., 1974; Rhoads et al., 1978b; Yingst and Rhoads, 1978; Boyer, 1980; Nowell et al., 1981; Eckman et al., 1981). Sediments can also become destabilized by organisms relative to azoic or bound conditions (Rhoads and Young, 1970; Southard et al., 1971; D. K. Young, 1971; R. A. Young and Southard, 1978; Rhoads et al., 1978b; Yingst and Rhoads, 1978; Boyer, 1980; Nowell et al., 1981; Eckman et al., 1981; Grant et al., 1982).

It is the purpose of this chapter to focus on those physical organism–sediment relations that are of potential value in predicting the physical properties of sediments and the transport fates of fine-grained sediments and their associated contaminants. We will limit our discussion primarily to macrofaunal effects on subtidal muds. The geochemical aspects of this problem are covered in Chapter 2 of this volume. We also direct the reader to an excellent recent review of biodepositional and bioturbational effects on the distribution of pollutants in sediments (Lee and Swartz, 1980). We will not cover biogenic rate processes here (see Aller, 1978, and Lee and Swartz, 1980). Rather, our contribution will focus on how organism–sediment relations develop during ecological succession of the seafloor.* Generalizations will be made about how early and late successional stages influence the physical properties of sediments—some of the cause-and-effect relationships are well known, while others remain speculative. The successional paradigm for interpreting organism–sediment relations is new, and, therefore, poorly tested. New kinds of measurements and observational techniques will be required to explore and test this model. We will offer a few promising techniques and approaches which we feel are necessary for conducting future work.

2. Ecological Succession and Organism–Sediment Relations

Until recently, methods of describing bottom communities and their relationships to sediments followed, more or less, the procedures outlined

---

* The term *succession* has many connotations in the literature (see McIntosh, 1980, for a discussion of the history of the successional concept). We define *primary succession* as the predictable appearance of macrobenthic invertebrates belonging to specific functional types following a benthic disturbance. These invertebrates interact with sediment in specific ways. Because functional types are the biological units of interest for this study, our definition does not demand a sequential appearance of particular invertebrate species or genera.

by C. G. J. Petersen (1913) and popularized by G. Thorson (1957). The classical Danish method of defining benthic communities consists of identifying characterizing species. Characterizing species are those that are temporally and spatially persistent, dominate in biomass, and have relatively long life-spans. These criteria select for "equilibrium" species, which live on parts of the seafloor where physical disturbance is uncommon. Early organism–sediment work was also done from this "steady-state" point of view. In addition, much of this early work was limited to the sedimentary effects of individuals of a species or single-species populations (e.g., Crozier, 1918; Dapples, 1942; Rhoads, 1963; Gordon, 1966). A notable exception to this generalization is the early German work in

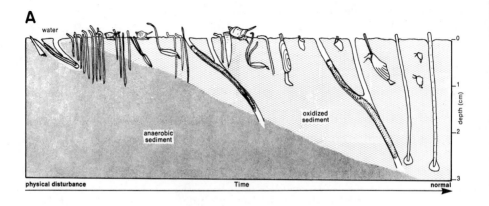

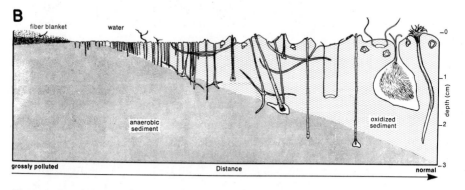

Figure 1. Development of infaunal assemblages over time following a major disturbance (A) and along chronic pollution gradients (B). Pioneering species (left side) tend to be tubicolous or otherwise sedentary organisms that live near the sediment surface and feed near the surface or from the water column. Particle bioturbation rates are low. High-order successional stages (right side) tend to be dominated by bioturbating infauna that feed at depth within the sediment. Particle mixing by bioturbation is intensive. Figure from Rhoads et al. (1978a). The pollution gradient diagram (B) is modified from Pearson and Rosenberg (1976).

the North Sea, where the effects of whole faunas on sediments were considered (see Schafer, 1972).

A more ecologically realistic concept of benthic communities was proposed by Johnson (1971, 1972), who suggested that a benthic community is a temporal and spatial mosaic, "parts of which are at different levels of succession . . . in this view the community is a collection of relics of former disasters." This successional perspective has its origin in plant ecology as popularized by Clements (1916).

Subsequent experimental and field work by McCall (1977), Myers (1977a,b), Pearson and Rosenberg (1978), Rhoads et al. (1977), Wolff et al. (1977), Santos and Bloom (1980), and Santos and Simon (1980) helped to define the taxonomic and functional structure of these "temporal mosaics." This dynamic approach to community structure suggests that earlier work on subtidal organism–sediment relations be reevaluated in terms of a successional paradigm, and that future work test the validity of these successional dynamics.

Figure 1 depicts organism–sediment relations as they develop during succession of a subtidal granular bottom. Most of the information about this successional pattern comes from mud bottoms (e.g., McCall, 1977; Pearson and Rosenberg, 1978; Rhoads et al., 1977, 1978a), but major qualitative features of this succession are apparently also shared with sand-dwelling faunas (e.g., Myers, 1977a,b; Dauer and Simon, 1975).

2.1. Pioneering Stages

Pioneering stages usually occur near shore, above the mean shoaling depth of storm waves. Disturbance assemblages may, however, be found at any water depth—even in abyssal trenches (Jumars and Hessler, 1976)—and in any sediment type. The controlling factors are related to the intensity and frequency of disturbance* (Osman and Whitlatch, 1978).

Small, opportunistic, tube-dwelling polychaetes are the first faunal components to colonize a new or newly disturbed bottom. Their aggregations may reach densities of 10^5 m^{-2} within a few days to weeks after a disturbance (McCall, 1977; Rhoads et al., 1978a). Tubicolous amphipods appear shortly after the polychaete acme.

Most pioneering species feed near the sediment surface or from the water column. Tube walls serve to isolate the colonizing organisms from

---

* The example of disturbance used in our review is the physical reworking of the seafloor by waves and currents. This is only one category of disturbance that can restructure benthic communities. For a comprehensive review of the physical, biological, and chemical perturbations that can affect macrofaunal succession we refer the reader to Pearson and Rosenberg (1978) and Dayton and Oliver (1980).

the sediment by controlling the rate of diffusion of ambient pore water solutes into the tube environment (Aller, 1980).

An organism colonizing a newly disturbed bottom or new sediment is faced with formidable physiological problems. The pore water chemistry will be unpredictable, especially regarding the vertical concentration gradients of pore water metabolites, metal sulfides, dissolved oxygen,* and organic decomposition products. Reduced compounds present at high concentrations in the subsurface sediment diffuse across the tube walls and enter the tube interior. Aller (1980) has shown that organisms occupying closely spaced, small-diameter tubes efficiently share the work (measured as tube irrigation) required to maintain a constant and low concentration of solutes (e.g., NH_4) in their tubes. The unpredictable chemical and trophic environment below the sediment–water interface may also explain why most pioneering species feed at, or above, the sediment–water interface.

Most pioneering species have adaptive strategies that have been described as r-selected, while many infaunal deposit-feeding species share K-selected attributes (McCall, 1977; Rhoads et al., 1978a). Vermeij (1978) has pointed out that the often-cited opportunistic-to-equilibrium (or r-to-K) end-member classification of benthic marine communities does not represent the full range of possible adaptive types. He recognizes a third end-member called stress-tolerant, which is representative of species that predominantly inhabit physiologically stressful areas such as the intertidal zone. At this time, the organism–sediment successional model (Fig. 1) does not consider this stress-tolerant group of species. However, it is interesting to note that many intertidal species such as *Ensis directus* (bivalvia) make brief appearances in subtidal pioneering stages (McCall, 1977).

In summary, the sedimentary effects of pioneering species are limited to the near-surface region of the bottom (<2 cm) and they include the following:

1. Construction of dense tube aggregations, which may affect microtopography and bottom roughness on a scale dictated by tube diameter, tube height, and tube spacing.

---

* Mapping the depth of the *redox potential discontinuity* (RPD) below the surface of marine sediments has proven useful for assessing the affect of organic pollution on the benthic ecosystem (Pearson and Stanley, 1979; Rhoads and Germano, 1982). The cause-and-effect relationships are complex. In areas of the seafloor experiencing organic enrichment, biological oxygen demand and chemical oxygen demand are highly correlated with organic decomposition. This oxygen demand causes the $Eh = 0$ horizon to rebound toward the sediment surface. The enrichment is also associated with the appearance of pioneering species, which are relatively ineffective in exchanging ambient pore water with overlying water; hence the RPD remains near the sediment surface.

2. Fluid bioturbation, which pumps water into and out of the bottom through vertically oriented tubes. Particle bioturbation, although present, is of subordinate importance.
3. Surface deposit feeding and suspension feeding, which cover the surface of the bottom with fecal pellets, especially the fusiform pellets of opportunistic polychaetes.

2.2. Equilibrium Stages

High-order successional stages are found where major physical disturbances such as intense annual storms rarely affect the bottom. Examples have been described for the Maldanid–*Nucula*–*Syndosmya* (polychaete–bivalve) community in the Clyde Sea (H. B. Moore, 1931), The *Nucula*–*Nephtys* (bivalve–polychaete) assemblage in Buzzards Bay (Sanders, 1958; Rhoads and Young, 1970), the *Molpadia*–*Euchone* (holothurian–polychaete) community of Cape Cod Bay (Rhoads and Young, 1971; D. K. Young and Rhoads, 1971), the *Maldane*–*Amphiura* (polychaete–ophiuroid) community of Ria de Muros, Spain (Tenore et al., 1982), and the *Lumbrineris*–*Alpheus*–*Diolodonta* (polychaete–shrimp–bivalve) community in Kingston Harbor, Jamaica (Wade, 1972). A late successional stage is dominated by infaunal deposit feeders, many of which are represented by "head-down" conveyor belt feeders (*sensu* Rhoads, 1974). Some of these species are tubicolous (e.g., the polychaete family Maldanidae) but many others are free-living. Equilibrium assemblages are associated with a deeply oxygenated sediment surface where the RPD commonly reaches depths of over 10 cm. Feeding is concentrated at, but not limited to, the RPD. The RPD is in fact related to the feeding depth.* The redox zone appears to be a region of high microorganism productivity (see Yingst and Rhoads, 1980). Both vertical particle mixing and pore water exchange by respiratory pumping are important bioturbational processes.

Organism–sediment relations are complex and well developed in equilibrium systems. In summary, the sedimentary effects of equilibrium species in shallow water environments appear to be as follows:

1. Transfer of both water and particles over vertical distances of up to 10–20 cm takes place.

---

* The depth of the redox boundary can be seen as a brown oxidized layer at the surface of the sediment, or locally around tube and burrow walls. In sediments high in iron content, this colored layer may not, in many cases, reflect the presence of free molecular oxygen in sediment pore water. Once oxidized, ferric iron may persist long after the associated pore water has been depleted of dissolved oxygen (Revsbech et al., 1979).

2. Intensive particle mixing produces homogeneously mixed fabrics and many of the particles at and below the sediment surface may be in the form of fecal pellets.
3. Head-down feeding produces void spaces (feeding pockets) at depth within the bottom.
4. The RPD is located over 2 cm into the bottom and commonly to depths of 10–20 cm.
5. Surface microtopography may be featureless and planar if tidal resuspension "smoothes" over biologically produced features at the sediment surface, or, in the absence of the smoothing effect, the surface may be covered with numerous feeding pits and fecal or excavation mounds.

In some cases pioneering and equilibrium species may coexist in the same sediment, if physical disturbance involves only near-surface sediment. Large and deeply burrowed infaunal species may not be affected by such small-scale disturbances (e.g., Woodin, 1978). The pioneers may arrive as settling larvae, stimulated by chemical cues associated with the disturbed sediment (e.g., Gray, 1974). One of us (DCR) has seen mixing of low- and high-order successional species in Long Island Sound by the passive wash-in of opportunists from adjacent habitats, but this mixing represents a transient state. Pioneering species may be excluded from equilibrium assemblages by competitive exclusion mechanisms such as trophic group amensalism (Rhoads and Young, 1970). Thayer (1979) has described this phenomenon as biological "bulldozing."

3. Ecological Succession and Geotechnical Properties: A Laboratory Experiment

If sedimentary properties are influenced by the stage of succession, one should be able to observe changes in geotechnical parameters by experimentally introducing the different successional stages into the same sediments and observing the effects of dominant species on geotechnical properties. We conducted such an experiment using sediments and organisms from Long Island Sound (Table I).

A study site called FOAM* is located in 14 m of water approximately 1 mile from shore on a mud bottom. This station is maintained in a low-order successional stage by seasonal resuspension of the seafloor. Several

---

* Station FOAM is located near McCall's (1975, 1977) site A. The FOAM site has been the location of intensive geochemical investigations. The initials are, appropriately, an acronym for Friends of Anoxic Muds (Goldhaber et al., 1977).

Table I. Experimental Design of the Successional Geotechnical Experiment

	Sediment source (tank no.)									
	Station FOAM[a]			Station NWC[b]						
Introduced macrofauna	1	2	3	4	5	6	7	8	9	10
None (control)	X				X					X
FOAM ambient fauna			X[c]					X[d]		
NWC ambient fauna		X[e]							X[f]	
250 Nucula annulata only				X						
49 Yoldia limatula only					X					
25 Nephtys incisa only							X			

[a] 85% Silt–clay, mean grain size 0.0125 ± 0.0074 mm (N = 10) (top 10 cm).
[b] 96% Silt–clay, mean grain size 0.0046 ± 0.0005 mm (N = 20) (top 10 cm).
[c] 6 Nephtys incisa and 4 Clymenella torquata.
[d] 3 Nephtys incisa, 184 Nucula annulata, 20 Cylichnella canaliculata, 1 Cylichnella oryza, plus numerous spionids not retained on the 1.0-mm sieve.
[e] 15 Nephtys incisa, 39 Nucula annulata, 1 Cylichnella canaliculata, 1 Nassarius, plus numerous spionids not retained on the 1.0-mm sieve.
[f] 3 Nephtys incisa, 107 Nucula annulata, 20 Cylichnella canaliculata, 1 Nassarius trivittatus, 1 Pinnixia sayana, 1 Sipunculida goldfingia(?).

centimeters of sediment are resuspended during storms (McCall, 1978). The distribution of species at FOAM is patchy, and their abundance and diversity are unpredictable. This high degree of variance arises from the short life spans and high population turnover rates of pioneering species. At best, one may predict that, from the total list of opportunistic species known for Long Island Sound, a small subset of these species will dominate the faunal assemblage at any one time.

Figure 2A shows structural evidence of storm erosion and resuspension of FOAM sediments. A sediment profile photograph (Rhoads and Cande, 1971; Rhoads and Germano, 1982) also shows the density of small tubicolous polychaetes typically present at FOAM (Fig. 2B). In some years, the bivalve *Mulinia lateralis* may arrive as an early pioneer in great abundance. In this case, tubicolous polychaetes are excluded from dense *M. lateralis* patches (Fig. 2C). McCall (1977) has described the faunal dynamics near this sampling station.

Station NWC, located in 20 m of water near the center of the sound, is below the mean shoaling depth of storm waves and supports an equilibrium infaunal community (McCall, 1977; Rhoads et al., 1977). Three species persistently dominate this station in both biomass and number: the errant polychaete *Nephtys incisa* and two protobranch bivalves, *Yoldia limatula* and *Nucula annulata*. Figure 3A shows the relatively homogenous sedimentary fabric produced by deep and intensive bioturbation at this station. The sediment surface may be covered with a layer of organic–mineral aggregates (Fig. 3B).

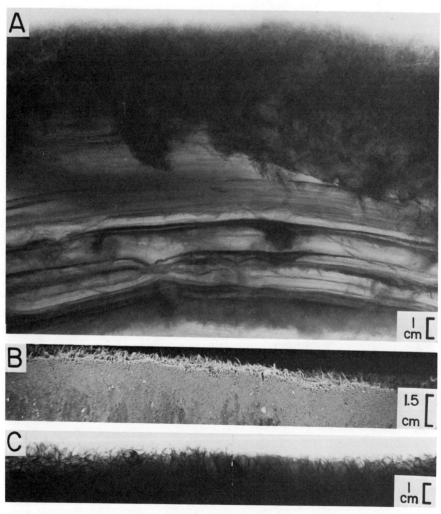

Figure 2. Sedimentary fabrics and biogenic structures in muddy sediments representing a physically disturbed bottom. (A) X-radiograph of station FOAM, showing current laminated sediment lying below a bioturbated interval. Note the unconformity separating these two fabrics. (B) *In situ* sediment profile photograph showing dense populations of pioneering tubicolous polychaetes typically found at station FOAM. (C) X-radiograph of a dense population of the pioneering mactrid bivalve *Mulinia lateralis* also sometimes found at station FOAM.

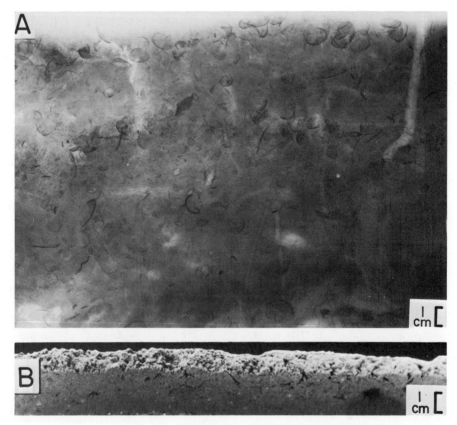

Figure 3. Sedimentary fabrics and biogenic structures in muddy sediments representing a bottom infrequently disturbed by storm. (A) X-radiograph showing the bioturbated fabric typical of station NWC. Note the burrow-mottled structure and random orientation of shells within the sediment. (B) *In situ* sediment profile photograph of organic–mineral aggregates at the sediment surface at station NWC. Note the absence of near-surface-dwelling benthos. Some of the aggregates are fecal pellets; others represent pieces of cohesive surface sediment that have been broken and moved by particle bioturbation.

3.1. Methods

The successional/geotechnical experiment was set up by bringing defaunated sediment from stations FOAM and NWC into the laboratory. Large macrofauna were removed by passing sediment through a 1-mm mesh sieve. Fourteen liters of the sieved mud were placed into each of 10 Plexiglas tanks, each measuring 36 cm long, 27 cm wide, and 30 cm deep. After settling and compaction, the depth of sediment in each tank

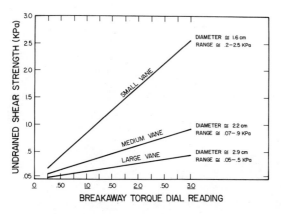

Figure 4. The relationship between shear vane breakaway torque and the undrained shear strengths of our experimental muds. Three vane sizes were used to measure shear strength over the range <0.5–2.5 kPa (1 kPa ≈ 10 g·cm^{-2}).

was approximately 15 cm. Ten liters of seawater covered the sediment in each tank and was circulated at 30 liters·hr^{-1} between the test tanks and a temperature-controlled, 148-liter reservoir. Water temperatures were regulated during the experiment to match the changing ambient water temperature in Long Island Sound (15–22°C during the experimental period). Eighteen days after sediment had been placed in the test tanks, benthic macrofauna were introduced into the tanks as outlined in Table 1.

Undrained sediment shear strength and sediment water content were measured in each tank over the course of the experiment (57 days duration). Vertical profiles of these two variables were measured at intervals of 2.0–2.5 cm to depths of about 12 cm. Water contents were determined by sectioning small cores (2.5-cm diameter) and weighing each section (wet weight). The samples were then dried at 105°C for 3 days. The sections were weighed, and percent water and water content determined. Shear vane measurements were made with a modified torque screwdriver with specially constructed vanes. The three vane sizes are shown in Fig. 4, which relates their breakaway torque to undrained shear strength in muds.

The successional/geotechnical experiment began on July 7, 1980, and terminated on September 2, 1980. Geotechnical measurements were made 13 times over the 57-day period. We report here only data for the beginning and the end of the experiment (day 57). Because of the sediment treatments involved in setting up this experiment (sieving and remolding) and the relatively short term of the experiment, we cannot directly compare field and laboratory values of water content and undrained shear strength from FOAM and NWC sediments. The standards of comparison in this experiment are the controls which lack macrofauna ≥1 mm in size. This experiment addresses the problem of detecting a change in geotechnical properties relative to the controls. The magnitude of the change can be used to estimate qualitatively the relative effects of pioneering and equilibrium species on geotechnical properties.

3.2. Results

3.2.1. FOAM Sediments

Vertical profiles of undrained shear strength for tank 3 (containing the ambient FOAM fauna) and tank 2 (containing the ambient NWC fauna) are shown in Figs. 5A and B, respectively. The data are plotted as the difference between values measured in the two experimental tanks and values (from the same depth intervals) in the control tank (1). Horizontal bars graphed to the right of the 0 percentage difference datum represent an increase in shear strength relative to the control; values plotted to the left represent decreases.

FOAM sediments containing near-surface-living pioneer species experience a 12% increase in shear strength in the 0- to 2.5-cm depth interval relative to the control. In comparison, a 50% decrease in shear strength was measured for this same depth interval with the presence of subsurface deposit feeders. The 2.5- to 5.0-cm depth interval in both tanks shows a decrease in shear strength. This decrease is most dramatic in tank 3 (90%). Radiographs of this sediment show the presence of methane gas voids at depth, which might explain this reduction in shear strength. Methane forms after pore water sulfate has been depleted. The supply of interstitial sulfate would be expected to be lower in the less intensively bioturbated

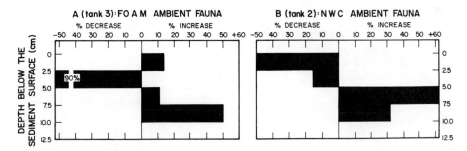

Figure 5. The effects of pioneering and equilibrium species on undrained sediment (mud) shear strengths. (A) Changes in undrained shear strengths of experimental FOAM muds populated with a pioneering macrofaunal benthic assemblage (FOAM). (B) Changes in undrained shear strengths of experimental FOAM muds populated with station NWC equilibrium species. In A and B, plotted values of percent change in shear strengths are relative to a control tank containing FOAM sediment without macrofauna. Shear strength changes shown are for 57 days after introduction of macrofauna. See Table I.

tank (Fig. 5A) compared with the more intensively bioturbated tank (Fig. 5B). Below 5 cm, both tanks show an increase in shear strength relative to the control. This depth interval is well below the living depths of the introduced faunas. Possible "far-field" effects of benthos on sediments will be discussed later.

No difference in sediment water content was observed between the experimental tanks. Surface values ranged from 60% to 61% and declined to about 51% at 15 cm. The shapes of the curves were also similar.

3.2.2. NWC Sediments

Vertical profiles of undrained sediment shear strengths in NWC sediments, containing both Pioneering and equilibrium species, are given in Figs. 6A–D. In this experiment two controls were run. The average profile

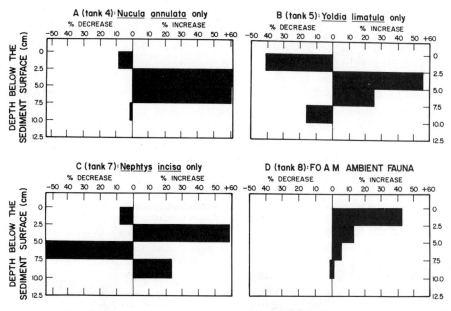

Figure 6. The effects of pioneering and equilibrium species on undrained sediment (mud) shear strengths. (A) Changes in undrained shear strengths of experimental NWC muds populated with 250 bivalves (*Nucula annulata*). (B) Changes in undrained shear strengths of experimental NWC muds populated with 49 bivalves (*Yoldia limatula*). (C) Changes in undrained shear strengths of experimental NWC muds populated with 25 errant polychaete worms (*Nephtys incisa*). (D) Changes in undrained shear strengths of experimenal NWC muds populated with the pioneering FOAM fauna. In A–D, plotted values of percent change in shear strengths are relative to the mean of two control tanks containing NWC sediment without macrofauna. Shear strength changes shown are for 57 days after introduction of macrofauna. See Table I.

change in shear strength (at day 57) is used as a datum to compare measured changes in the experimental tanks. Tank 4 (Fig. 6A) contains the late successional or equilibrium-stage bivalve *Nucula annulata*. This protobranch bivalve has a small but measurable effect of decreasing shear strengths in the interval 0–2.5 cm. *Yoldia limatula*, another equilibrium-stage protobranch, has a relatively greater effect in decreasing near-surface shear strengths (tank 5, Fig. 6B). *Yoldia* lives below *N. annulata* and can burrow to 3–4 cm depending on its size. Below the living depths of these two clams, shear strengths increase relative to the controls.

The burrowing depth of the errant polychaete *Nephtys incisa* depends on its length. In tank 7 (Fig. 6C), most of the burrow systems are limited to the upper 3–4 cm, but a few large specimens penetrated to 8.5 cm. The effect of *N. incisa* on sediment shear strength above 5 cm is comparable to that of *N. annulata* (Fig. 6A). The increase in sediment shear strength below the burrow-penetrated area may represent another "far-field" effect. The decrease in shear strength in the interval 5.0–7.5 cm may be related to the presence of methane gas voids or local influence of a large *N. incisa* burrow.

Tank 8 (Fig. 6D) shows that the ambient FOAM fauna increases sediment shear strength from the surface to 7.5 cm, but that below this depth there is no difference between the test and control tanks. Although FOAM and NWC sediments are somewhat different in organic content and grain size, the addition of pioneering species to these two types of muds results in near-surface increases in shear strength (tank 3, Fig. 5A, and tank 8, Fig. 6B).

Water content profiles of tanks shown in Figs. 6A–D were not significantly different from one another, nor did they differ from those reported from the FOAM sediment experiments.

3.3. Conclusions

The geotechnical experiment has shown that the successional stage can have an important effect on sediment shear strengths. Within the 0- to 2.5-cm surface interval, particle advection and burrow excavation associated with equilibrium species "dilates," "fluffs," or otherwise decreases the shear strength of the surficial sediment. Although not measured in this experiment, equilibrium species are known to pelletize sediment intensively (Rhoads and Young, 1970; D. K. Young, 1971). This pelletization effectively increases the modal grain size of the sediment, reducing the number of particle-to-particle contracts, and increases sediment porosity.

Pioneering species, as described earlier, live and feed near the sediment surface and are not effective in mixing sediment. When produced, pellets are deposited on the sediment surface. The numerous spionid polychaetes present in the FOAM assemblage pump water into and out of the sediment through their tubes. This irrigation may stimulate near-surface populations of microorganisms and meiofauna.* The apparent increase in sediment shear strengths associated with this assemblage may be related to mucus binding of particles (e.g., Martin and Waksman, 1940; Frankel and Meade, 1973). The presence of the tubes also increases sediment resistance to horizontal shearing forces.

The most interesting results of these experiments are the most difficult to interpret. Every experimental tank showed an effect below the living depths of the introduced macrofauna which we have called a "far-field" effect. We are cautious in our interpretation of these apparent effects. One would need to repeat this experiment with more control tanks to better evaluate sources of variance in the data. However, we believe that many of the observed "far-field" phenomena may represent real biological effects. Rowe (1974) has shown that burrowing anemones can affect sediment shear strengths as far as 20 cm away from the burrow and tube. Aller (1978) has shown that microbial ATP and production and consumption of decomposition products in the sediment are affected well below the sediment actually occupied by *Y. limatula*. In this case, the far-field effect is related to how the clam influences the rate of exchange of pore water constituents such as ammonium and sulfate. Bioturbation, especially respiratory pumping, affects the rates of chemical reactions as well as the concentration gradients of dissolved compounds in pore water.

Another far-field effect is related to the depth of the RPD below the sediment surface. Respiratory pumping can maintain the RPD several millimeters or centimeters below the surface. The RPD usually lies below the zone of living macrofauna, and is a region of intensive microbial production (Yingst and Rhoads, 1980). It is also possible that the chemical and physical gradients set up by macrofaunal reworking affect both the distribution and activities of interstitial meiofauna. This size class of organisms can also influence the physical properties of sediment (e.g., Cullen, 1973; Boyer, 1980).

A possible far-field effect is related to the density or size stratification of detrital particles at the base of the reworked zone. Silt- or sand-sized particles with specific gravities ≥2.0 are reworked through the zone of bioturbation and become concentrated at the base of this zone as a graded

---

* Total meiofaunal and microbial adenosine triphosphate (ATP) at the sediment surface at FOAM is more than twice that at the sediment surface at station NWC (Aller and Yingst, 1980).

bed (van Straaten, 1952; Rhoads and Stanley, 1965; Rhoads and Young, 1970; Cadee, 1979).

An unexpected result of this experiment was that the water content profiles did not track measured changes in sediment shear strength. We offer two possible explanations. First, sediment water content may be insensitive to the biological activity present in our tanks, at least on the time scale of 57 days. FOAM sediments in the field have summer water contents (percent water) of between 50% and 60% within the upper 10 cm of the sediment surface (Aller et al., 1980), while NWC sediments normally have water contents $\geq 70\%$ all year long (Rhoads et al., 1977). Alternatively, our procedures for storing water content cores (7°C for approximately 1 week) may have resulted in sediment compaction, which "erased" differences in water content.

More geotechnical experiments like the one described here are needed to evaluate fully the relationship between succession and sediment geotechnical properties. Separate experiments must be done for muddy and sandy substrata. Sandy sediments may behave very differently from muds under the same bioturbational regimes, as sands tend to compact rather than "dilate" when mixed (Webb, 1969; Myers, 1977a; Powell, 1977).

4. Sediment Transport and Biogenic Activity

Several biogenic structures and biologically mediated changes in sedimentary properties are of potential importance for determining substratum stability: (1) production of bed roughness; (2) increase in modal grain size through the formation of fecal pellets and the secretion of mucopolysaccharides, which coat pellets and particles and bind them to one another; and (3) change in sediment packing, shear strength, and water content. Each of the above factors will be discussed and then related to the successional paradigm outlined earlier.

4.1. Bed Roughness

Organisms produce pits and depressions on the seafloor related to burrowing and foraging activity. Benthos also produce elevations such as mounds from burrow excavation, tracking, and feeding (Frey, 1975). The length scales of these features range from meters (ray pits) to decimeters (shrimp mud mounds) in diameter, and on down to submillimeter scales, in the case of fecal pellets deposited on the sediment surface.

Not all bioturbational activity results in an increase in bed roughness on scales of centimeters. If most of the species are small and mobile, the surface may be homogeneously reworked, resulting in a very "smooth" and flat bottom. An example of this is described for the mud facies in central Buzzards Bay (Rhoads and Young, 1970). However, if the reworking species are large and relatively sedentary, each individual may generate considerable topographic relief by its localized feeding activity. Examples of this local topographic effect are described for central Cape Cod Bay, Massachusetts, related to the presence of mounds made by the holothurian *Molpadia oolitica* (Rhoads and Young, 1971), and for burrowing shrimp mounds in Discovery Bay, Jamaica (Aller and Dodge, 1974).

Another commonly encountered and qualitatively important biological roughness element consists of tubes that project a few millimeters above the bottom. These tubes may be constructed of parchmentlike organic material or mucus-cemented sediment grains. Polychaetes and amphipod crustaceans commonly form dense aggregations of tubes in physically disturbed habitats. The spatial scales of the aggregations range from several thousand square meters (e.g., Fager, 1964) to approximately 10^5 m^{-2} (e.g., Eckman, 1979). Tube populations may develop very quickly following a disturbance through larval settlement, reaching densities of 10^5 m^{-2} within a few days (McCall, 1977). These tube fields are usually short-lived phenomena lasting 2 years or less, as the tube-forming species are subject to mortality from intensive predation (Fager, 1964) or competitive amensalistic interactions with burrowing species (e.g., Myers, 1977a,b). As we will see later, individual tubes themselves may enhance bottom erosion, resulting in eventual washout of the population (Eckman et al., 1981). On the other hand, some tube fields and their associated benthic communities may be relatively permanent, with sediment-stabilizing species coexisting with sediment-destabilizing species (Wilson, 1979).

Inferences about the influence of tube fields on substratum stability are largely based on the association of dense tube aggregations with increased topographic elevation of the bottom, a decrease in mean grain size, an increase in the quantity of organic-rich detritus settling between tubes (Mills, 1967, 1969; Myers, 1977a; M. Lynch and Harrison, 1970; Woodin, 1976; Featherstone and Risk, 1977; Bailey-Brock, 1979), or absence of traction bed forms within tube fields (Fager, 1964). Indirect inferences about habitat stability have also been based on diversity patterns. Tube patches have a higher abundance and diversity of non-tube-constructing species living within them than are found on the adjacent seafloor lacking tubes (Woodin, 1978; Wilson, 1979; Bailey-Brock, 1979; Eckman, 1979).

An extensive fluids literature (see review by Wooding, 1973) exists on roughness elements; however, few flume experiments have been conducted on the effects of animal tubes or other biologically produced roughness elements on substratum stability. Rhoads et al. (1978b) did flume work with the capitellid polychaete *Heteromastus filiformis* and showed that, 3 days after a population was introduced into a mud, the critical erosion velocity was increased by 80% relative to the critical velocity required to entrain tube-free sediment. The mechanism of sediment stabilization in this experiment is difficult to interpret, because the authors also demonstrated, in a separate set of experiments, that bacterial mucus films can increase entrainment velocities by up to 60%. The *H. filiformis* study did not separate tube effects from microorganism effects. This study also reported an apparent decrease in critical erosion velocity for muds in central Long Island Sound associated with the appearance of a dense set of the polychaete *Owenia filiformis*. Again, the correlation between the presence of tubes and sediment stabilization does not allow one to identify a simple cause-and-effect relationship because the effects of microbial binding were not measured.

While there is only one experimental study of the natural tube problem *per se* (Eckman et al., 1981), there are many experimental studies of the effects of roughness spacing on flow resistance. The original study was carried out by Schlichting (1936) (cf. Morris, 1955; Sayre and Albertson, 1963; Lettau, 1969; Raupach, 1981). Of particular interest is a study by Nowell and Church (1979), for they presented not only resistance diagrams but also detailed velocity field and turbulent flux profiles. They investigated the effect that different densities of boundary roughness elements have on the shape of the velocity, kinetic energy, and turbulent energy dissipation profiles measured over the total boundary layer. The roughness elements used in their experiments were Lego® construction blocks. The density of the blocks was expressed as a ratio of the plan area of N roughness elements (blocks) to the total bed area of the flume channel.

At low densities of blocks (less than 1 unit area of blocks per 22 unit areas of flume bottom), each block acts as an isolated element. Turbulent vortices are shed from individual roughness elements and the turbulent energy is transferred to the bed. The kinetic energy and energy dissipation functions show the characteristic monotonic increases all the way to the wall, which is hydraulically rough. At intermediate densities (1/16–1/22), the profile of kinetic energy is approximately constant over a distance of 1–2 roughness heights. At the highest densities (1/8–1/12), maximum turbulent intensity and the rate of turbulent energy dissipation were elevated to near the tops of the roughness elements, which "protected" the bed within the tube field from higher-energy turbulence ("skimming flow" *sensu* Morris, 1955).

The Nowell and Church (1979) study has important implications for

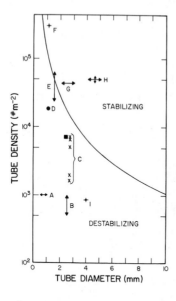

Figure 7. The relationship between tube diameter and tube spacing in affecting noncohesive bed stability. Tube values in the stabilizing field are proposed to initiate "skimming flow," which protects the bed from turbulence. Tube values in the destabilizing field cause bed scour by reattachment of wake turbulence. Modified from Eckman et al. (1981). (A) McCall and Fisher (1980) (oligochaetes); (B) Fager (1964) (*Owenia fusiformis*); (C) Eckman et al. (1981) (*O. fusiformis*); (D) McCall (1977) (*Streblospio benedicti*); (E) Bailey-Brock (1979) (*Chaetopterid* polychaetes); (F) Myers (1977a,b) (*Corophium insidiosum*); (G) Mills (1969) (*Ampelisca abdita*); (H) Fager (1964) (*Owenia fusiformis*); (I) Myers (1977a,b) (*Microdentopus gryllotolpa*).

predicting how the substratum might respond to the spacing and size of biological features projecting above the bed. Using the 1/12 ratio as a threshold roughness spacing for initiating skimming flow, Eckman et al. (1981) have shown that natural densities of tubicolous species span the inferred range of stabilizing densities (Fig. 7). Of particular interest is the prediction that densities of *Owenia*, reported by Fager (1964) to stabilize a sand bottom from wave surge, should destabilize the bed (i.e., give values for critical shear velocity lower than those predicted by the Shields curve*). Average densities reported by Fager (1964) actually destabilized sediment in the Eckman et al. (1981) flume study. The cause for disparity between Fager's field observations and the *Owenia* flume experiment is unknown, but Eckman et al. (1981) suggest that the difference between the field observations and their flume experiments is related to the presence of sediment-binding exudates from surface-living bacteria, diatoms, or a commensal anemone that is attached to the *Owenia* tube.

Changes in substratum elevation, sediment quality, and grain size, which are frequently cited as evidence for tube stabilization of the seafloor, must be interpreted with care. Even though the tube density falls within the destabilizing region of Fig. 7, the bed may be stabilized by exudates of chemautotrophic microorganisms or diatoms. The presence

---

* The *Shields* curve takes into account important fluid and sedimentary parameters to provide the ratio of dimensionless shear to a particle Reynolds number (i.e., ratio of entrainment to stabilizing forces). This curve allows one to compare data on measured grain sizes and fluid shear stresses regardless of flume design. Moreover, laboratory results can be compared with field measurements or estimates (see Madsen and Grant, 1976; Miller et al., 1977).

of tubes may be directly coupled with these exudates. Microorganism productivity is enhanced through the increased supply of dissolved nutrients pumped across the sediment–water interface through tubes (Webb, 1969, Aller and Yingst, 1978; Aller, 1980).

Tube fields that fall well within the stabilizing region of Fig. 7 may still be susceptible to scour and erosion, especially at the leading edge and sides of a patch where flow accelerations occur. In a rotational tidal current, all sides of a patch may be subject to scour. Most flume and field studies have not addressed the problem of form drag associated with tube patches. The question of whether "skimming flow" is actually induced by tube-dwelling organisms is still unanswered, as flume work has not yet been done with tube densities predicted to be stabilizing (Fig. 7).

Another biogenic factor that can armor or protect sediments from erosion is the presence of dense concentrations of shell material at the sediment–water interface. The quantity of shell at the sediment surface is related to both the sedimentation rate and the shell production rate.

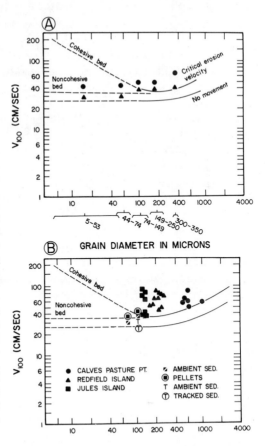

Figure 8. Competency diagrams showing the effects of biological activity on critical threshold velocities. (A) The maximum observed change in mean rolling velocity of five glass bead beds following exposure to microbial growth (replotted from Ullman, 1975). Flow data coverted to z = 100 cm and plotted on a competency diagram. Triangular symbols are initial (aseptic) bead values; circles are values after microbial binding. Microbial binding displaces threshold values above the quartz critical erosion velocity boundary. From Rhoads et al. (1978b). (B) Effects of fecal pellets, tracks, and exudate binding on natural sediments. Data on Calves Pasture Pt., Redfield Island, and Jules Island, Barnstable Harbor, Massachusetts, from Boyer (1980). Other data from Nowell et al. (1981).

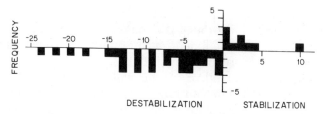

Figure 9. Change in critical threshold velocity for initiating bed movement in fine sandy intertidal sediments reworked by meiofauna. Horizontal scale is change in flume velocity (cm · sec^{-1}); vertical axis is measurement frequency. Of the 41 flume runs, 78% ($N = 32$) showed that particle bioturbation lowered the critical velocity. The change in critical velocity is relative to a recently eroded bed surface. From Boyer (1980).

Pioneering assemblages are highly productive systems, and opportunistic bivalve molluscs (e.g., species of the families Mactridae, Tellinidae, and Pandoridae) may appear in abundances of 10^3–10^4 m^{-2} (McCall, 1977; Rhoads et al., 1978a). These clam populations usually persist for less than 1 year. As the death assemblage accumulates, the disarticulated valves may be packed so densely that they armor or pave the interface (Fig. 2C).

Sediments may be destabilized by bed roughness produced by surficial tracks and trails (Eckman et al., 1979). Nowell et al. (1981) have conducted flume experiments to determine the threshold velocity for very fine sand tracked by a small (approximately 4 mm) clam, *Transenella tantilla*. The critical shear velocity was reduced by 20% from that of the untracked sediment (Fig. 8B). Boyer (1980) has also observed decreases in critical flume entrainment velocities related to the presence of tracks and trails of small macrofauna as well as meiofauna that fluff-up the interface (Fig. 9). Although macrofaunal and meiofaunal bioturbation can erase or otherwise reduce the topographic relief of tracks and trails (Cullen, 1973; MacIlvaine and Ross, 1979; Boyer, 1980), meiofaunal reworking produces a smaller-scale boundary roughness on the order of 1 mm or less (Grant et al., 1982; see comment in Nowell et al., 1981). In our flume experiments, we have found that ostracods, copepods, and foraminifera moving horizontally just below the sediment surface are effective in producing a characteristic "hummocky" relief on the surface. Grant et al. (1982) show that meiofaunal reworking of the bed may lower the critical shear velocity to near the predicted Shields values.

4.2. Pelletal Textures

Much of the potential food for detritus-feeding invertebrates is associated with sediments with high proportions of particles within the silt–clay size range (Sanders, 1958; Sanders et al., 1962; Purdy, 1964;

Whitlatch, 1976). The high surface-area-to-volume ratios of small particles provide a large expanse for the attachment and growth of microbial populations that produce mucopolysaccharide exudates (Zobell, 1943; D. L. Lynch and Cotnoir, 1956; Hobbie and Lee, 1980). The size fraction of particles selected in the feeding process varies among species. Many deposit feeders preferentially select particles larger than 10 μm (Whitlatch, 1976) and below 1 mm (Rhoads and Stanley, 1965). In addition to organic–mineral aggregates, relatively "clean" mineral particles may be ingested and stripped of adsorbed or absorbed organic films (Sulanowski, 1978; Pinck et al., 1954). Particle selection may also be based on grain density and surface texture (Self and Jumars, 1978; Taghon et al., 1978).

In addition to the size specificity of food particles, much of the potential food for detritus-feeding invertebrates is located within the upper few centimeters of the seafloor. In the intertidal zone, usable organic material may be present to a depth of 20 cm (Johnson, 1977; Sulanowski, 1978) and within the upper 2 cm of subtidal coastal sediments (Johnson, 1974, 1977). The above generalizations are based on detailed analyses of vertical profiles of particle morphology, organic content, and total protein and carbohydrate in marine muds and sands (Johnson, 1974, 1977; Whitlatch, 1977, 1980). Detritus feeding is therefore concentrated within, but not limited to, these depths.

With few exceptions, marine benthic invertebrates produce fecal pellets* and these pellets or castings are deposited at, or near, the sediment surface. This generalization also holds true for head-down conveyor belt feeders, which may ingest particles at depths of 20 cm below the surface. Ingested material is usually transported upward through the digestive tract and egested at the sediment surface. In equilibrium communities, these pellets may be subsequently advected downward by bioturbation and distributed throughout the bioturbation zone (Rhoads, 1967). Pioneering species are less efficient in vertical mixing; therefore, pellets remain near the sediment surface. Pelletization may result in a change in the grain size of surface sediments. In muds, this process commonly involves the addition of a very fine pelletal sand fraction to an unpelletized silt–clay matrix (Rhoads and Young, 1970). Because many detritivores preferentially select for particles within the silt–clay fraction, sand-sized fecal pellets are avoided until they break down and the constituent particles are repopulated by algae and/or microbes (Frankenberg and Smith, 1967; Fenchel, 1970; Hargraves, 1976; Levinton and Lopez, 1977; Levinton, 1980). Most fecal pellets are produced by the resident benthos,

---

* Pseudofeces (particles manipulated but not ingested by an organism) may also be ejected or deposited at the sediment surface. Pseudofeces are usually less consolidated than fecal pellets and castings.

but H. B. Moore (1931) describes pelletal muds from the Clyde Sea that are composed predominantly of zooplankton feces.

Jumars et al. (1981) have modeled the fates of sediments that contain deposit feeders. Their Markov model involves the probabilities of particle selection, pellet disintegration, and burial, and the probability of transport of pellets versus "free" unpelletized sediment. The authors show how the steady-state abundance of a given particle type comprising a pellet is a function of pellet durability and particle (feeding) selectivity. High densities of a given particle type in surface sediments result from packaging of particles into durable fecal pellets.

The distribution of pellets in the sediment column depends on their chance of burial and preservation. Burial may be brought about by a net input of "new" sediment or by downward advection by particle bioturbation.

Particle residence times are also affected by the feeding types present (Jumars et al., 1981): surface deposit feeders only, subsurface deposit feeders only, or both of these feeding types occurring together. Particle residence times are also sensitive to the relative erodibilities of pellets compared to unpelletized sediment.

The production of very fine sand-size pellets from silt–clay-size particles would, a priori, appear to favor selective transport of pellets relative to unpelletized sediment. The threshold shear velocity for entrainment of fine sand is lower than that for cohesive silt and clay. Flume observations in natural sediment containing organisms, however, show that pellets do not always have lower critical shear velocities than silt or clay, and may have higher critical shear velocities (Fig. 8) (Nowell et al., 1981). In nature, it is probably a rare phenomenon to have single, discrete particles of fine sand or smaller grain sizes move independently of other particles. Most particles below fine sand size are associated with organic–mineral aggregates (Rhoads, 1973; Johnson, 1974) held together with microbial mucus (Rhoads, 1973).

Existing competency curves are of limited use in predicting the stability of natural fine-grained sediments because their use is restricted to the particular apparatus or experimental location chosen. Moreover, hydraulic engineers avoid biological growth in their flume channels by adding growth inhibitors to the water. It is not surprising then, to expect major departures in threshold velocities of natural sediments from those predicted by "azoic" competency diagrams (Boyer, 1980; Nowell et al., 1981). However, biologically reworked intertidal sediments fall well within the values predicted from the Shields curve (Shields, 1936; Madsen and Grant, 1976; Grant et al., 1982).

Nowell et al. (1981) have conducted flume studies on the threshold velocities of polychaete and bivalve pellets and castings in natural and artificial sediment. Individual pellets, as well as mounds of fecal castings, were held together and to the bed by "elastic" mucus. These pellets and castings were abraded in situ by greater than critical shear velocities. Mucus cemented individual pellets and fecal mounds to the bottom (Fig. 8B). Boyer's (1980) flume observations of intertidal sediments support these observations. The most transportable feces are those produced by species that eject their pellets several centimeters into the water above the bed. Pseudofeces may also be forcefully pumped into the water column (Rhoads, 1963).

The transport of pellets appears to be largely controlled by the presence of binding microbial, meiofaunal, or diatom mucus (excluding algae). Below the plant "photosynthetic compensation depth," microbial and meiofaunal exudates are probably the major binding agents. Microbial binding of medium, sand-sized, to coarse, silt-sized, spherical glass beads has been shown to increase critical entrainment velocities by as much as 60%[*] in the Rhoads flume (Fig. 8A). The onset of a mucus binding effect, starting from initially sterile glass bead surfaces, was observed to be as short as 3 days in laboratory flume experiments (Rhoads et al., 1978b).

In nature, bacterial colonization of surfaces is enhanced by the adsorption of glycoproteinaceous films onto surfaces (Baier, 1973; Neihof and Loeb, 1973). Organic films can be adsorbed onto sterile and clean surfaces immersed in seawater within a period of 10 min (Goupil et al., 1973). Clay minerals bear net negative surface charges, promoting the rapid adsorption of mucopolysaccharide films (Khailov and Finenko, 1970; Neihof and Loeb, 1972). Microbial recolonization of newly produced pellets may take place within a few hours in the summer (J. Y. Yingst, personal communication). In the absence of intensive particle bioturbation, iron hydroxide encrustations may also serve to bind surface sediments (Parthenaides and Paswell, 1970). Subtidal carbonates, not subjected to bioturbation, may lithify in situ by precipitation of cements. This condition has been proposed to explain the abundant flat pebble (carbonate) conglomerates observed in Cambrian strata (Sepkoski, 1982).

In our flume work with Long Island Sound muds, we have rarely seen individual pellets roll along the bed. As velocity is increased, organic–mineral aggregates[†] (including pellets) that project above the bed start to vibrate and rock in place.

---

[*] Alcian blue staining of the glass beads showed the presence of carbohydrate-rich mucus at very small point contacts between grains. These contact films were barely visible under 50× magnification.

[†] Many of these aggregates may have their origins from the water column by the formation of particulates from dissolved organic matter, adsorbed onto bubble surfaces (micelle formation), or onto other surfaces in the water column (Riley, 1963; Krank, 1973, 1975). When the aggregates reach the seafloor they are mixed into the bottom and become part of the detrital food chain.

We have called this the *critical excitation velocity* (Rhoads et al., 1978b). This same phenomenon has been described by Nowell et al. (1981) for fecal mounds of polychaetes and by Boyer (1980) for polychaete and *Hydrobia* sp. (gastropod) pellets. The excitation velocity can be well above the predicted critical flume entrainment velocity for detached aggregates. The elastic mucus that binds most particles to one another and to the bed only fails when the flow rate is increased well above the excitation velocity.* As pieces of organic–mineral aggregates break away from the bed, they move downstream as traction or suspended load (depending on the density of the detached aggregate and the flow velocity relative to aggregate diameter). These aggregates can be quite large (1–2 mm) and include pellets and pellet aggregates, as well as irregularly shaped detrital particles. With sustained flow, strings of mucus and mucus-bound particles are observed on the surface of the bed; one end of the strand is attached to the bed while the free end vibrates and "flails" at a high frequency in the turbulent vortices. These filaments represent coiled sheets of mucus and are probably the "fibrous binding material" observed in broken sediment surfaces in flume experiments described by MacIlvaine and Ross (1979), Nowell et al. (1981), and Boyer (1980).

It is important for future experimental flume work to measure quantitatively the binding mucus at the sediment–water interface. This is a formidable problem. Mucopolysaccharides are a complex and poorly defined group of macromolecules. Quantitative separation from sediments is difficult, as mucopolysaccharides are closely associated with proteins and inorganic salts (Hobbie and Lee, 1980).

Another problem involves characterizing seasonal changes in mucus production and degradation. Yingst and Rhoads (1980) summarized earlier work on soil microbes: Soil microorganisms are known to be temperature-controlled in their production of polysaccharides (Martins and Craggs, 1946; Harris et al., 1966). Microorganism assemblages in soils produce a range of chemically different binding agents (Aspiras et al., 1971), which may differ in their biodegradability as well as in the physical properties that they confer on soil aggregates. The effect of environmental variables such as temperature on the binding of sediment particles by microorganisms may involve the production of agents of different biodegradability (Martins and Craggs, 1946). Aspiras et al. (1971) found that maximum structural stability of microbially bound solid aggregates was reached sooner at high temperatures, but was maintained longer at lower temperatures, once maximum stability was reached.

* Using microbially bound glass beads, Ullman (1975) found that the mucus bonds failed after the flume velocity was increased approximately 77% above the excitation velocity. Boyer (1980) also found that mucus-coated, intertidal sands had a "tear-away" velocity approximately 70% higher than the excitation velocity using the same flume.

Mucopolysaccharides also appear to be chiefly produced in microaerophilic environments (Mitchell and Nevo, 1964). Therefore, the production of mucus may vary with redox potential (Eh) and the position of the RPD in sediments. Even if one could make quantitative extractions of mucopolysaccharides from sediments, we question the value of such a parameter for predicting the bed response to fluid shear stress. The critical shear velocity may not be simply a linear function of the absolute quantity of mucus present. Once grain-to-grain contacts are bridged by an elastic film, additional mucus (increasing film thickness) may have only a relatively minor effect on binding strength. To develop a complete understanding of fine-grained sediment transport we must begin to understand the binding of sediments by mucus (Richards and Parks, 1976). However, we foresee serious analytical and methodological problems in characterizing biological adhesion.

Some pellets may move as discrete particles. Risk and Moffat (1977) describe the transport of both pellets and pseudofeces of the intertidal tellinid bivalve *Macoma baltica* from muddy sediments of the Minas Basin, Nova Scotia. Feces are picked up on the flood tide and transported shoreward. These findings are especially interesting, as Boyer (1980) commonly found that intertidal pellets composed of fine sand and exposed during low tide are tightly bound to the bottom by microbial and algal mucus.

When pellets or pellet aggregates are entrained, they may move as bed load. This can result from high settling velocities relative to their critical shear velocities (Nowell et al., 1981). If the whole bed is suspended into the water column by wave surge, the presence of fecal pellets can increase the settling velocity of pellet-bound mud by two orders of magnitude (Haven and Morales-Alamo, 1966; McCall, 1979).

4.3. Geotechnical Mass Properties

The mass or bulk geotechnical properties of sediments, such as density, water content, Atterberg limits (Lambe and Whitman, 1969), and macroscopic shear strength, are all influenced to a substantial degree by benthic organisms (Chapman, 1949; McMaster, 1962, 1967; Webb, 1969; Rhoads, 1970; Silva and Hollister, 1973; Richards and Parks, 1976; Bokuniewicz et al., 1975; Myers, 1977a,b; Boyer, 1980).

The ease with which free-living infaunal burrowers move through, and feed upon, sediments is known to be a function of sediment water content and state of compaction (Chapman, 1949). The limiting force with which burrowing infaunal metazoans can displace sediment is determined either by the internal coelomic pressures that can be generated by muscular contraction of their flexible and extensible body wall (Chapman and Newell, 1947; Clark, 1964) or by hydrostatic pressures exerted by blood or water vascular systems (Nicol, 1967). To facilitate burrowing and feeding, some metazoa, especially bivalves, also liquify the sediment by injecting water anteriorly into the bottom. This causes an instantaneous local increase in pore water pressure, and the liquid limit of the sediment is temporarily exceeded. At this instant, the organism moves forward into the liquified zone. In sands, this fluidized sediment represents a transient state, as overburden pressure soon causes the sediment to collapse on itself. In cohesive silts and clays, the sediment may remain dilated long after a burrowing organism has passed through or otherwise processed the sediment. This activity is evident from the presence of abandoned burrows and feeding voids. The mass properties of muds, therefore, reflect the cumulative burrowing history to a greater degree than do rapidly consolidating sands. If burrowing organisms are relatively active, a few individuals per unit area of the bottom can have a major impact on remolding and dilating muds.* The effect of burrowers on a specific sediment is related to the rate of burrowing versus the rate of consolidation. In the sense that muds compact less rapidly than cohesionless sands, the relative effect of burrowing organisms on the bulk density of muds would be expected to be greater than on sands.

Burrowing in muds has the effect of increasing sediment water content (Harrison et al., 1964; Harrison and Wass, 1965; Rhoads, 1970, 1973; Rhoads and Young, 1970). Intensive particle bioturbation, characteristic of equilibrium communities, is associated with fine-grained sediment water contents that are greater than 60%, and commonly over 70% (Table II). Burrowing also decreases near-surface sediment compaction (Bokuniewicz et al., 1975) and decreases undrained shear strength, as shown in the geotechnical experiments described earlier (Section 3.2). Because many pioneering species are not efficient in mixing particles vertically, these bottom types may compact and have water contents of less than 60%. However, generalizations are difficult to make because the water content will also depend on the frequency and intensity of physical resuspension by currents. Such a "fluid" pioneering bottom has been described for the Ria Arosa, Spain, by Tenore et al. (1982). These mud bottoms are apparently resuspended quite frequently (more than once a month) and have a water content of over 60% in the upper 12 cm.

Rowe (1974) made in situ sediment shear strength measurements in a highly burrowed mud in Buzzards Bay, Massachusetts, and found average shear strengths to be 0.98 kPa (10 g·cm^{-2}). Located on this burrowed

---

* Burrowing may operate together with pelletization to cause sediment dilation. The presence of pellets can decrease packing through the production of interpellet void space.

Table II. Near-Surface Water Contents of Muds Bioturbated by Errant Infaunal Deposit Feeders

Location	Water content (percent)[a]	Depth interval (cm)	Reference
Buzzards Bay, Massachusetts, USA	>60	0–1	Rhoads and Young (1970)
Cape Cod Bay, Massachusetts, USA	>60	0–3	Rhoads and Young (1971)
Long Island Sound, Connecticut, USA	>70	0–1	Rhoads et al. (1977)
Ria Muros, Galicia, Spain	~70	0–2	Tenore et al. (1982)
Clyde Sea, Scotland	>80	0–6	H. B. Moore (1931)

[a] Weight of water expressed as a percentage of the initial weight of solids plus water.

bottom are sedentary anemones (*Ceriantheopsis americanus*). Shear strengths of the sediment within 20 cm of the anemones increase to 1.83 kPa (18.7 g·cm^{-2}). The mechanisms by which *C. americanus* and other macrofauna increase sediment shear strengths, or otherwise promote or permit compaction, are poorly understood. In some cases, the cause and effect may be obvious, i.e., walled tube structures within the bottom obviously increase the breakaway torque of a shear vane apparatus or impede the vertical penetration of the bottom by a penetrometer or fall cone. Moreover, large tube dwellers may laterally compact sediments several millimeters away from the immediate tube (e.g., Aller and Yingst, 1978). In the case of *C. americanus*, the mucus shed from the oral crown may have spread over the sediment surface and caused particle-to-particle adhesion.

Other stabilization mechanisms may be less obvious, such as the interactions between macrofauna, meiofauna, and microorganisms. For example, pore water irrigation is associated with enhanced production of microbial populations and their mucopolysaccharide exudates (Webb, 1969). These viscous and elastic-binding mucus secretions, generated by bacteria as well as by macro- and meiofauna, may fill intergranular pore spaces (Frankel and Meade, 1973). Because many geotechnical properties are related to water content, the inferred range of influence of benthic organisms on the rheological properties of sediments is shown in Fig. 10.

In order to evaluate how biological changes in mass properties might influence sediment transport, we have examined much of the literature on the erosion of cohesive sediments. This literature does not present a consensus on the relationships between measured geotechnical properties and critical threshold velocities (e.g., Partheniades, 1965; Einstein and Krone, 1962; Gularte et al., 1980). Richards and Parks (1976) state that

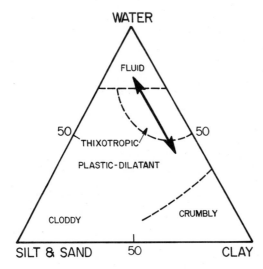

Figure 10. Mass geotechnical properties of sediments related to grain size and water content. In clayey or silt-rich sediments, the properties of the sediment can be significantly affected by the successional stage populating the sediment. The inferred range of effects is shown by the large double arrow. Pioneering species may facilitate the compaction of muds (plastic–dilatant field), but this depends on the frequency and intensity of bottom resuspension by currents. High-order successional species tend to enhance the water content of the bottom through intensive particle bioturbation (thixotropic–fluid field). Diagram modified from Boswell (1961).

"at present there is not a satisfactory way to predict the critical (fluid) shear stress of a cohesive bed surface or erosion rates as a function of excess strain." In part, the problem is procedural. Intercomparison of data is difficult owing to the diversity of methods used to measure geotechnical properties, the range of properties measured, the variety of open and closed channel flumes used, the different criteria applied to identify initiation of motion and mass erosion, and the different manipulation histories and mineralogical compositions of test sediments.

Partheniades (1965), in his review of the literature on cohesive sediment erosion, concludes that shear strength and Atterberg limits are poor predictors of cohesive sediment erodibility. Because incipient bed erosion is related to how fluid shear stress is transferred to the sediment–water interface, one might expect that subsurface geotechnical properties might prove to be uncoupled from processes at the immediate sediment–water interface.*

It is clear that most geotechnical measurements are made on vertical and horizontal scales that are too coarse to address adequately the problem of a natural sediment–water interface. As we have seen in Figs. 2 and 3, and described in earlier sections, the biological and physical structure of the sediment surface must be measured and described on the scale of a millimeter or less.

Partheniades (1965) emphasizes the importance of physiochemical properties of the interface, particularly the configuration and strength of

---

* Subsurface mass properties might be more appropriately described for determining the extent of massive bed erosion.

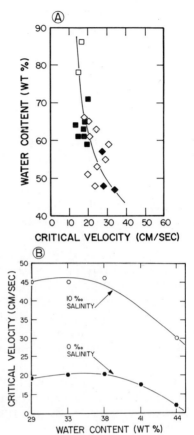

Figure 11. The relationship among sediment water content (weight of water expressed as a percentage initial weight of solids and water), salinity, and critical entrainment velocity. (A) Data from fine-grained shallow and deep-water muds. Velocity given for z = 100 cm. □, From Postma (1967); ◇, from Migniot (1968); ■, from Lonsdale and Southard (1974). Redrawn from Southard (1974). (B) Fine-grained sediment (grundite) erosion as a function of sediment water content and ambient water salinity. Velocity given for closed flume channel. Much of the flocculation between clay particles takes place at salinities ≤10‰. From Gularte (1978).

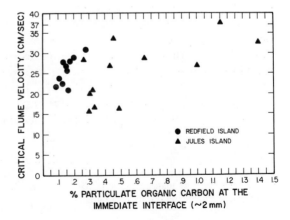

Figure 12. The critical velocity for bed motion in sandy intertidal sediments containing different concentrations of particulate organic carbon at the sediment surface. From Boyer (1980).

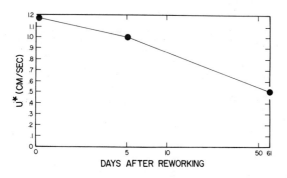

Figure 13. Change in critical velocity ($U^*_{\text{crit.}}$) of muddy sediments from Buzzards Bay, Massachusetts, subjected to bioturbation over a period of 61 days. Data plotted from Young and Southard (1978).

aggregates located at the sediment surface. Although his experiments were done with azoic* clay-rich sediments, his perspective is very relevant to the conditions present on a natural interface composed of mucus-bound organic–mineral aggregates and fecal pellets. Gularte et al. (1980) also suggest that physiochemical interparticle forces at the sediment surface are extremely important for determining the erosional response of azoic sediments composed of clay minerals.

Some workers have found reasonably good inverse correlations between erosion resistance and sediment water content (Postma, 1967; Southard et al., 1971; Lonsdale and Southard, 1974). Void space, salinity, and sediment shear strength have also been found to be important (Einsele et al., 1974; Gularte et al., 1980) (Fig. 11). Critical flume erosion velocity has been found to be correlated with the organic carbon content of intertidal sediments (Boyer, 1980) (Fig. 12). R. A. Young and Southard (1978) found that the threshold shear velocity for particle movement increased as the Leco carbon content of experimental sediments was increased over the range of 0.8–2.0 weight percent. [Organic matter content is correlated with water content ($r = 0.62$) in muddy Long Island Sound sediments (Yingst and Rhoads, 1978).]

We previously noted that intensive burrowing and particle bioturbation can produce surface water contents in muds of greater than 60% (Table II), and that bioturbation decreases sediment shear strengths. Particle bioturbation would than appear to lower critical shear velocities of muds. R. A. Young and Southard (1978) have experimentally shown that $U^*_{\text{crit.}}$ (critical bottom shear velocity) can be substantially decreased in sediments reworked by errant deposit feeders (Fig. 13). Grant et al. (1982)

---

* Insufficient data are given in most engineering studies to determine whether or not bacterial or plant activity was excluded from the sediments under examination. By *azoic* we mean that no micro- or macroorganisms were deliberately introduced into, or maintained in, the sediment during the experiment.

indicate that small (<1 mm) macrofaunal and meiofaunal invertebrates reworking intertidal fine sands may lower the critical shear velocity of previously stabilized sediments.

The flume measurements of Rhoads et al. (1978b) suggest a seasonal change in subtidal mud erodibility in central Long Island Sound over a 2½-year period (Fig. 14). The critical rolling velocity for organic–mineral aggregates ranged from 16 cm · sec^{-1} to 28 cm · sec^{-1} (velocities calculated for a depth off the bottom of 1 m). The seafloor was most easily eroded in spring and early summer, and most resistant to erosion in winter. These

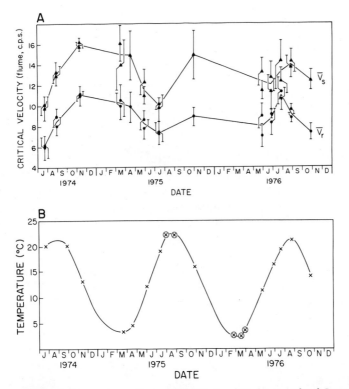

Figure 14. Seasonal changes in seafloor erodibility in central Long Island Sound over a 2½-year period. Measurements made from diver-obtained box cores that were inserted into a laboratory flume. (A) Seasonal change in mean critical rolling velocity and mean suspension velocity at station NWC, central Long Island Sound. Vertical bars represent one standard deviation. Curves fitted to the average of two replicate samples except for April and October, 1975, and October, 1976, when data points were determined from single box cores. Data points are significantly different at the 0.95 level if separated by ≥0.8 velocity units on the y axis. (B) Seasonal change in bottom sediment temperature at station NWC. Circled data points are times when bottom temperatures were taken without sampling with flume cores. From Rhoads et al. (1978b).

flume observations are supported by seasonal field measurements of near-bottom turbidity in Long Island Sound (Rhoads and Boyer, 1983).

R. A. Young and Southard (1978) made in situ Seaflume (a flume deployed on the seafloor; R. A. Young, 1977) measurements in Buzzards Bay, Massachusetts, over the period August through February, 1977. The station selected for study appears to be similar to the Long Island Sound bottom studied by Rhoads et al. (1978b). Young and Southard found $U^*_{\text{crit.}}$ to vary by a factor of two at measurement sites located only a few meters apart. They attribute this variance to spatial patchiness in unmeasured physical or biological parameters rather than bulk geotechnical properties. No systematic change in $U^*_{\text{crit.}}$ was detected over the time span of the study. The area of Buzzards Bay where Young and Southard made their measurements was severely affected in 1969 by an oil spill, which killed off many bioturbating species. As of this writing, the equilibrium fauna has still not completely recovered. This fact, plus the relatively short period of measurement, may explain the spatial patchiness in the results of the Buzzards Bay study.

5. A Qualitative Predictive Model

There have been two earlier attempts to place organism–sediment relationships into a time-dependent successional framework (Webb et al., 1976; Rhoads et al., 1977). We have updated these earlier models in Table III.

The underlying assumptions in Table III are that we are relating sedimentary properties to end-member series of succession on a subtidal muddy seafloor. A late pioneering stage may be associated with physical properties of the seafloor that are between the extreme end-member cases outlined in the table.* These transition states are poorly studied. A second assumption is that the described biological processes are operating at high rates. The effects of seasonality are not considered. In environments with a large temperature range, rates of biological activities may change by a factor of two or more with each 10°C change in temperature. This means that the rate of particle bioturbation by equilibrium species will be dramatically reduced in winter. The physical properties of sediments in late winter may therefore approach "azoic" values. Fluid bioturbation or ir-

---

* The time involved in a succession such as the one outlined in Table III and in Fig. 1 depends on many factors, and is poorly known. Our best estimate is that a primary succession, starting at a macrofaunal free state, will take tens of years to develop into a persistent equilibrium community. This estimate assumes that the system remains undisturbed in the course of its recovery.

rigation will similarly be decreased in winter. The temperature effect on benthic metabolism is important for explaining seasonal changes in pore water chemistry, and rates at which nutrients are pumped into, and out of, the bottom (Rhoads et al., 1977; Aller, 1978, 1980; Nixon et al., 1980; Propp et al., 1980; Zeitschel, 1980). If nutrient fluxes are coupled to benthic microbial and mucopolysaccharide production, one might expect a strong seasonal control on sediment cohesion. These interacting factors may be the main reasons for the observed seasonal changes in critical flume entrainment velocities described for Long Island Sound sediments (Rhoads et al., 1978b).

Seasonal changes in pelagic primary production and sedimentation of these products to the seafloor affect metabolic rates of benthos (Hargrave, 1980). This pulse of organic matter could temporarily increase U^*_{crit} (R. A. Young and Southard, 1978). However, U^*_{crit} may decrease if the organic matter is in the form of large aggregates (Riley, 1963), which are not firmly bound to the bottom. Seasonal changes in benthic primary productivity may be important in determining temporal patterns in the critical shear velocity of intertidal sediments (Grant et al., 1982).

Recruitment patterns of the benthos can also affect rate-dependent processes, as seen in Table III. Recruitment is often correlated with changes in water temperature and primary production cycles. During recruitment, the density of organisms on the bottom increases by orders of magnitude over long-term, mean abundances. This is especially true of pioneering stages (Rhoads et al., 1978a). The recruited organisms are all of meiofaunal size (<0.5 mm) and occur in a thin zone near the sediment–water interface (Mare, 1942; Yingst, 1978). During the recruitment period there is intensive meiofaunal processing of the sediment surface (Cullen, 1973). Competition between permanent and temporary meiofauna results in high mortality, and abundances of total meiofauna may drop dramatically over a short period of time (Bell and Coull, 1980). In this regard, the muddy seafloor in Long Island Sound is most easily eroded in the spring, a period of intensive benthic recruitment (Rhoads et al., 1978b).

We now return to the original intention of this paper: to propose hypotheses about the transport fates of fine-grained sediments and their contaminants. The organism–sediment–fluid interactions outlined in Table III predict that particulate and dissolved pollutants will experience different transport fates depending on the ecological conditions of the seafloor with which they came into contact. For example, a suspended particulate contaminant passing over a pioneering assemblage has a relatively high probability of being intercepted by the numerous suspension and surface deposit feeders present. Once the pollutant has arrived at the sediment surface, it also has a higher probability of being ingested and

Table III. Geotechnical and Transport Properties of Nearshore Marine Muds as Influenced by Faunal Successional Stage[a]

Physical sedimentary properties	Pioneering stages		Equilibrium stages	
	Observations	Interpretations	Observations	Interpretations
1. Shear strength	Increased in upper 0- to 2-cm interval relative to macrofauna-free muds (3.2).	May be related to pioneering species pumping water across the sediment surface, stimulating surface primary production and microbial growth at the sediment surface (mucopolysaccharide binding) (4.2–4.3). Slow rate of particle advection allows sediment to compact.	Shear strengths in the upper 0- to 2-cm interval are decreased relative to macrofauna-free muds (3.2).	Intensive particle bioturbation breaks up mucus-bound surface, producing a "granular" texture at the sediment surface (4.2). Sediment is maintained in a dilated state (3.2).
2. Water content (and density/void ratio correlates)	Depends on frequency and intensity of bottom resuspension by water turbulence.	Low water contents (<60%) related to low rate of particle advection. However, frequent bottom resuspension may keep sediment near the liquid limit (4.3).	Commonly >60% in the upper 0- to 2-cm interval of muds (4.3).	Related to intensive particle reworking and presence of intergranular or interpelletal void space as well as presence of open burrows (4.3).

Effects of Marine Benthos on Sediment Physical Properties

3. Biogenic texture	Surface pelletized, especially by polychaetes (4.2).	Most particle feeding is done from the water column or sediment surface. Pellets are deposited at the surface and are not advected downward (4.2).	Pellets found distributed throughout the actively bioturbated depth interval (4.2).	Most pellets are deposited near the sediment surface but many are advected downward by biogenic mixing (4.2).
4. Surface relief	Patches of densely packed tubes. Patches may be associated with topographic elevations composed of organic-rich fine-grained sediment (2.1–4.1).	Sediment accretion within dense patches (4.1).	Relief depends on species present and their modes of reworking. If species are laterally mobile, relief may be small. If the species are sedentary, the surface may be covered with feeding pits and excavation or fecal and pseudofecal mounds (4.1).	High water content and low shear strength of sediment do not permit topographic relief to develop on a large scale. Relief limited to activities and structures of individual sedentary organisms (2.2–4.1).
5. Sediment transport	Based on observations 1, 2, and 4 above, we predict an increase in $U^*_{crit.}$ at high population densities. At low densities, bottom may be destabilized relative to a macrofauna-free mud (4.1).	Related to mucopolysaccharide binding (4.2–4.3) and/or "skimming" flow (4.1).	Based on observations 1 and 2 above $U^*_{crit.}$ decreased relative to macrofauna-free mud (4.3).	Related to the presence of low-density organic mineral aggregates at the sediment surface. Seasonal changes in erodibility may be related to efficiency of mucopolysaccharide binding and rates of bioturbation (4.3).

[a] Numbers in parentheses refer to sections of the text in which observations and interpretations are discussed in more detail.

incorporated into fecal pellets. The pellets will be deposited at the surface and may be bound there by mucopolysaccharides or trapped in stable eddies between densely packed tubes. Dissolved pollutants can come into contact with anaerobic subsurface sediment by being pumped downward 1 or 2 cm through tubes and diffusing through tube walls into the ambient sediment. The residence time of pollutants in a pioneering stage depends on the frequency and intensity of major physical disturbances, which are associated with massive mortality and washout of both sediments and fauna. Solid-phase pollutants transported in suspension over an equilibrium community are less likely to be actively filtered or otherwise intercepted by feeding benthos because most of the infauna feed below the sediment surface. However, once a particle settles to the bottom, there is a higher probability that the contaminant will be advected downward, ingested, and incorporated into the deposit in the form of an aggregate. It is also likely that some of the particulate pollutants will reappear at the interface and be resuspended and dispersed by bottom currents with low mean critical shear velocities (e.g., $U^*_{\text{crit.}} = 2.0$ for 42-μm particles in Long Island Sound). The resuspended particles may scavenge other dissolved pollutants from the water column (e.g., Aller, 1978; Aller et al., 1980). Dissolved pollutants can be actively drawn into the bottom to depths of 10–20 cm by the respiratory pumping of burrowing or tube-dwelling species. Muds occupied by equilibrium species are chemically reactive owing to intensive particle and fluid bioturbation, and the potential for particle resuspension by weak bottom currents is increased (Aller, 1980). Moreover, the mean residence times of pollutants in sediments populated by equilibrium communities are expected to be longer than in areas of the seafloor dominated by pioneering assemblages.

6. Recommendations for Future Work

In order to test further how geotechnical properties of sediments are related to benthic successional history, it is necessary to make in situ observations of organism–sediment relations. Conventional means of bottom sampling disturb or destroy important structures near the sediment–water interface.* Free-fall samplers are preceeded by a pressure wave, which washes away small tubes, pelletal layers, and fecal mounds. Existing methods of faunal analysis, such as grab sampling and sieving, are similarly inadequate because important information about the vertical

---

* The only ship-deployed sampler that we have seen that recovers a nearly undisturbed sediment–water interface sample is a spade box core that has been fitted with internal vertical partitions. These partitions dampen out water motion over the bottom as the sample is recovered (e.g., Hessler and Jumars, 1974).

stratification of infauna is lost (Rhoads et al., 1981). Similarly, information about the dispersion patterns of tubes and feeding structures, surface pellet layers, the depth of the RPD, and the depth of subsurface feeding voids is also lost.

The Rhoads–Cande Profile Camera was developed specifically for the purpose of observing organism–sediment relations in situ (Rhoads and Cande, 1971). The ability of this photographic technique to identify successional stages accurately has been documented by Bosworth et al. (1980) and Rhoads and Germano (1982). The profile camera technique has been used successfully for over 10 years. We have also found sediment radiography (e.g., Figs. 2A and 3A) to provide useful information about sedimentary processes related to the successional paradigm (Tenore et al., 1982).

A newly developed and promising remote sensing technique is ultrasound imaging (Orr and Rhoads, 1982). Our initial laboratory work has been done with a 1.6-MHz backscattering system. Small-scale (approximately 1-cm) structures can be imaged acoustically at the surface of sediments; more importantly, subsurface structures can be seen to a depth of 10 cm. The ultrasound technique can be adapted to a towed sled vehicle pulled over the seafloor. Such a system might allow one to map rapidly the distribution of successional stages by providing acoustic information on sedimentary and biological structures at and below the sediment–water interface. At the Woods Hole Oceanographic Institution, A. J. Williams and W. D. Grant (personal communication) have constructed a 1-MHz acoustic profiler capable of measuring bed form topography to ≥1 mm.

High-frequency sound may have potential for remotely measuring and detecting changes in mass geotechnical properties (Richardson and Young, 1981). This technique requires inserting an acoustic probe vertically into the surface of the sediment. Sound is propagated horizontally and the difference in travel times along two different path lengths is measured. Vertical profiles in acoustic properties of the sediments can be obtained by sequentially moving the probe downward. This study has shown that bioturbated muds have lower impedance, a lower Rayleigh reflection coefficient, lower bulk and shear moduli, and a lower shear wave velocity propagation than nonbiotubated muds (see Richardson and Young, 1981).

Future work must involve measurement of geotechnical properties on smaller scales than has been done in the past. For problems relating to sediment transport, these measurements may have to be made within 1 or 2 mm of the sediment–water interface. We are not aware of any existing geotechnical instruments that can make measurements on this scale.

In situ measurement of critical shear stresses and shear velocity at the ocean floor is also desirable. In this regard, the Seaflume concept of R. A. Young (1977) is probably the most appropriate means for determining the in situ critical shear velocity of natural sediments populated by organisms. In order to better understand complex organism–sediment–fluid interactions, more laboratory flume work is required. Recent advances in forward and backscattering laser–doppler velocimetry technology open up a new approach for accurately measuring water flow and suspended solids within a flume channel. Vertical velocity profiles may be made to within a fraction of a millimeter of the bed without interfering with flow or organisms (Grant et al., 1982). Once again, high-resolution profiling is necessary in organism–sediment work because many biological structures and their effects are of small scale.

The influence of tubes, pelletal layers, and mucopolysaccharide binders on sediment stability must be investigated under highly controlled laboratory conditions. Of particular importance is the characterization of mucus binders. This is both a biochemical and a mechanical problem. No one has yet measured the elasticity or breaking strength of these compounds, nor are their kinetics or the conditions of their breakdown completely known.

It is clear from our earlier discussion of successional processes and metabolic rate factors that the geotechnical and transport properties of sediments are temporally and spatially dynamic. A sampling program designed to address the problem of seafloor stability should take into account important biological events. In environments that experience a large annual range in temperature and solar radiation, sampling should be extensive in the spring (when rising temperatures and increasing day length stimulate recruitment and primary production). Sampling in late summer and early autumn will characterize biological processes working at their maximum rates. Early winter is associated with declining metabolic rates and sediment processing. Late winter sampling may closely approximate "azoic" sediment conditions, as benthic organisms, although present, are metabolically dormant.

Thermally uniform deep-water environments may also experience subdued seasonality. Cycles in surface water production result in the sedimentation of detritus to the seafloor in the form of zooplankton fecal pellets (Honjo, 1976). This pulse of detritus may be associated with an increase in benthic activity. In shallow-water tropical environments, periodic changes in nearshore salinity and water turbulence during monsoon periods can similarly cause major restructuring of the benthic community.

An areal sampling plan for benthic environs can be more equitably constructed if the physical disturbance history of the seafloor is known or can be estimated. One need not know this history precisely. Estimates may be made on the basis of a reconnaissance survey of sedimentary fabrics obtained in X-radiographs of box cores or with an interface camera survey (Rhoads et al., 1981; Rhoads and Germano, 1982). The geologist or sedimentologist need not be highly proficient in the identification of benthic species to make judgments about successional history. If the geologist can recognize the functional aspects of benthic communities as outlined in Fig. 1, this information is sufficient to recognize end-member disturbance regimes. Once this organism–sediment information is recorded on a hydrographic chart, the selection of stations for measuring sedimentary or biological properties is facilitated.

ACKNOWLEDGMENTS. This work was supported by NSF grant OCE 7826211 and EPA Cooperative Agreement 807331-01 to Donald C. Rhoads. J. Germano and D. Muschenheim procured cores of the seafloor while scuba diving. J. Germano identified the macrofauna in the microcosm experiments. M. Pimer of Professional Sea Services and the R/V Gale provided support at sea. D. Muschenheim and D. Kramer aided in the construction and fabrication of the experimental set-up and in the measurement program. A. Goodhue constructed the experimental tanks. C. Phelps of Draft-A-Line helped with the figures; W. Sacco provided the finished prints. B. Gorecki typed the original manuscript. Grain size information was provided by D. K. Young and M. Richardson of NORDA, Bay St. Louis, Mississippi. R. Gularte of Levingston Marine Corporation, Annapolis, Maryland, built and calibrated the shear vane device for our use. We also thank the many people who have reviewed the manuscript, especially J. Germano, P. Jumars, A. Nowell, J. Eckman, G. Taghon, W. Grant, and the editors, P. L. McCall and M. J. S. Tevesz.

References

Aller, R. C., 1977, The influence of macrobenthos on chemical diagenesis of marine sediments, Ph.D. dissertation, Department of Geology and Geophysics, Yale University, New Haven, Connecticut.

Aller, R. C., 1978, Experimental studies of changes produced by deposit feeders on pore water, sediment, and overlying water chemistry, Am. J. Sci. **217**:1185–1234.

Aller, R. C., 1980, Relationships of tube-dwelling benthos with sediment and overlying water chemistry, in: Marine Benthic Dynamics (K. R. Tenore and B. C. Coull, eds.), pp. 285–308, University of South Carolina Press, Columbia.

Aller, R. C., and Cochran, J. K., 1976, ^{234}Th/^{238}U disequilibrium in nearshore sediment: Particle reworking and diagenetic time scales, Earth Planet. Sci. Lett. **20**:37–50.

Aller, R. C., and Dodge, R. E., 1974, Animal–sediment relations in a tropical lagoon, Discovery Bay, Jamaica, J. Mar. Res. **32**:209–232.

Aller, R. C., and Yingst, J. Y., 1978, Biogeochemistry of tube-dwellings: A study of the sedentary polychaete Amphitrite ornata (Leidy), J. Mar. Res. **36**:201–254.

Aller, R. C., and Yingst, J. Y., 1980, Relationships between microbial distributions and the anaerobic decomposition of organic matter in surface sediments of Long Island Sound, U.S.A., *Mar. Biol.* **56**:29–42.

Aller, R. C., Benninger, L. K., and Cochran, J. K., 1980, Tracking particle associated processes in nearshore environments by use of ^{234}Th/^{238}U disequilibrium, *Earth. Planet. Sci. Lett.* **47**:161–175.

Aspiras, R. B., Allen, O. N., Harris, R. F., and Chesters, G., 1971, The role of micro-organisms in the stabilization of soil aggregates, *Soil Biol. Biochem.* **3**:347–353.

Atkinson, R. J. A., and Pullin, R. S. V., 1976, The red band-fish, *Cepola rubescens* L. at Lundy, *Rep. Lundy Fld. Soc.* **27**:1–6.

Baier, R. E., 1973, Influence of the initial surface condition of materials in bioadhesion, in: *Proceedings of the Third International Congress on Corrosion and Fouling* (R. F. Acker, B. F. Brown, J. R. DePalma, and W. D. Iverson, eds.), pp. 633–639, National Bureau of Standards, Washington, D.C.

Bailey-Brock, J. H., 1979, Sediment trapping by chaetopterid polychaetes on a Hawaiian fringing reef, *J. Mar. Res.* **37**:643–656.

Bell, S. S., and Coull, B. C., 1980, Experimental evidence for a model of juvenile macrofauna–meiofauna interactions, in: *Marine Benthic Dynamics* (K. R. Tenore and B. C. Coull, eds.), pp. 179–192, University of South Carolina Press, Columbia.

Berger, W. H., and Heath, G. R., 1968, Vertical mixing in pelagic sediments, *J. Mar. Res.* **26**:134–143.

Bokuniewicz, H. J., Gordon, R. B., and Rhoads, D. C., 1975, Mechanical properties of the sediment–water interface, *Mar. Geol.* **18**:263–278.

Boswell, P. G. H., 1961, *Muddy Sediments: Some Geotechnical Studies for Geologists, Engineers, and Soil Scientists*, Heffer, Cambridge.

Bosworth, W. S., Germano, J., Hartzband, D. J., McCusker, A. J., and Rhoads, D. C., 1980, Use of benthic sediment-profile photography in dredging impact analysis and monitoring, *Ninth World Dredging Conference*, 29–31 October, 1980, Vancouver, British Columbia, Canada.

Boyer, L. F., 1980, Production and preservation of surface traces in the intertidal zone, Ph.D. dissertation, Department of the Geophysical Sciences, The University of Chicago, Chicago, Illinois.

Cadee, G. C., 1979, Sediment reworking by the polychaete *Heteromastus filiformis* on a tidal flat in the Dutch Wadden Sea, *Neth. J. Sea Res.* **13**:441–456.

Carney, R. S., 1981, Bioturbation and biodeposition, in: *Principles of Benthic Marine Paleoecology* (A. J. Boucot, ed.), pp. 357–400, Academic Press, New York.

Chapman, G. A., 1949, The thixotropy and dilatancy of a marine soil, *J. Mar. Biol. Assoc. U.K.* **28**:123–140.

Chapman, G. A., and Newell, G. E., 1947, The role of the body-fluid in relation to movement in soft bodied invertebrates. i. The burrowing of *Arenicola*, *Proc. R. Soc. London Ser. B* **134**:431–455.

Clark, R. B., 1964, *Dynamics of Metazoan Evolution*, Clarendon Press, Oxford, 313 pp.

Clements, F. E., 1916, Plant succession: An analysis of the development of vegetation, Carnegie Institute, Washington, Publication 242, 512 pp.

Cool, D. O., 1971, Depressions in shallow marine sediments made by benthic fishes, *J. Sediment. Petrol.* **41**:577–578.

Crozier, W., 1918, The amount of bottom material ingested by holothurians (*Stichopus*), *J. Exp. Zool.* **26**:379–389.

Cullen, D. J., 1973, Bioturbation of superficial marine sediments by interstitial meiobenthos, *Nature* **242**:323–324.

Dapples, E. C., 1942, The effect of macro-organisms upon near shore marine sediments, *J. Sediment. Petrol.* **12**:118–126.

Dauer, D. M., and Simon, J. L., 1975, Repopulation of the polychaete fauna of an intertidal habitat following natural defaunation: Species equilibrium, *Oecologia (Berlin)* **22:**99–117.
Dayton, P. K., and Oliver, J. S., 1980, An evaluation of experimental analyses of population and community patterns in benthic marine environments, in: *Marine Benthic Dynamics* (K. R. Tenore and B. C. Coull, eds.), pp. 93–120, University of South Carolina Press, Columbia.
Eckman, J. E., 1979, Small-scale patterns and processes in a soft substratum, intertidal community, *J. Mar. Res.* **37:**437–457.
Eckman, J. E., Nowell, A. R. M., and Jumars, P. A., 1979, The influence of animal motility on sediment entrainment, *Eos* **60:**847 (abs.).
Eckman, J. E., Nowell, A. R. M., and Jumars, P. A., 1981, Sediment destabilization by animal tubes, *J. Mar. Res.* **39:**361–374.
Einsele, G., Overbeck, R., Schwarz, H. U., and Unsold, G., 1974, Mass physical properties, sliding and erodibility of experimentally deposited and differently consolidated clayey muds, *Sedimentology* **21:**339–372.
Einstein, H. R., and Krone, R. B., 1962, Experiments to determine modes of cohesive sediment transport in salt water, *J. Geophys. Res.* **67:**1451–1461.
Fager, E. W., 1964, Marine sediments; effects of a tube-building polychaete, *Science* **143:**356–359.
Featherstone, R. P., and Risk, M. J., 1977, Effects of tube-building polychaetes on intertidal sediments of the Minas Basin, Bay of Fundy, *J. Sediment. Petrol.* **47:**446–450.
Fenchel, T., 1970, Studies on the decomposition of organic detritus derived from the turtle grass *Thalassia testudinium*, *Limnol. Oceanogr.* **15:**14–20.
Frankel, L., and Meade, D. J., 1973, Mucilaginous matrix of some estuarine sands of Connecticut, *J. Sediment. Petrol.* **43:**1090–1095.
Frankenberg, D., and Smith, K. L., Jr., 1967, Coprophagy in marine animals, *Limnol. Oceanogr.* **12:**443–450.
Frey, R. W. (ed.), 1975, *The Study of Trace Fossils*, Springer-Verlag, New York.
Ginsburg, R. N., and Lowenstam, H. A., 1958, The influence of marine bottom communities on the depositional environment of sediments, *J. Geol.* **66:**310–318.
Goldhaber, M. B., Aller, R. C., Cochran, J. K., Rosenfeld, J. K., Martens, C. S., and Berner, R. A., 1977, Sulfate reduction, diffusion, and bioturbation in Long Island Sound sediments: Report of the FOAM group, *Am. J. Sci.* **277:**193–237.
Gordon, D. C., Jr., 1966, The effects of the deposit feeding polychaete *Pectinaria gouldii* on the intertidal sediments of Barnstable Harbor, *Limnol. Oceanogr.* **11:**327–332.
Goupil, D. W., DePalma, V. A., and Baier, R. E., 1973, Prospects for non-toxic fouling-resistant paints, in: *Proceedings of the 9th Marine Technology Society Conference*, pp. 445–458, Marine Technology Society, Washington, D.C.
Grant, W. D., Boyer, L. F., and Sanford, L. P., 1982, The effect of biological processes on the initiation of sediment motion in non-cohesive sediment, *J. Mar. Res.* **40:** (in press).
Gray, J. S., 1974, Animal–sediment relationships, *Oceanogr. Mar. Biol. Annu. Rev.* **12:**223–261.
Gularte, R. C., 1978, The erosion of cohesive marine sediments as a rate process, Ph.D. dissertation, University of Rhode Island, Kingston, Rhode Island.
Gularte, R. C., Kelley, W. E., and Nacci, V. A., 1980, Erosion of cohesive sediments as a rate process, *Ocean Eng.* **7:**539–551.
Hanor, J. S., and Marshall, N. F., 1971, Mixing of sediment by organisms, in: *Trace Fossils: A Field Guide to Selected Localities in Pennsylvanian, Permian, Cretaceous, and Tertiary Rocks of Texas and Related Papers* (B. F. Perkins, ed.), pp. 127–135, Louisiana State University Press, Miscellaneous Publication 71-1, Baton Rouge.
Hargrave, B. T., 1976, The central role of invertebrate faeces in sediment decomposition, in:

The role of Terrestrial and Aquatic Organisms in Decomposition Processes (J. M. Andersen and A. Macfadyen, eds.), pp. 301–321, Blackwell, Oxford.

Hargrave, B. T., 1980, Factors affecting the flux of organic matter to sediments in a marine bay, in: *Marine Benthic Dynamics* K. R. Tenore and B. C. Coull, eds.), pp. 243–264, University of South Carolina Press, Columbia.

Harris, R. F., Chesters, G., and Allen, O. N., 1966, Dynamics of soil aggregation, *Adv. Agron.* **18**:107–169.

Harrison, W., and Wass, M. L., 1965, Frequencies of infaunal invertebrates related to water content of Chesapeake Bay sediments, *Southeast. Geol.* **6**:177–187.

Harrison, W., Lynch, M. P., and Altschaefel, A. G., 1964, Sediments of lower Chesapeake Bay with emphasis on mass properties, *J. Sediment. Petrol.* **34**:727–755.

Haven, D. S., and Morales-Alamo, R., 1966, Aspects of biodeposition by oysters and other invertebrate filter feeders, *Limnol. Oceanogr.* **11**:487–498.

Haven, D. S., and Morales-Alamo, R., 1972, Biodeposition as a factor in sedimentation of fine suspended solids in estuaries, *Geol. Soc. Am. Mem.* **133**:121–130.

Hessler, R. R., and Jumars, P. A., 1974, Abyssal community analysis from replicate box cores in the central No. Pacific, *Deep Sea Res.* **21**:185–209.

Hobbie, J. E., and Lee, C., 1980, Microbial production of extracellular material: Importance in benthic ecology, in: *Marine Benthic Dynamics* (K. R. Tenore and B. C. Coull, eds.), pp. 341–346, University of South Carolina Press, Columbia.

Holland, A. F., Zingmark, R. G., and Dean, J. M., 1974, Quantitative evidence concerning the stabilization of sediments by marine benthic diatoms, *Mar. Biol.* **27**:191–196.

Honjo, S., 1976, Coccoliths: Production, transportation and sedimentation, *Mar. Micropaleontol.* **1**:65–79.

Johnson, R. G., 1971, Animal–sediment relations in shallow water benthic communities, *Mar. Geol.* **11**:93–104.

Johnson, R. G., 1972, Conceptual models of benthic marine communities, in: *Models in Paleobiology* (T. J. M. Schopf, ed.), pp. 149–159, Freeman and Cooper, San Francisco.

Johnson, R. G., 1974, Particulate matter at the sediment–water interface in coastal environments, *J. Mar. Res.* **33**:313–330.

Johnson, R. G., 1977, Vertical variation in particulate matter in the upper twenty centimenters of marine sediments, *J. Mar. Res.* **35**:273–282.

Jumars, P. A., and Hessler, R. R., 1976, Hadal community structure: Implications from the Aleutian Trench, *J. Mar. Res.* **34**:547–560.

Jumars, P. A., Nowell, A. R. M., and Self, R. L. F., 1981, A simple model of flow–sediment–organism interactions, *Mar. Geol.* **42**:155–172.

Khailov, K. M., and Finenko, Z. Z., 1970, Organic macromolecular compounds dissolved in sea-water and their inclusion into food chains, in: *Marine Food Chains* (J. H. Steele, ed.), pp. 6–18, University of California Press, Berkeley.

Krank, K., 1973, Flocculation of suspended sediment in the sea, *Nature* **246**:348–350.

Krank, K., 1975, Sediment deposition from flocculated suspensions, *Sedimentology* **22**:111–123.

Lambe, T. W., and Whitman, R. V., 1969, *Soil Mechanics*, John Wiley and Sons, New York, 553 pp.

Lee, H., II, and Swartz, R. C., 1980, Biological processes affecting the distribution of pollutants in marine bioturbation, in: *Contaminants and Sediments* (R. A. Baker, ed.), Volume 2, pp. 555–605, Science Publishers, Ann Arbor, Michigan.

Lettau, P., 1969, Note on aerodynamic roughness parameter estimation, *J. Appl. Meteorol.* **8**:828–832.

Levinton, J. S., 1980, Particle feeding by deposit-feeders: Models, data, and a prospectus, in: *Marine Benthic Dynamics* (K. R. Tenore and B. C. Coull, eds.), pp. 423–441, University of South Carolina Press, Columbia.

Levinton, J. S., and Lopez, G. R., 1977, A model of renewable resources and limitation of deposit-feeding benthic populations, *Oecologia (Berlin)* **31:**117–190.

Lonsdale, P., and Southard, J. B., 1974, Experimental erosion of North Pacific red clay, *Mar. Geol.* **17:**M51–M60.

Lynch, D. L., and Cotnoir, L. J., Jr., 1956, The influence of clay minerals on the breakdown of certain organic substrates, *Soil Sci. Soc. Am. Proc.* **20:**367–370.

Lynch, M., and Harrison, W., 1970, Sedimentation caused by a tube-building amphipod, *J. Sediment. Petrol.* **40:**434–435.

McCall, P. L., 1975, The influence of disturbance on community patterns and adaptive strategies of the infaunal benthos of central Long Island Sound, Ph.D. dissertation, Yale University, New Haven, Connecticut.

McCall, P. L., 1977, Community patterns and adaptive strategies of the infaunal benthos of Long Island Sound, *J. Mar. Res.* **35:**221–266.

McCall, P. L., 1978, Spatial–temporal distributions of Long Island Sound infauna: The role of bottom disturbance in a nearshore marine habitat, in: *Estuarine Interactions* (M. L. Wiley, ed.), pp. 191–219, Academic Press, New York.

McCall, P. L., 1979, The effects of deposit-feeding oligochaetes on particle size and settling velocity of Lake Eerie sediments, *J. Sediment. Petrol.* **49:**813–818.

McCall, P. L., and Fisher, J. B., 1980, Effects of tubificid oligochaetes on physical and chemical properties of Lake Erie sediments, in: *Aquatic Oligochaete Biology* (R. O. Brinkhurst and D. G. Cook, eds.), pp. 253–317, Plenum Press, New York.

MacIlvaine, J. C., and Ross, D. A., 1979, Sedimentary processes on the continental slope of New England, *J. Sediment. Petrol.* **49:**565–574.

McIntosh, R. P., 1980, The relationship between succession and the recovery process in marine sediments, in: *The Recovery Process in Damaged Ecosystems* (J. Cairns, Jr., ed.), pp. 11–62, Ann Arbor Scientific Publishers, Ann Arbor, Michigan.

McMaster, R. L., 1962, Seasonal variability in compactness in marine sediments: A laboratory study, *Geol. Soc. Am. Bull.* **73:**643–646.

McMaster, R. L., 1967, Compactness variability of estuarine sediments: An *in situ* study, in: *Estuaries* (G. Lauff, ed.), pp. 261–267, American Association for the Advancement of Science, Washington, D.C.

Madsen, O. S., and Grant, W. D., 1976, Sediment transport in the coastal environment, Ralph M. Parsons Laboratory for Water Resources and Hydrodynamics, Report 209, Massachusetts Institute of Technology, Cambridge, Massachusetts, 105 pp.

Mare, M. F., 1942, A study of the marine benthic community with special reference to the microorganisms, *J. Mar. Biol. Assoc. U. K.* **25:**517–554.

Marshall, N., and Lucas, K., 1970, Preliminary observations on the properties of bottom sediments with an without eelgrass, *Zostera marina*, *Proc. Natl. Shellfish Assoc.* **60:**107–111.

Martin, J. P., and Waksman, S. A., 1940, Influence of microorganisms on soil aggregation and erosion, *Soil Sci.* **50:**29–47.

Martins, J. P., and Craggs, B. A., 1946, Influence of temperature and moisture on the soil-aggregating effect of organic residues, *J. Am. Soc. Agron.* **38:**322–339.

Migniot, C., 1968, Etude des propriétés physiques de différents sediments très fins et de leur comportement sous des actions hydrodynamiques, *La Houille Blanche* **23:**591–620.

Miller, M. C., McCave, I. N., and Komar, P. D., 1977, Threshold of sediment motion under unidirectional currents, *Sedimentology* **24:**507–527.

Mills, E. L., 1967, The biology of an ampeliscid amphipod crustacean sibling species pair, *J. Fish. Res. Board Can.* **24:**305–355.

Mills, E. L., 1969, The community concept in marine zoology, with comments on continua and instability in some marine communities: A review, *J. Fish. Res. Board Can.* **26:**1415–1428.

Mitchell, R., and Nevo, Z., 1964, Effects of bacterial polysaccharide accumulation on infiltration of water through sand, *Appl. Microbiol.* **12**:219–223.

Moore, D. G., and Scruton, P. C., 1957, Minor internal structure of some recent unconsolidated sediments, *Bull. Am. Assoc. Petrol. Geol.* **41**:2723–2751.

Moore, H. B., 1931, The muds of the Clyde Sea area. III. Chemical and physical conditions; rate and nature of sedimentation; and fauna, *J. Mar. Biol. Assoc. U. K.* **17**:325–358.

Moore, H. B., 1939, Faecal pellets in relation to marine deposits, in: *Recent Marine Sediments* (P. Trask, ed.), pp. 516–523, American Association of Petroleum Geologists/Dover Press, New York.

Morris, H. M., 1955, A new concept of flow in rough conduits, *Trans. Am. Soc. Civ. Eng.* **120**:373–398.

Myers, A. C., 1977a, Sediment processing in a marine subtidal sandy bottom community. I. Physical aspects, *J. Mar. Res.* **35**:609–632.

Myers, A. C., 1977b, Sediment processing in a marine subtidal sandy bottom community. II. Biological consequences, *J. Mar. Res.* **35**:633–647.

Myers, A. C., 1979, Summer and winter burrows of a mantis shrimp, *Squilla empusa*, in Narragansett Bay, Rhode Island (U.S.A.), *Estuarine Coastal Mar. Sci.* **8**:87–98.

Neihof, R. A., and Loeb, G. I., 1972, The surface charge of particulate matter in seawater, *Limnol. Oceanorgr.* **17**:7–16.

Neihof, R. A., and Loeb, G. I., 1973, Molecular fouling surfaces in seawater, in: *Proceedings of the Third International Congress on Marine Corrosion and Fouling* (R. F. Aker, B. F. Brown, J. R. dePalma, and W. P. Iverson, eds.), pp. 710–718, National Bureau of Standards, Gaithersburg, Virginia.

Neuman, A. C., Gebelein, C. P., and Scoffin, T. P., 1970, The composition, structure, and erodibility of subtidal mats, Abaco, Bahamas, *J. Sediment. Petrol.* **40**:274–297.

Nicol, J. A. C., 1967, *The Biology of Marine Animals*, John Wiley and Sons, New York.

Nixon, S. W., Kelly, J. R., Furnas, B. N., Oviatt, C. A., and Hale, S. S., 1980, Phosphorus regeneration and the metabolism of coastal marine bottom communities, in: *Marine Benthic Dynamics* (K. R. Tenore and B. C. Coull eds.), pp. 219–242, University of South Carolina Press, Columbia.

Nowell, A. R. M., and Church, M., 1979, Turbulent flow in a depth-limited boundary layer, *J. Geophys. Res.* **84**:4816–4824.

Nowell, A. R. M., Jumars, P. A., and Eckman, J. E., 1981, Effects of biological activity on the entrainment of marine sediments, *Mar. Geol.* **42**:133–153.

Orr, M., and Rhoads, D. C., 1982, Acoustic imaging of structures and macrofauna in the upper 10 cm of sediments using a megahertz backscattering system, *Mar. Geol.* (in press).

Osman, R. W., and Whitlatch, R. B., 1978, Patterns of species diversity: Fact or artifact? *Paleobiology* **4**:41–54.

Partheniades, E., 1965, Erosion and deposition of cohesive soils, *Proc. Am. Soc. Civ. Eng., J. Hydraul. Div.* **91**:105–139.

Parthenaides, E., and Paswell, R. E., 1970, Erodibility of channels with cohesive boundary, *Proc. Am. Soc. Civ. Eng., J. Hydraul. Div.* **96**:755–771.

Pearson, T. H., and Rosenberg, R., 1978, Macrobenthic succession in relation to organic enrichment and pollution of the marine environment, *Oceanogr. Mar. Biol. Annu. Rev.* **16**:229–311.

Pearson, T. H., and Stanley, S. O., 1979, Comparative measurement of the redox potential of marine sediments as a rapid means of assessing the effect of organic pollution, *Mar. Biol.* **53**:371–379.

Pemberton, G. S., Risk, M. J., and Buckley, D. E., 1976, Supershrimp: Deep bioturbation in the strait of Canso, Nova Scotia, *Science* **192**:790–791.

Petersen, C. G. J., 1913, Valuation of the sea. II. The animal communities of the sea bottom and their importance for marine zoogeography, *Rep. Danish Biol. Stat.* **21**:1–44.

Phelps, D. K., 1966, Partitioning of the stable elements, Fe, Zn, Se, Sm, within a benthic community, Anasco Bay, Puerto Rico, in: *Radioecological Concentration Processes* (B. Aberg and F. P. Hungate, eds.), pp. 721–734, Pergamon Press, New York.

Pinck, L. A., Dyal, R. S., and Allison, F. E., 1954, Protein–montmorillonite complexes, their preparation and the effects of soil micro-organisms on their decomposition, *Soil Sci.* **78**:109–118.

Postma, H., 1967, Sediment transport environment, in: *Estuaries* (G. H. Lauff, ed.), pp. 158–179, American Association for the Advancement of Science, Washington, D.C.

Powell, E. N., 1977, Particle size selection and sediment reworking in a funnel feeder, *Leptosynapta tenuis* (Holothuroidea, Synoptidae), *Int. Rev. Ges. Hydrobiol.* **62**:385–403.

Propp, M. V., Tarasoff, V. G., Cherbadgi, I. I., and Lootnik, N. V., 1980, Benthic–pelagic oxygen and nutrient exchange in a coastal region of the Sea of Japan, in: *Marine Benthic Dynamics* (K. R. Tenore and B. C. Coull, eds.), pp. 265–284, University of South Carolina Press, Columbia.

Purdy, E., 1964, Sediments as substrates, in: *Approaches to Paleoecology* (J. Imbrie and N. Newell, eds.), pp. 238–271, John Wiley and Sons, New York.

Raupach, T., 1981, A wind tunnel study of turbulent flow close to regularly arrayed rough surfaces, *Boundary Layer Meteorol.* **18**:373–397.

Revsbech, N. P., Jorgensen, B. B., and Blackburn, T. H., 1979, Oxygen in the sea bottom measured with a microelectrode, *Science* **207**:1355–1356.

Rhoads, D. C., 1963, Rates of sediment reworking by *Yoldia limatula* in Buzzards Bay, Massachusetts and Long Island Sound, *J. Sediment. Petrol.* **33**:723–727.

Rhoads, D. C., 1967, Biogenic reworking of intertidal and subtidal sediments in Barnstable Harbor and Buzzards Bay, Massachusetts, *J. Geol.* **75**:461–476.

Rhoads, D. C., 1970, Mass properties, stability and ecology of marine muds related to burrowing activity, in: *Trace Fossils* (T. P. Crimes and J. C. Harper, eds.), pp. 391–406, Seel House Press, Liverpool.

Rhoads, D. C., 1973, The influence of deposit-feeding benthos on water turbidity and nutrient recycling, *Am. J. Sci.* **273**:1–22.

Rhoads, D. C., 1974, Organism–sediment relations on the muddy sea floor, *Oceanogr. Mar. Biol. Annu. Rev.* **12**:263–300.

Rhoads, D. C., and Boyer, L. F., 1983, Seasonal patterns in sediment resuspension in central Long Island Sound, U.S.A., (in preparation).

Rhoads, D. C., and Cande, S., 1971, Sediment profile camera for *in situ* study of organism–sediment relations, *Limnol. Oceanogr.* **16**:110–114.

Rhoads, D. C., and Germano, J., 1982, Characterization of benthic processes using sediment-profile imaging: An efficient method of Remote Ecological Monitoring of the seafloor (REMOTS system), *Mar. Ecol. Prog. Ser.* (in press).

Rhoads, D. C., and Stanley, D. J., 1964, Biogenic graded bedding, *J. Sediment. Petrol.* **35**:956–963.

Rhoads, D. C., and Young, D. K., 1970, The influence of deposit-feeding organisms on sediment stability and community trophic structure, *J. Mar. Res.* **28**:150–178.

Rhoads, D. C., and Young, D. K., 1971, Animal–sediment relations in Cape Cod Bay, Massachusetts. II. Reworking by *Molpadia oolitica* (Holothuroidea), *Mar. Biol.* **11**:255–261.

Rhoads, D. C., Aller, R. C., and Goldhaber, M., 1977, The influence of colonizing macrobenthos on physical properties and chemical diagenesis of the estuarine seafloor, in: *Ecology of Marine Benthos* (B. C. Coull, ed.), pp. 113–138, University of South Carolina Press, Columbia.

Rhoads, D. C., McCall, P. L., and Yingst, J. Y., 1978a, Disturbance and production on the estuarine seafloor, *Am. Sci.* **66**:577–586.

Rhoads, D. C., Yingst, J. Y., and Ullman, W., 1978b, Seafloor stability in central Long Island Sound. Part I. Temporal changes in erodibility of fine-grained sediment, in: *Estuarine Interactions* (M. L. Wiley, ed.), pp. 221–224, Academic Press, New York.

Rhoads, D. C., Germano, J., and Boyer, L. F., 1981, Sediment-profile imaging: An efficient method of Remote Ecological Monitoring of the Seafloor (REMOTS system), *Oceans* **1**:561–566, Publication No. 81CH1685-7, IEEE, Piscataway, New Jersey.

Richards, A. F., and Parks, J. M., 1976, Marine geotechnology: Average sediment properties, selected literature and review of consolidation, stability and bioturbation–geotechnical interactions in the benthic boundary layer, in: *The Benthic Boundary Layer* (I. N. McCave, ed.), pp. 157–181, Plenum Press, New York.

Richardson, M. D., and Young D. K., 1981, Geoacoustic models and bioturbation, *Mar. Geol.* **38**:205–218.

Riley, G. A., 1963, Organic aggregates in seawater and the dynamics of their formation and utilization, *Limnol. Oceanogr.* **8**:372–381.

Risk, M. J., and Moffat, J. S., 1977, Sedimentological significance of fecal pellets of *Macoma baltica* in the Minas Basin, Bay of Fundy, *J. Sediment. Petrol.* **47**:1425–1436.

Rowe, G. T., 1974, The effects of the benthic fauna on the physical properties of deep-sea sediments, in: *Deep-Sea Sediments: Physical and Mechanical Properties* (A. L. Inderbitzen, ed.), pp. 381–400, Plenum Press, New York.

Sanders, H. L., 1958, Benthic studies in Buzzards, Bay. I. Animal–sediment relationships, *Limnol. Oceanogr.* **3**:245–258.

Sanders, H. L., Goudsmit, E. L., Mills, E. L., and Hampson, G. E., 1962, A study of the intertidal fauna of Barnstable Harbor, Massachusetts, *Limnol. Oceanogr.* **7**:63–70.

Santos, S. L., and Bloom, S. A., 1980, Stability in an annually defaunated estuarine soft-bottom community, *Oecologia (Berlin)* **46**:290–294.

Santos, S. L., and Simon, J. L., 1980, Marine soft-bottom community establishment following annual defaunation: Larval or adult recruitment? *Mar. Ecol. Prog. Ser.* **2**:235–241.

Sayre, W. W., and Albertson, M. L., 1963, Roughness spacing in rigid open channels, *Trans. Am. Soc. Civ. Eng.* **128**:343–372.

Schafer, W., 1972, *Ecology and Paleoecology of Marine Environments* (I. Oertel and G. Y. Craig, translators), University of Chicago Press, Chicago, Illinois.

Schlichting, H., 1936, Experimentelle Untersuchungen zum Rauhigkeitsproblem, *Ing. Arch.* **7**:1–34.

Schwartz, A., 1932, Der tierische Einfluss auf die Meeressedimente, *Senckenbergiana* **14**:118–172.

Scoffin, T. P., 1970, The trapping and binding of subtidal carbonate sediments by marine vegetation in Bimini Lagoon, Bahamas, *J. Sediment. Petrol.* **40**:249–273.

Self, R. F. L., and Jumars, P. A., 1978, New resource axes for deposit feeders, *J. Mar. Res.* **36**:627–641.

Sepkoski, J. J., 1982, Flat-pebble conglomerates, storm deposits, and the Cambrian bottom fauna, in: *Cyclic and Event Stratification* (G. Einsele and A. Seilacher, eds.), Springer-Verlag, New York (in press).

Shields, A., 1936, Application of similarity principles and turbulent research to bed-load movement (*Mitteilungen der Preussicher Versuchsanstalt für Wasserbau und Schiffen*, Berlin), in: W. M. Keck Laboratory of Hydraulics and Water Resources, Report 167 (W. P. Ott and J. C. van Uchelen, translators), California Institute of Technology, Pasadena, California.

Silva, A. J., and Hollister, C. D., 1973, Geotechnical properties of ocean sediments recovered with giant piston corer. I. Gulf of Maine, *J. Geophys. Res.* **78**:3597–3616.

Southard, J. B., 1974, Erodibility of fine abyssal sediment, in: *Deep-Sea Sediments: Physical and Mechanical Properties* (A. L. Inderbitzen, ed.), pp. 367–379, Plenum Press, New York.
Southard, J. B., Young R. A., and Hollister, C. D., 1971, Experimental erosion of calcareous ooze, *J. Geophys. Res.* **76**:5903–5909.
Sulanowski, J. S. K., 1978, Field study of relationship between organic matter and sedimentary particles, Ph.D. dissertation, The University of Chicago, Chicago, Illinois.
Swartz, R. C., and Lee, H., III, 1980, Biological processes affecting the distribution of pollutants in marine sediments. Part I. Accumulation, trophic transfer, biodegradation and migration, in: *Contaminants and Sediments*, Volume 2 (R. A. Baker, ed.), pp. 533–553, Ann Arbor Science Publishers, Ann Arbor, Michigan.
Taghon, G. L., Self, R. F. L., and Jumars, P. A., 1978, Predicting particle selection by deposit feeders: A model and its implications, *Limnol. Oceanogr.* **23**:752–759.
Tenore, K. R., Boyer, L. F., Corral, J., Garcia-Fernandez, C., Gonzalez, N., Gurrian, E. G., Hanson, R. B., Iglesias, J., Krom, M., Lopez-Jamar, E., McClain, J., Pamatmat, M., Perez, A., Rhoads, D. C., Rodriguez, R. M., Santiago, G., Tietjen, J., Westrich, J., and Windom, H. L., 1982, Coastal upwelling in the Rias Bajas, N.W. Spain: Contrasting the benthic regimes of the Ria de Arosa and de Muros, *J. Mar. Res.* **40** (in press).
Thayer, C. W., 1979, Biological bulldozing and the evolution of marine benthic communities, *Science* **203**:458–461.
Thorson, G., 1957, Bottom communities, in: *Treatise on Marine Ecology and Paleoecology*, Volume I: Ecology (J. W. Hedgpeth, ed.), pp. 461–534, Geological Society of America Memoir 67, Geological Society of America, New York.
Ullman, W., 1975, Stabilization of the sediment—water interface by the presence of the extracellular products of microorganisms, Senior thesis, Department of Geology and Geophysics, Yale University, New Haven, Connecticut.
van Straaten, L. M. J. U., 1952, Biogene textures and the formation of shell beds in the Dutch Wadden Sea. I–II, *Koninkl. Nederl. Akad. Wet. Proc. Ser. B* **55**:500–516.
Vermeij, G. J., 1978, *Biogeography and Adaptation Patterns of Marine Life*, Harvard University Press, Cambridge, Massachusetts.
Wade, B., 1972, A description of a highly diverse soft-bottom community in Kingston Harbour, Jamaica, *Mar. Biol.* **13**:57–69.
Webb, J. E., 1969, Biologically significant properties of submerged marine sands, *Proc. R. Soc. London Ser. B* **174**:355–402.
Webb, J. E., Djorges, D. J., Gray, J. S., Hessler, R. R., van Andel, Tj. H., Werner, F., Wolff, T., Zijlstra, J. J., and Rhoads, D. C., 1976, Organism–sediment relationships (Working Group Reports—Group E), in: *The Benthic Boundary Layer* (I. N. McCave, ed.), pp. 273–295, Plenum Press, New York.
Whitlatch, R. B., 1976, Seasonal changes in the community structure of the macrobenthos inhabiting the intertidal sand and mud flats of Barnstable Harbor, Massachusetts, Ph.D. dissertation, The University of Chicago, Chicago, Illinois.
Whitlatch, R. B., 1977, Seasonal changes in the community structure of the macrobenthos inhabiting the intertidal sand and mud flats of Barnstable Harbor, Massachusetts, *Biol. Bull.* **152**:275–294.
Whitlatch, R. B., 1980, Patterns of resource utilization and coexistence in marine intertidal deposit-feeding communities, *J. Mar. Res.* **38**:743–765.
Wilson, W. H., 1979, Community structure and species diversity of the sedimentary reefs constructed by *Petaloproctus socialis* (polychaeta: Maldanidae), *J. Mar. Res.* **37**:623–641.
Wolff, W. J., Sandee, A. J. J., and DeWolf, L., 1977, The development of a benthic ecosystem, *Hydrobiologia* **52**:107–115.

Woodin, S. A., 1976, Adult–larval interactions in dense infaunal assemblages: Patterns of abundance, *J. Mar. Res.* **34**:24–41.

Woodin, S. A., 1978, Refuges, disturbance, and community structure: A marine soft-bottom example, *Ecology* **59**:274–284.

Wooding, R. A., 1973, Drag due to regular arrays of roughness element geometry, *Boundary Layer Meteorol.* **5**:285–308.

Yingst, J. Y., 1978, Patterns of micro- and meiofaunal abundance in marine sediments, measured with the adenosine triphosphate assay, *Mar. Biol.* **47**:41–54.

Yingst, J. Y., and Rhoads, D. C., 1978, Seafloor stability in central Long Island Sound. Part II. Biological interactions and their potential importance for seafloor erodibility, in: *Estuarine Interactions* (M. N. Wiley, ed.), pp. 245–260, Academic Press, New York.

Yingst, J. Y., and Rhoads, D. C., 1980, The role of bioturbation in the enhancement of microbial turnover rates in marine sediments, in: *Marine Benthic Dynamics* (K. R. Tenore and B. C. Coull, eds.), pp. 407–421, University of South Carolina Press, Columbia.

Young, D. K., 1971, Effects of infauna on the sediment and seston of a subtidal environment, *Vie Milieu (Supplement)* **22**:557–571.

Young, D. K., and Rhoads, D. C., 1971, Animal–sediment relations in Cape Cod Bay, Massachusetts. I. A transect study, *Mar. Biol.* **11**:242–254.

Young, R. A., 1977, Seaflume: A device for in-situ studies of threshold erosion velocity and erosional behavior of undisturbed marine muds, *Mar. Geol.* **23**:M11–M18.

Young, R. A., and Southard, J. B., 1978, Erosion of fine-grained marine sediments: Sea-floor and laboratory experiments, *Geol. Soc. Am. Bull.* **89**:663–672.

Zeitzschel, B., 1980, Sediment–water interactions in nutrient dynamics, in: *Marine Benthic Dynamics* (K. R. Tenore and B. C. Coull, eds.), pp. 195–218, University of South Carolina Press, Columbia.

Zobell, C. E., 1943, The effect of solid surfaces on bacterial activity, *J. Bacteriol.* **46**:38–59.

Chapter 2

The Effects of Macrobenthos on Chemical Properties of Marine Sediment and Overlying Water

ROBERT C. ALLER

1. Introduction	53
2. Diagenetic Reactions	54
3. Reactive Particle Redistribution	56
3.1. Homogeneous Reworking	56
3.2. Selective Reworking	67
4. Solute Transport	71
4.1. Apparent Diffusion	71
4.2. Biogenic Advection	75
4.3. Average Diffusion Geometry	77
5. Macrofaunal Influence on Sediment–Water Exchange Rates	83
6. Reaction Rates	87
7. Chemistry of the Burrow Habitat	89
8. Spatial and Temporal Patterns in Sediment Chemistry	92
9. Summary	93
Appendix: Solutions to Model Equations	94
References	96

1. Introduction

The composition of any environment or object is determined by a particular balance between material transport processes and chemical reactions within and around it. In the case of marine sedimentary deposits, the dominant agents of mass transport are often large bottom-dwelling animals that move particles and fluids during feeding, burrowing, tube construction, and irrigation. Such biogenic material transport has major direct

---

ROBERT C. ALLER • Department of the Geophysical Sciences, University of Chicago, Chicago, Illinois 60637.

and indirect effects on the composition of sediments and their overlying waters. In this chapter I review some of what is presently known about these effects, their implications for both chemical and biological properties of a deposit, and how they can be conceptualized in quantitative models.

2. Diagenetic Reactions

Many of the most important reactions taking place in sediments are associated with the decomposition of organic matter and other biogenic components such as $CaCO_3$ and SiO_2-nH_2O (Berner, 1976a). These reactions influence pH and oxidation–reduction potential and cause the depletion or buildup of characteristic reactants or products in both the fluid and solid phases of a deposit (Baas-Beckling et al., 1960; Thorstenson, 1970; Ben-Yaakov, 1973; Goldhaber and Kaplan, 1974). Because of

Table I. Idealized Decomposition Reactions

1. Aerobic respiration

$(CH_2O)_x(NH_3)_y(H_3PO_4)_z + (x + 2y)O_2 \rightarrow xCO_2 + (x + y)H_2O + yHNO_3 + zH_3PO_4$

2. Nitrate reduction

$5(CH_2O)_x(NH_3)_y(H_3PO_4)_z + 4xNO_3^- \rightarrow xCO_2 + 3xH_2O$
$\qquad + xHCO_3^- + 2xN_2 + 5yNH_3 + 5zH_3PO_4$

3. Manganese reduction

$(CH_2O)_x(NH_3)_y(H_3PO_4)_z + 2xMnO_2 + 3xCO_2 + xH_2O \rightarrow 4xHCO_3^-$
$\qquad + 2xMn^{2+} + yNH_3 + zH_3PO_4$

4. Iron reduction

$(CH_2O)_x(NH_3)_y(H_3PO_4)_z + 4xFe(OH)_3 + 7xCO_2 \rightarrow 8xHCO_3^-$
$\qquad + 3xH_2O + 4xFe^{2+} + yNH_3 + zH_3PO_4$

5. Sulfate reduction

$2(CH_2O)_x(NH_3)_y(H_3PO_4)_z + xSO_4^{2-} \rightarrow 2xHCO_3^- + xH_2S + yNH_3 + 2zH_3PO_4$

6. Methane production

$2(CH_2O)_x(NH_3)_y(H_3PO_4)_z \rightarrow xCO_2 + xCH_4 + 2yNH_3 + 2zH_3PO_4$

7. Fermentation (generalized)

$12(CH_2O)_x(NH_3)_y(H_3PO_4)_z \rightarrow xCH_3CH_2COOH + xCH_3COOH$
$\qquad + 2xCH_3CH_2OH + 3xCO_2 + xH_2 + 12yNH_3 + 12zH_3PO_4$

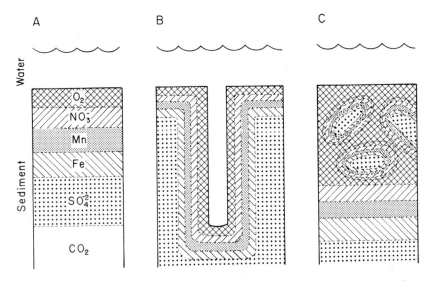

Figure 1. (A) Classically assumed vertical zonation of electron acceptor use in sediments. (B) Reaction zonation around irrigated burrow microenvironment. (C) Reaction geometries associated with fecal pellet microenvironments.

the central role of decomposition reactions in controlling the diagenesis of sediments as well as their importance in biogeochemical cycles, a subset of such reactions will be used here to illustrate the effects of macrofaunal activity on sediment chemistry.

Idealized representations of major organic matter decomposition pathways are listed in Table I (after Richards, 1965; Stumm and Morgan, 1970; Claypool and Kaplan, 1974; Froelich et al., 1979). Stoichiometries are variable but tend overall to $x \sim 106$, $y \sim 16$, and $z \sim 1$ (Redfield, 1934; Grill and Richards, 1964; Sholkovitz, 1973). The metabolic energy yields available to an organism for each mole of carbon utilized during decomposition depend in part on the electron acceptor involved and decrease in the order $O_2 > NO_3^- \gtrsim MnO_2 > FeOOH > SO_4^{2-} > CO_2$. This fact, together with the competitive exclusion principle of ecology, implies that reactions (1)–(6) should occur successively in association with changes in the types of bacteria that mediate them (Stumm and Morgan, 1970; Claypool and Kaplan, 1974). Observations have generally supported these inferences (Froelich et al., 1979). In contrast, fermentation reactions like reaction (7) are apparently temporally and spatially ubiquitous.

The concept of biogeochemical successions, together with the view that sedimentary deposits are laterally homogeneous and essentially one-dimensional bodies, has led to the dogma that decomposition reactions are vertically stratified below the sediment–water interface (Fig. 1A). This dogma has guided both sampling strategies and interpretations of sediment properties.

Macrobenthos influence this hypothetical reaction distribution and its maintenance in four major ways:

1. Material is translocated continuously between reaction zones during feeding, burrowing, and tube construction.
2. Burrow and fecal pellet formation alters reaction and solute diffusion geometries, creating a mosaic of biogeochemical microenvironments rather than a vertically stratified distribution (Figs. 1B,C).
3. New reactive organic substrates in the form of mucus secretions may be introduced into the deposit independent of sedimentation processes.
4. Feeding and mechanical disturbance may influence microbial populations that mediate reactions.

Macrobenthos can also directly alter sedimentary minerals during passage of material through their guts, but this class of effects is not considered here (Dapples, 1942; Pryor, 1975; L. S. Hammond, 1981).

The extent to which these effects are realized in any given case depends on the functional groups of animals present, their abundance, certain taxonomic peculiarities, and the size of individuals. Functional groups are defined on the basis of particular combinations of feeding type, life habit, and mobility, i.e., the specific ways in which animals interact with the seafloor (Brenchley, 1978). In the subsequent discussion emphasis will be placed on deposit-feeding benthos of varied life habits and mobilities.

3. Reactive Particle Redistribution

3.1. Homogeneous Reworking

Organic and other particles are constantly supplied to the sea floor, where they accumulate and undergo reactions such as those listed in Table I. Within a single type of sedimentary deposit and in the absence of disturbance by macrobenthos, spatial variation in the distribution of any physical or chemical property is predominantly in the vertical dimension and is determined by a balance between the rate of upward accretion of material and chemical reactions. For example, the concentration, $\hat{C}$, of a reactive organic matter substrate subject to microbial degradation is formally given by (after Berner, 1980):

$$\frac{\partial \hat{C}}{\partial t} = -\frac{\partial}{\partial x}(\omega \hat{C}) + R \qquad (1)$$

where $\hat{C}$ is the mass per volume of sediment; t is time; x is the vertical dimension, with its origin at the sediment–water interface, positive into the deposit; ω is the sedimentation or accretion rate; and R is the reaction rate. In practice, compaction is usually ignored and a steady-state distribution assumed, so that:

$$0 = -\omega \left(\frac{\partial \hat{C}}{\partial x}\right) + R \qquad (2)$$

This equation can be solved by defining an appropriate reaction rate. As a first approximation, organic matter decomposition is often assumed to follow first-order kinetics with respect to reactant concentration, so that:

$$R = -k\hat{C} \qquad (3)$$

where k is the first-order reaction rate constant. Models employing this approximation have been successful in many cases, particularly for describing SO_4^{2-} reduction. Complications are that k is an approximation to more complex Michaelis–Menton kinetics for multiple substrates, the value of k is temperature-dependent, and over large depth intervals k can vary significantly (Jørgenson, 1978; Berner, 1980).

The simplest influence macrobenthic activity has on this otherwise stratightforward set of processes is the nonselective or homogeneous mixing of sedimentary particles during feeding, burrowing, and construction activities. Particles are biologically transported within the deposit at a different rate and style than net accretion, ω. There is no strict physical analogue for such mixing even in the simple nonselective case.

On short time scales particles can move in streams around deposit-feeding individuals (Fig. 2A). If only one kind of animal of a given size is present, particle transport often occurs as advective loops between the sediment–water interface and the depth of feeding (Rhoads, 1974; Aller and Dodge, 1974; Amiard-Triquet, 1974; Fisher et al., 1980). Particles are also subject to local agitation and dispersion during burrowing or manipulation by appendages and mouth parts (Cullen, 1973; Fisher et al., 1980). Even with only one type of animal, natural variation in individual size either within a given population or with time following larval settlement results in various loop sizes (Whitlatch, 1974) (Fig. 2B). The

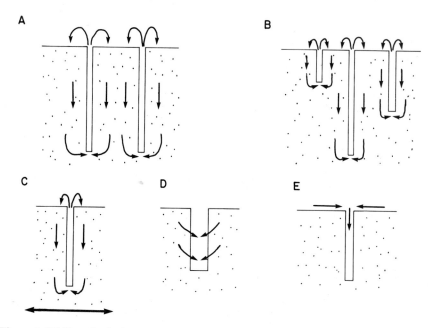

Figure 2. (A) Hypothetical transport paths of particles around individual subsurface deposit feeders of uniform size. (B) Multiple transport loop scales around organisms of varied size. (C) Particle transport loops are subject to lateral motion depending on mobility of deposit feeders. (D) Infilling of biogenic cavities results in particle transport independent of feeding activity. (E) Vertical dimension of feeding loops goes to zero in case of surface deposit feeders and in some cases may result in injection below the interface of surface-derived fecal material.

lateral mobility and patchiness of infauna in addition to instantaneous advective events such as burrow infilling further complicate particle motions (Hanor and Marshall, 1971; Benninger et al., 1979) (Figs. 2C,D).

In the face of such complexity, it is often assumed that when a sufficient number and variety of small-scale mixing events occur within a sediment interval they can be treated as random particle eddies and quantified in a mixing coefficient analogous to the physical eddy diffusion coefficient in a fluid (Goldberg and Koide, 1962; Guinasso and Schink, 1975). A major inconsistency in this analogy is that the path length of particle motion is often both specific in orientation and large in size scale with respect to chemical or physical property gradients in the sediment. For instance, reduced sulfide-rich material from depth may be placed directly at the oxidized sediment–water interface without mixing with material in between. Despite such conceptual inaccuracies, some insight into the consequences of reworking as compared to simple sedimentation can be gained from the analogy.

Ignoring lateral variation in sediment properties and vertical compaction, the distribution of a sediment component in the bioturbated zone is now described by:

$$\frac{\partial \hat{C}}{\partial t} = \frac{\partial}{\partial x}\left[D_B \left(\frac{\partial \hat{C}}{\partial x}\right)\right] - \omega \left(\frac{\partial \hat{C}}{\partial x}\right) + R \qquad (4)$$

where D_B is the particle mixing coefficient. All other variables are the same as for equation (1).

Because animal abundance and activity vary with depth, so does D_B (Mare, 1942; Myers, 1977; Guinasso and Schink, 1975). This variation can be accommodated either by assigning a continuous functional form to D_B, such as an exponential decrease with depth (Schink and Guinasso, 1977; Santschi et al., 1980; Olsen et al., 1981), or by assuming that D_B is constant over one or more empirically defined sediment intervals (Guinasso and Schink, 1975; Benninger et al., 1979; Nittrouer et al., 1979) (Fig. 3). Although this latter approach may appear completely unrealistic, reworking is sometimes dominated by one or two species feeding at well-defined horizons or burrowing within restricted intervals (Rhoads, 1974). The kinds of data available to evaluate values of D_B are often limited enough that in any case only an average coefficient acting over an interval can be determined.

Because both large and small "particle eddies" occur owing to variation in size and functional groups of animals, it could also be reasoned that D_B may actually increase with depth in much the way that eddy diffusion coefficients have length scale dependence in oceanic dispersion processes (Okubo, 1971). This possibility is not being considered here in part because of the imperfections in the diffusion analogy to bioturbation

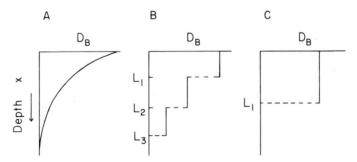

Figure 3. Commonly assumed depth dependences of the particle transport coefficient D_B (A) Exponentially decreasing magnitude with depth. (B) Multiple step-function or composite layer model. (C) Single mixed layer overlying uninhabited zone.

Table II. Isotopes Commonly Used in Evaluation of Particle Reworking Rates

Isotope	Half-life	Input type
^{234}Th	24.1 days	Continuous
^{7}Be	44 days	Continuous
^{228}Th	1.9 yr	Continuous
^{210}Pb	22 yr	Continuous
^{137}Cs	30.1 yr	Pulse
^{32}Si	275 yr	Continuous
^{14}C	5730 yr	Continuous
^{239,240}Pu	2.4×10^4 yr, 6540 yr	Pulse

and in part because experimental data to test such dependence are not available.

Values of D_B are estimated from the distribution within a deposit of particle-associated components having both well-defined inputs to the deposit and known reaction rates. These may be stable tracers such as colored particles, tektites, or foraminifera, all of which enter a deposit in pulses, or naturally occurring radioactive tracers that are supplied continuously during deposition and decay away. Man-made tracers such as the several isotopes of Pu and Cs are also used. The advantage of the naturally occurring radioactive tracers is that continual supply often allows assumption of a steady-state distribution ($\partial \hat{C}/\partial t = 0$) and easy evaluation of equations like (4).

The most commonly used isotopes in the evaluation of particle transport, biogenic or otherwise, are listed in Table II. Because of their varied half-lives, these isotopes are useful over a range of different depth and time scales depending on biological activity and net sedimentation rate. For example, consider the simple case of an inhabited and uniformly mixed zone of sediment overlying an uninhabited interval, as depicted in Fig. 3C. The steady-state vertical distribution of a continuously supplied isotope, i, irreversibly adsorbed to particles is, for zone 1 ($0 \leq x \leq L$):

$$\frac{\partial A_i}{\partial t} = 0 = D_B \left(\frac{\partial^2 A_i}{\partial x^2}\right) - \omega \left(\frac{\partial A_i}{\partial x}\right) - \lambda_i A_i \tag{5a}$$

and, for zone 2 ($x \geq L$):

$$\frac{\partial A_i}{\partial t} = 0 = -\omega \left(\frac{\partial A_i}{\partial x}\right) - \lambda_i A_i \tag{5b}$$

with the following boundary conditions:

1. $x = 0$, $J_0 = D_B (\partial A_i/\partial x) - w A_i$
2. $x = L$, $A_{i_{Zone\,1}} = A_{i_{Zone\,2}}$
3. $x = L$, $\partial A_i/\partial x = 0$
4. $x \to \infty$, $A_i = 0$

where A_i is the activity of isotope i; λ_i is the decay constant of isotope i; and J_0 is the constant flux of constituent A_i supplied to the sediment at $x = 0$. These boundary conditions require continuity of material fluxes between vertical zones and eventual decay of the tracer at depth. Neither compaction nor loss of volume during reaction is considered. The solutions to these equations are given in the Appendix for easy reference.

The kinds of vertical property distributions predicted by such assumptions are illustrated in Fig. 4 for ^{234}Th and ^{210}Pb. Cases in which D_B

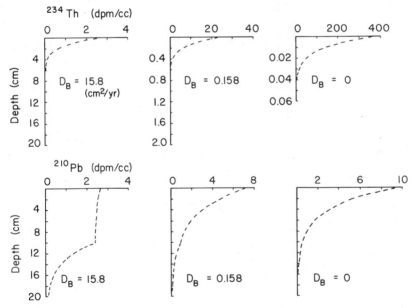

Figure 4. Predicted vertical profiles for ^{234}Th (upper row) and ^{210}Pb (lower row) assuming single layer mixed at rates $D_B = 15.8$ cm^2/yr (0.5×10^{-6} cm^2/sec); 0.158 cm^2/yr (0.5×10^{-8} cm^2/sec); and 0 cm^2/yr with a sedimentation rate $\omega = 0.1$ cm/yr. The flux of ^{234}Th and ^{210}Pb to the surface is fixed at 40 dpm/cm^2 per yr and 1 dpm/cm^2 per yr respectively. Note changes in activity and depth scales in the three cases. These nuclides have decay time scales in the same range as reactive organic matter.

Table III. Representative Particle Reworking Coefficients from Marine Environments

Location	Environment	Water depth (m)	Sediment type	Mixed layer interval (cm)	Sedimentation rate (cm/yr)	D_B (cm²/yr)	Method	Reference
Long Island Sound	Estuarine	8–35	Clay–silt–sand	0–5	<0.1–0.3	19 ± 16 (0.3–50)	^{234}Th	(1)
		13–15	Clay–silt	0–3	<0.1–0.3	3–9	^{7}Be	(2)
		15	Silt–clay	3–15	<0.1–0.3	0.6–0.9	^{210}Pb, ^{239,240}Pu	(3)
Narragansett Bay	Estuarine	10	Silt–clay	0–9	0.01	5–32	^{234}Th, ^{210}Pb, ^{239,240}Pu	(4)
				10–30	0.01	0.3–2.5	^{234}Th, ^{210}Pb, ^{239,240}Pu	(4)
Hudson River	Estuarine	—	Silt–clay	0–8	1–3	0.25–1	^{137}Cs, ^{239,240}Pu	(5)
New York Bight	Inner shelf	23	Silt–sand	0–5	—	16	^{234}Th	(6)
	Midshelf	69–87	Silt–clay	0–8	0.01–0.14	3.8–11	^{234}Th, ^{210}Pb, ^{239,240}Pu	(4)
				8–40	0.01–0.14	1.7–2	^{234}Th, ^{210}Pb, ^{239,240}Pu	(4)
	Slope	850–1800	Silt–clay	0–15	0.01–0.14	0.01–1.3	^{234}Th, ^{210}Pb, ^{239,240}Pu	(4)
Northwestern Pacific, Washington	Midshelf	20–200	Silt	0–10	~0.3	10 ± 12 (0.75–44)	^{210}Pb	(7)

Location	Setting	Depth	Sediment type	Depth range (cm)		Value	Tracer	Ref.
Southeastern Atlantic, Amazon	Outer shelf	100–200	Silt–sand	0–10	~0.2	5.5 ± 5.4 (0.28–19)	^{210}Pb	(7)
	Slope	200–2000	Clay–silt–sand	0–10	~0.1	2.7 ± 3.6 (0.25–9.8)	^{210}Pb	(7)
	Shelf	42	Silt–clay	0.5	—	30	^{234}Th	(8)
Mediterranean Sea	Pelagic	1000	—	0–12	—	0.38	^{239,240}Pu	(9)
North Atlantic	Pelagic	2705	Calcareous ooze	0–10	0.003	0.19	^{210}Pb	(10)
		4810	—	0–8	—	0.22	^{239,240}Pu	(9)
		1410	—	0–9	—	0.25	^{239,240}Pu	(9)
Southwestern Atlantic	Pelagic	4910	Calcareous ooze	—	—	0.063	^{210}Pb	(10)
		4920	—	0–4	—	0.14	Pu	(9)
		1345	—	0–6	—	0.10	Pu	(9)
Antarctic	Pelagic	4080–4730	Siliceous ooze	0–8	0.0004–0.001	0.03–0.25	^{210}Pb	(10)
Northern Pacific, equatorial	Pelagic	4640–5050	Red clay, siliceous ooze	0–7	0.00014–0.00031	0.25–0.44	^{210}Pb	(10)
Western Pacific, equatorial	Pelagic	1598	Calcareous ooze	0–8	0.001	0.12	^{210}Pb	(11)

[a] References: (1) Aller et al. (1980), (2) Krishnaswami et al. (1980), (3) Benninger et al. (1979), (4) Santschi et al. (1980), (5) Olsen et al. (1981), (6) Cochran and Aller (1979), (7) Carpenter et al. (1982), (8) DeMaster et al. (1980), (9) Guinasso and Schink (1975), (10) Turekian et al. (1978), (11) Peng et al. (1979).

is fixed at either 0.5×10^{-6} cm^2/sec (15.8 cm^2/yr), 0.5×10^{-8} cm^2/sec (0.158 cm^2/yr), or 0 are shown for comparison. The nonzero values of D_B, and the other values chosen ($\omega = 0.1$ cm/yr, $L = 10$ cm, $J_{Th} = 40$ dpm/cm^2 per yr, $J_{Pb} = 1$ dpm/cm^2 per yr), are of realistic magnitudes for estuarine and shelf sediments (Table III) (Aller et al., 1980; Benninger et al., 1979; Nittrouer et al., 1979; Carpenter et al., 1982). When L is greater than the depth at which A_i becomes small, as in the case of ^{234}Th, its exact value does not affect the tracer distribution.

These examples illustrate that, in the range of sedimentation rates usually encountered, biogenic reworking and the decay rate of an isotope largely determine both the penetration patterns of a reactive component in a deposit and the absolute quantity of a constituent decomposing at any given depth. Sedimentation rate may be ignored entirely in any given case if $\omega^2 \ll 4D_B\lambda_i$. In the present example reactant supply or flux at the sediment–water interface is held constant, so that only the vertical distribution of reaction rate and not the integrated quantity of decomposing material is affected. The concentration at the sediment–water interface is thus lowered by reworking while that at depth undergoes a relative increase.

In most sedimentary basins, physical processes such as resuspension and current transport act to buffer the interface concentration of a sediment component by lateral exchange and homogenization of surface material. As a result, spatial variability in biogenic reworking together with resuspension within a basin can cause a transfer of reactive particles from regions of relatively low biogenic reworking into regions of high reworking through a process of particle exchange during lateral and vertical mixing (Aller and Cochran, 1976). Exchange increases the total amount, or inventory, of a reactant in a deposit in areas of high biological activity. This effect is shown quantitatively in Fig. 5, in which the interface boundary condition (1) in model equations (5) has been changed to a fixed concentration, x, $= 0$, $A = A_0$, to simulate surface buffering by resuspension. The ^{234}Th inventory increases with increased values of D_B. In the real world, the actual effect of reworking on surface concentrations (Fig. 4), relative vertical distributions, and inventory (Fig. 5) will vary between the two extremes illustrated by the constant flux and constant surface concentration models (Figs. 4 and 5). For example, data from Long Island Sound, an estuarine basin along northeast North America, indicate a general increase in ^{234}Th inventories with D_B, although there is considerable variation owing to localized peculiarities in supply, physical transport, and biological reworking styles (Fig. 6A). ^{239,240}Pu inventories have also been shown to increase in areas of increased biological reworking, as reflected by depth of Pu penetration (Santschi et al., 1980) (Fig. 6B).

Because the decomposition of organic matter can be approximately

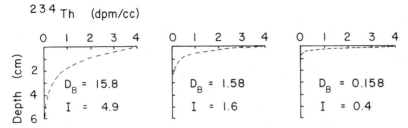

Figure 5. Predicted vertical distributions of ^{234}Th assuming a single mixed layer (Fig. 3C) and varied rates of mixing. In this case the surface activity is assumed to be buffered by resuspension and supply from surrounding areas at a value of 4 dpm/cc. As a result the inventory (I) varies with mixing rate from 4.9 dpm/cm^2 to 0.4 dpm/cm^2 when D_B varies from 15.8 to 0.158 cm^2/yr.

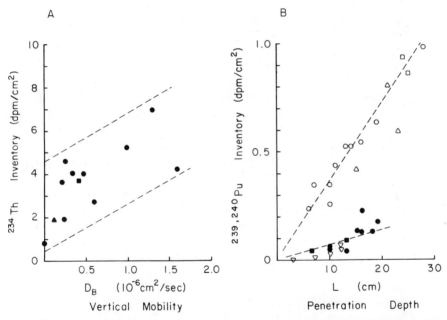

Figure 6. (A) ^{234}Th inventories in sediments from Long Island Sound as a function of D_B. Inventory increases with increases in mixing rate, illustrating capture of resuspended particles during reworking. After Aller et al. (1980). (B) Variation in ^{239,240}Pu inventories as a function of Pu penetration depth into sediment. Penetration depth varies with intensity of biogenic or physical reworking. Nearshore stations (depth <100 m): ○, Buzzards Bay, Gayhead (from Livingston and Bowen, 1979); △, Narragansett Bay; □, New York Bight. Offshore stations (depth >100 m): ●, Wilkinson Basin, Georges Bank (from Livingston and Bowen, 1979); ■, New York Bight; ▽, Atlantic deep sea (from Noshkin and Bowen, 1973). After Santschi et al. (1980).

described by first-order decay kinetics [equation (3)], the model distributions derived for particle-associated nuclides are directly applicable to the distribution of metabolizable organic matter or other reactive material in a deposit. The effective first-order decomposition rate constants, k, typical of organic components in surface sediments are in fact comparable to λ_i values for ^{234}Th and ^{210}Pb (e.g., Grill and Richards, 1964; Berner, 1981, Jørgenson, 1977a, 1978; Nixon et al., 1980). Disregarding the effects of microenvironments and solute transport, which will be discussed later, these considerations and previous qualitative observations by many investigators (Rhoads, 1974) indicate that a larger proportion of organic matter is decomposed at depth in a biogenically reworked deposit than in unreworked deposits of comparable material supply. It is likely that, as a result, a correspondingly larger proportion is remineralized by anaerobic rather than aerobic pathways. This can be seen qualitatively by simply comparing the downward flux, J_L, of reactive material out of the bioturbated zone at $x = L$ in the mixed and unmixed cases: $J_L = \omega A(L)$, where $A(L)$ mixed $> A(L)$ unmixed. Reduced products of anaerobic decomposition may not permanently build up in a deposit, however, because particles can be subject to reoxidation before final burial.

Organic matter subduction during reworking together with resuspension must result in an increased inventory of reactive organic matter in areas inhabited by deposit feeders (Fig. 7) through the same combination of physical and biological reworking processes outlined previously for ^{234}Th and Pu. Because deposit feeders utilize detritus and associated bacteria for food, the resulting increase in food supply may itself encourage increases in animal size and abundance, further stimulating localized capture of resuspended reactive material. Biogenically mixed patches should therefore act as biogeochemical hotspots in sedimentary basins for this reason alone, other factors being equal. This also raises the interesting possibility of competition for food between laterally separated patches of deposit feeders that might not otherwise interact. The intensity of such competition would be determined by both the nature of resuspension in the water column and the reworking styles of the benthic infauna.

Average values of D_B have been determined in a variety of nearshore, shelf, and deep-sea environments. Some of these values, the techniques used, and the depth interval over which they were determined are listed in Table III. Although there can be substantial variability within any region, or with depth in a deposit, in general there is a decrease in biological reworking from shallow ($D_B \sim 10^{-6}$cm^2/sec) to deep-sea ($D_B \sim 10^{-8}$cm^2/sec) environments. This is consistent with the decreased metabolic activity generally observed in deeper water (Smith, 1978; Zeitschel, 1980). High D_Bs calculated for slope canyon deposits probably reflect

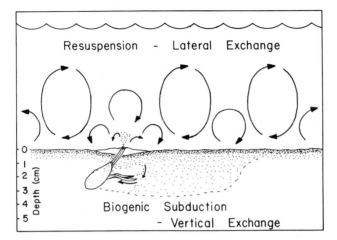

Figure 7. Sketch of how vertical exchange of particles within the sediment during biogenic reworking and lateral exchange of particles during resuspension can result in capture of a reactive component into biogenically reworked patches at the expense of surrounding areas. No net sedimentation takes place during exchange. After Aller and Cochran (1976).

physical rather than biological transport (Carpenter et al., 1982). In almost all cases, biogenic particle reworking is on the average slower than solute transport by molecular diffusion. Particle reworking per se is therefore unlikely to control solute transport directly in most deposits. Indirect effects are discussed later in this chapter.

3.2. Selective Reworking

Although the concept of homogeneous, random mixing as reflected in the mixing coefficient, D_B, is of great practical application and is useful for quantifying and illustrating several fundamental effects of reworking, it can be a misleading description of the processes taking place. During feeding and burrow construction, animals are capable of selecting particles on the basis of particle position in the sediment, size, shape, surface texture, and density (Dapples, 1942; Fager, 1964; Rhoads, 1974; Fenchel et al., 1975; Self and Jumars, 1978; Jumars et al., 1981; Khripounoff and Sibuet, 1980; Levinton, 1980). This selection may result in lateral and vertical segregation of different classes of particles depending on the particle sizes available, the functional kinds of animals present, and their age–size distributions.

Because many chemical properties of sediments correlate with grain size, any spatial segregation of particle types will result in a corresponding variability in chemical composition and reaction rates (Fig. 8). For ex-

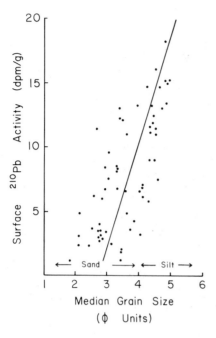

Figure 8. Many reactive metals and organic components associate preferentially with fine-grained sediments. In this case, ^{210}Pb activity in surface sediments is shown to increase as sediment particle size decreases in Washington Shelf sediments off northwest North America. After Nittrouer et al. (1979).

ample, in the case of aerobic reaction rates, oxygen consumption by bacteria on particles is directly proportional to particle surface area (Hargrave, 1972; Hargrave and Phillips, 1977). Thus, the oxygen flux to any surface is approximately constant (Fig. 9A). Because of differences in surface area, fine-grained sediments should be exponentially more reactive on a mass or volume basis than coarser-grained material when surface-dependent processes such as remineralization reactions are involved (Figure 9B). Reaction rate distribution can therefore be altered by selective feeding in ways not accounted for by a general nonselective mixing coefficient D_B.

In some cases where relatively fine material is preferentially ingested at depth and defecated at the sediment–water interface, a biogenic stratigraphy or graded bedding is formed (van Straaten, 1952; Rhoads and Stanley, 1965; Cadée, 1976; Baumfalk, 1979) (Fig. 10). As a result, reaction rates and properties associated with particle surface area are increased near the sediment–water interface. Because of proximity of solute release to overlying water, this should lower the buildup of metabolites in the sediment, as discussed later. It will also accentuate interaction with overlying water by resuspension of small partcles known to be important in adsorption and scavenging reactions in the water column (Aston and Chester, 1973; Turekian et al., 1980).

Other groups of deposit feeders may egest relatively coarse particles at the interface or defecate fine material preferentially at depth (Whitlatch, 1974; Powell, 1977). Small particles placed at the sediment surface by

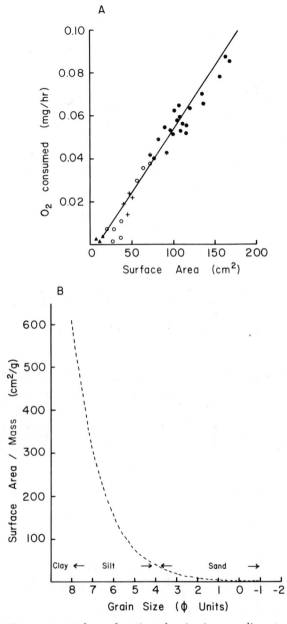

Figure 9. Reaction rates are often a function of grain size or sediment surface area. (A) O_2 consumption by aerobic bacteria on particles and other oxidation reactions as a function of surface area. The constant slope of the relation demonstrates that the oxygen flux to a surface is constant. After Hargrave and Phillips (1977). (B) The theoretical relation between surface area per mass of sediment and particle diameter or grain size. Particles are assumed to be simple spheres of density 2.5 g/cc. Actual clay particles have considerably greater surface areas per mass, ranging from ~10–30 m^2/g (Kaolinite) to 40–100 m^2/g (Smectite). Fine-grained material is therefore much more reactive than a comparable mass of coarse material with respect to surface-dependent processes such as organic matter decomposition.

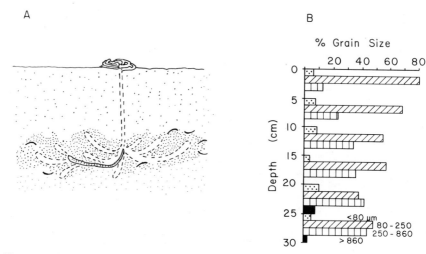

Figure 10. Segregation of particle sizes often occurs during feeding. (A) Subsurface feeding by the lugworm *Arenicola* and surface defecation of ingested material can result in pockets of coarse material at depth. After Rijken (1979). (B) Average vertical grain size distribution resulting from feeding by *Arenicola* in tidal flats of Dutch Wadden Sea. After Baumfalk (1979).

one species inhabiting an area can also be preferentially subducted by another or nonspecifically infill vacated burrows. For instance, selective subduction occurs in some areas of Barnstable Harbor, Massachusetts, inhabited by both the maldanid polychaete *Clymenella torquata* and the terebellid *Amphitrite ornata* (Aller and Yingst, 1978) (Fig. 11). *Amphitrite*

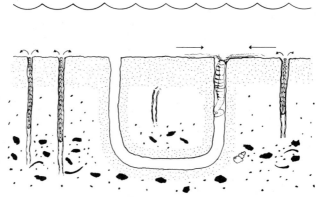

Figure 11. Small particles brought to the interface by the feeding activities of some organisms may be subducted by others during both burrow construction and feeding. In this case the interaction between the subsurface feeder *Clymenella torquata* and the surface deposit feeder *Amphitrite ornata* results in biogenic graded bedding interrupted by fine-grained shafts at depth in areas of Barnstable Harbor, Massachusetts.

builds its tubewall from small particles defecated at the interface by *Clymenella* and subducts them as a distinct shaft. The specific microenvironmental geometry of particle redistribution, and therefore reaction rates, is important in determining solute buildup patterns and sediment solute exchange with overlying water.

4. Solute Transport

The solutes consumed or generated in sediments by diagenetic reactions are subject to transport by molecular diffusion, advection, and mixing processes caused by physical or biological activity. In the absence of biological or physical disturbances, solute distributions can be quantitatively described by transport–reaction equations conceptually equivalent to those used previously for the solid phase:

$$\frac{\partial C}{\partial t} = D_s \left(\frac{\partial^2 C}{\partial x^2}\right) - \omega \left(\frac{\partial C}{\partial x}\right) + R \qquad (6)$$

where C is the solute concentration in pore water; D_s is the bulk sediment diffusion coefficient; and R is the solute reaction rate per liter of pore water. Other variables are defined as for equation (1). Compaction and adsorption reactions are again ignored for ease of illustration. As before it is assumed that variations in physical and chemical properties are predominantly vertical and can be modeled by one-dimensional equations. Solutes diffuse along concentration gradients maintained by the composition of an upper well-stirred water column reservoir and reactions within the sediment pile. No independent pore water flow is allowed, so that advective transport is equivalent to net sedimentation, ω, and steady-state conditions are usually assumed. Models of this type have been successfully applied in a wide range of environments, particularly anoxic basins and sediments below the bioturbated zone (Lerman, 1979; Berner, 1980).

4.1. Apparent Diffusion

The construction and irrigation of burrows by macroinfauna complicate this otherwise straightforward one-dimensional conceptualization of transport–reaction. In the bioturbated zone solutes can diffuse along lateral concentration gradients into biogenic cavities as well as vertically toward the sediment–water interface. Solute diffusion geometry is therefore three-dimensional and time-dependent.

Three classes of models have been proposed to quantify these processes. One class retains the one-dimensional formalism and assumes that macrofaunal activities can be accounted for in an increased apparent diffusion coefficient of solutres (D. E. Hammond et al., 1975; Vanderborght et al., 1977; Goldhaber et al., 1977; Aller, 1978). Like the particle mixing coefficient, D_B, this effective solute diffusion coefficient, D'_s, is analogous to eddy transport coefficients used to describe physical mixing in the overlying water column. In the strictest analogy, a single D'_s should therefore apply to all solutes over a defined sediment depth interval, transport in the underlying sediment being determined by molecular diffusion. Because apparent coefficients are determined empirically, they include the effects of coupling between particle reworking and solute transport that result from reversible adsorption (Schink and Guinasso, 1978). Reported values of D'_s obtained from modeling pore water profiles range from $\sim 10^{-5}$ to $\sim 10^{-4}$ cm^2/sec (Vanderborght et al., 1977; Goldhaber et al., 1977; Aller, 1978; Korosec, 1979; Filipek and Owen, 1980). These are 10–100 times higher than typical bulk sediment molecular diffusion coefficients for the same deposits; D_ss vary from $\sim 10^{-6}$ to $\sim 10^{-5}$ cm^2/sec depending on sediment structure and temperature (Li and Gregory, 1974).

The effect of a mixed layer on solute depletion or buildup is illustrated for NH_4^+ and SO_4^{2-} (produced or consumed during sulfate reduction) (Table I) as follows. In this example, the composite layer model equations used are, for Zone I (mixed) ($0 \leq x \leq L_1$):

$$\frac{\partial C_1}{\partial t} = 0 = D_1 \left(\frac{\partial^2 C_1}{\partial x^2} \right) + R_0 e^{-\alpha x} + R_1 \qquad (7a)$$

and, for Zone II (unmixed) ($L_1 \leq x \leq L_2$):

$$\frac{\partial C_2}{\partial t} = 0 = D_2 \left(\frac{\partial^2 C_2}{\partial x^2} \right) + R_0 e^{-\alpha x} + R_1 \qquad (7b)$$

where R_0, R_1, and α are constants and with the following boundary conditions:

1. $x = 0$, $C = C_T$
2. $D_1(\partial C_1/\partial x) = D_2(\partial C_2/\partial x)$, $x = L_1$
3. $C_1 = C_2$, $x = L_1$
4. $\partial C_2/\partial x = 0$, $x = L_2$

Compaction is ignored, advection is taken as unimportant relative to diffusion over the depth interval considered, and a steady state is assumed for simplicity. Adsorption terms cancel at steady state (Berner, 1976b). The solution is given in the Appendix for easy reference. Boundary condition (4) requires an impermeable layer or saturation condition at depth L_2, as would occur in either an experimental aquarium setup or deep in a deposit.

Reaction rates typical of terrigenous sediments from Mud Bay, South Carolina, at T = 29°C are used in the calculations (Aller, 1980a). These are $R_{SO_4} = -0.383\exp(-0.36x) - 0.061$ and $R_{NH_4^+} = 0.267\exp(-0.61x) + 0.0081$ mM/day. Estimates of molecular diffusion coefficients at T = 29°C are made from the relation $D_s \sim \phi^2 D$ where ϕ is porosity and D is the free solution diffusion coefficient (Li and Gregory, 1974; Lerman, 1978; Ullman and Aller, 1982). No viscosity correction for seawater density is made. Model predicted profiles for $C_T = 0.2$ μM, $L_1 = 5$ and 10 cm, $D_2(SO_4^{2-}) = 0.717$, and $D_2(NH_4^+) = 1.33$ cm²/day ($\phi = 0.851$) are plotted in Fig. 12. These are calculated to show the effect on a constituent for variations in apparent diffusion coefficients relative to the molecular diffusion coefficient of that ion. The values of D_1 in the examples are therefore not the same for both NH_4^+ and SO_4^{2-}. Identical values would

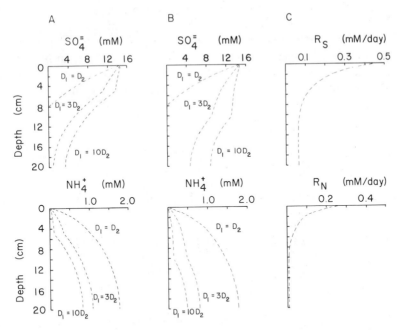

Figure 12. Vertical pore water profiles predicted for SO_4^{2-} and NH_4^+ for mixed layers of 5 cm (A) and 10 cm (B) with varied ratios of apparent diffusion coefficient (D_1) to molecular diffusion coefficients (D_2) in underlying sediment. The assumed consumption rate, R_S, for SO_4^{2-} and production rate, R_N, for NH_4^+ in each case are given in C.

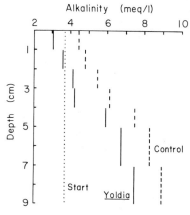

Figure 13. Vertical profiles of pore water alkalinity (~HCO_3^-) in experimental tanks with and without *Yoldia limatula*, a mobile, subsurface deposit-feeding bivalve (illustrated in Fig. 7) after 45 days. The starting profile was vertically homogeneous. In this case, a composite layer model in which a surface mixed layer overlies an unmixed layer, such as that shown in Fig. 12, works well in explaining the effects of macrofauna on the profile. Part of the decrease in alkalinity near the interface is caused by sulfide oxidation in both tanks. After Aller (1978).

be required for a conceptually consistent application of the mixing model in an actual deposit. This does not mean that solutes cannot be modeled individually in this way, but such modeling becomes progressively more arbitrary.

Several important qualitative conclusions can be drawn from these examples. Even a relatively thin mixed layer can radically change the buildup of reaction products or depletion of reactants both within and below the mixed zone compared with the molecular diffusion case. This effect is accentuated by the fact that mixing occurs near the sediment–water interface where reactions, at least in these cases, are most rapid. Any

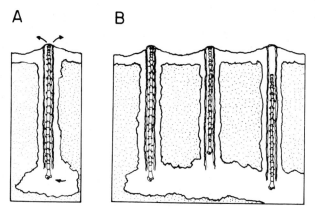

Figure 14. Inverted oxidation–reduction stratigraphies can result from selective feeding and irrigation at distinct horizons in a deposit. For example, when individual *Clymenella torquata* (A) occur in patches (B) the zone of feeding can become a distinct oxidized plane overlain by reduced material.

constituent subject to diffusion should be influenced in this way. Concentration changes for solutes having relatively small molecular diffusion coefficients, for instance, dissolved organic matter, will be particularly great below the zone of elevated transport.

The use of apparent transport coefficients works well when animals are highly mobile and restrict their activities to very near the sediment–water interface (Aller, 1978) (Fig. 13). As the zone of animal influence increases in thickness, a discrepancy between model and measured solute profiles usually arises. The deeper the zone of bioturbation, for example, the larger D'_s must be to explain solute distributions at depth. The result is that solute profiles near the sediment–water interface become less accurately simulated at the expense of modeling deeper regions. This kind of model is also not capable of simulating the maxima or minima that are sometimes observed in concentration profiles.

4.2. Biogenic Advection

A portion of the overlying water drawn into burrows during irrigation may be directly exchanged with surrounding pore waters, particularly in permeable sands. Direct exchange also occurs during active burrowing or tube construction. Some infauna, such as the lugworm *Arenicola marina*, may use permeable sediments as a filter for respiratory currents to augment food supply either by capture of suspended material or by stimulating local populations of bacteria (Krüger, 1959; Hylleberg, 1975). When animals are closely spaced and their burrows or tubes are relatively uniformly oriented, distinct horizons within the sediment column can become preferentially irrigated by such transport mechanisms. For instance, beds of the maldanid worm *C. torquata*, a conveyor belt deposit feeder that inhabits poorly sorted sands, often produce a second oxidized layer at their feeding depth within the sediment column owing to both the depletion of organic matter by selective feeding on fine-grained material and irrigation of the resulting highly permeable, coarse lag layer (Fig. 14) (Rhoads, 1974).

Such observations imply that solutes can be biogenically advected between relatively well-mixed and distinct reservoirs within the sediment and the overlying water column (Grundmanis and Murray, 1977; D. E. Hammond and Fuller, 1979; McCaffrey et al., 1980). Models constructed on this basis assign biogenic exchange of pore water to an apparent advection velocity rather than a random diffusive mixing process. These types of models have not been used to quantify solute profiles directly but have been employed to qualitatively explain inverted oxidation–reduction stratigraphies and pore water maxima or minima

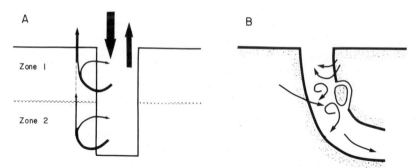

Figure 15. (A) A portion of overlying water advected into burrows during irrigation may in turn be advected into and through surrounding permeable sediment. (B) Collapsing burrows also result in rapid advective exchange of local pore water with overlying water.

(Grundmanis and Murray, 1977) (Fig. 15), or used as a mass balance construct to account for observed fluxes across the sediment–water interface (D. E. Hammond and Fuller, 1979; Korosec, 1979; McCaffrey et al., 1980). Apparent advection velocities of 0.3–4 cm/day have been calculated for estuarine and shelf muds on the basis of in situ ^{222}Rn profiles or laboratory tracer experiments with ^{22}Na (D. E. Hammond and Fuller, 1979; Luedtke and Bender, 1979; McCaffrey et al., 1980; Smethie et al., 1981). These apparent rates compare well with actual irrigation velocities of individual burrows and in that sense are of reasonable magnitude (Mangum, 1964; Mangum et al., 1968; Hoffman and Mangum, 1970; Coyer and Mangum, 1973; Aller and Yingst, 1978).

The concept of advective transport implies that bulk flow of pore water, usually in some specific direction, takes place in a deposit. Although such flow represents a component of biogenic irrigation activity, the model formalization of pore water solute transport in this way can lead to serious misconceptions. Many tube dwellings are lined by polysaccharide membranes that are permeable to diffusive transport of solutes but not advective flow (Foster-Smith, 1978; R. C. Aller, unpublished). This can be readily demonstrated by removing a section of burrow lining or tube and noting that in general it is extremely hard to push water through the wall; however, surrounding sediment is often oxidized, implying diffusive permeability. For this reason the major length of a lined burrow or tube is unlikely to be involved in advective exchange.

Muddy sediments are themselves impermeable to advective flow. It seems likely that much of the advective component of pore water exchange in muds is due not to the activities of oriented sedentary organisms but to either mobile benthos or periodic localized collapse of burrow walls. Such transport is not generally directional or specific to a particular interval of sediment as implied by the previously described advection models. In contrast to muds, permeable sands are probably subject to considerable pore water flow, biogenic or otherwise (Riedl et al., 1972). Selection between application of the two end-member models of apparent diffusion or apparent advection, neither of which is very realistic, should therefore depend not only on the functional types of animals present but also on sediment mass properties.

4.3. Average Diffusion Geometry

An alternative to incorporating the effects of macrobenthos in empirically elevated transport coefficients can be derived by noting that the bioturbated zone is not a homogeneous body but rather one permeated by cylinders of water. These cylinders are the burrows or tubes of infauna and are usually irrigated by their inhabitants. As a result, solute concentrations within the burrows are often close to those of the overlying water (Aller and Yingst, 1978; R. C. Aller and J. Y. Yingst, in preparation). Interstitial solutes generated within the sediment can diffuse along gradients either toward the sediment–water interface or into nearby burrows. The converse is true for solutes whose source is overlying water.

The geometry of solute diffusion is therefore altered from the one-dimensional case by the presence of burrows. The exact geometry is three-dimensionally complex and time-dependent. It is possible to simplify this complexity for purposes of modeling by postulating that at any instant the sediment is a mosaic of microenvironments represented on the average by a central irrigated tube or burrow and its immediately surrounding sediment (Aller, 1980a). Knowing the composition of this average microenvironment is the same as knowing the composition of the bulk deposit.

In the simplest imaginary case all animals inhabiting a deposit are vertically oriented tube dwellers of equal size and uniform spacing. The average microenvironment in this instance can be approximated by a hollow cylinder or annulus of sediment, the inner boundary of which is an inhabited tube and the outer boundary of which is half the distance between adjacent tubes (Fig. 16). The bioturbated zone as a whole consists of many such hollow cylinders packed together as a unit. Only the volume of sediment at the center of any three adjacent cylinders is not assigned to a given tube. This represents ~9% of the sediment volume and can be ignored for reasons that will become obvious. A major advantage of assuming cylindrical symmetry is that it simplifies a three-dimensional problem mathematically into two dimensions while retaining a relatively realistic solute diffusion geometry. It is also the natural symmetry of the system as most macrobenthos are themselves cylindrical in shape.

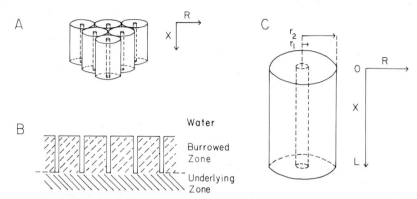

Figure 16. (A) Sketch of upper region of a deposit idealized as packed hollow cylinders of sediment. (B) Vertical cross-section of a deposit having the idealized diffusion geometry of A. (C) The simple hollow cylinder of sediment imagined as the average microenvironment within the bioturbated zone. The dimensions shown are those used in the transport–reaction model described in the text (Aller, 1980a–c).

When populations of infauna of variable sizes and spacing are present it is still possible to utilize the cylinder microenvironment concept, but the dimensions represent a weighted average of those actually present in a deposit. In this sense, the final cylinder size and orientation used in modeling are fictitious. Steady-state solute distributions within any cylinder are given by:

$$\frac{\partial C}{\partial t} = 0 = D_s \left(\frac{\partial^2 C}{\partial x^2} \right) + D_s \left\{ \frac{\partial}{\partial r} [r(\partial C/\partial r)] \right\} + R \qquad (8)$$

where r is the radial distance from the tube axis. Advection owing to sedimentation is ignored because of the small diffusion scales involved. Boundary conditions on the cylinder are taken as:

1. $x = 0, r = r_1, C = C_T$
2. $r = r_2, \partial C/\partial r = 0$
3. $x = L, \partial C/\partial x = B$

These conditions prescribe that (1) concentrations at the sediment–water interface and in tubes are the same and constant, (2) concentrations go through a maximum or minimum half way between any two tubes, and (3) the flux of solutes at the base of a cylinder matches that from the underlying, unburrowed zone. Like the cylinder microenvironment itself, these boundary conditions are approximations. Burrow waters may not be identical in solute concentration with overlying waters, the lining of some burrows hinders diffusion, and the outer boundary of a cylinder is not tangent at all points to adjacent cylinders. The solution is given in the Appendix for $R = R_0\exp(-\alpha x) + R_1 + k(C_{eq} - C)$.

Despite these approximations and idealizations, the cylinder microenvironment produces good agreement with actual data. This is shown in Fig. 17, which illustrates the solute profiles predicted for a site in Mud Bay, South Carolina, along with measured profiles (Aller, 1980a). Predicted vertical profiles are obtained by integrating the solution to equation (8) over finite vertical depth intervals comparable to sampling intervals. The reaction rates for NH_4^+ and SO_4^{2-} are the same as used previously in Fig. 12. Si is assumed to be produced by a reaction rate of the form $R = k(C_{eq} - C)$, where k and C_{eq} are constants. Measured values of B in these profiles are -0.1, 0.011, and 0.060 mM/cm for SO_4^{2-}, NH_4^+, and $SI(OH)_4$, respectively. The model profiles are generated in practice by fixing r_1 and L and varying r_2 until the best fit is obtained. An average microenvironment in this example was initially estimated from the average burrow size ($r_1 \sim 0.05$ cm), burrowing depth ($L \sim 15$ cm), and abundance ($r_2 \sim 2.5-5$ cm; $N = 50-500/m^2$) of *Heteromastus filiformis*, a polychaete that numerically dominates the macroinfauna of Mud Bay. The final value of $r_2 \sim 2.1$ cm ($N = 722/m^2$), necessary to obtain a good fit, is not unrealistic, and its higher than estimated value reflects animal mobility, error in estimating abundances, and the presence of other benthos besides *Heteromastus*. Once the size of the average microenviron-

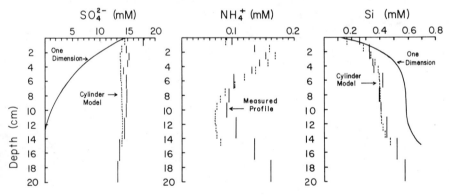

Figure 17. SO_4^{2-}, NH_4^+, and Si concentration profiles in pore water from Mud Bay, South Carolina, used to illustrate behavior of the cylinder microenvironment model. Solid vertical bars, measured concentrations; dashed vertical bars, cylinder model profiles; solid continuous curves, one-dimensional model profiles having the same diffusion, reaction, and boundary constants used in the cylinder model. The NH_4^+ profile predicted by the one-dimensional model is off the scale and is not plotted. After Aller (1980a). Compare with the composite layer models in Fig. 12, which use the same reaction rate distributions for SO_4^{2-} and NH_4^+.

ment was estimated from the NH_4^+ profile, which is very sensitive to choice of geometry, the Si profile was determined by adjustment of the first-order reaction rate k ($= 0.2$/day); C_{eq} ($= 577$ µM) was estimated from asymptotic concentrations in longer gravity cores. Molecular diffusion coefficients are used and assumed to be isotropic.

The buildup of pore water solutes depends strongly on the size and abundance of infauna. This is shown in Fig. 18, in which the mean pore water concentration in the bioturbated zone is plotted as a function of either r_1 or r_2 for both NH_4^+ and $Si(OH)_4$. Reaction rates are those used previously. It is assumed that solid-phase SiO_2 is not limiting. These plots illustrate that concentration depends strongly on animal abundance and to a lesser but still large extent on organism size.

The model demonstrates quantitatively that pore water concentrations can be expected to vary in space to the same degree as do animal populations. These graphs also show that the buildup of pore water constituents subject to zeroth order reactions, such as NH_4^+, is more affected

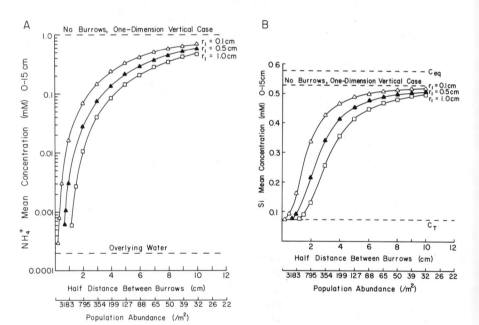

Figure 18. (A) Expected average concentration of NH_4^+ in 0- to 15-cm interval as a function of burrow spacing or abundance (r_2) and fixed burrow size (r_1). Equivalent population abundances (N) per m² are also indicated. Concentrations are bounded above by the average concentration predicted by the one-dimensional model and below by the overlying water value. Reaction rate as in Fig. 12C. (B) Expected average concentration of Si in pore water from 0–15 cm as function of burrow size and abundance. The apparent solubility, C_{eq}, is also plotted.

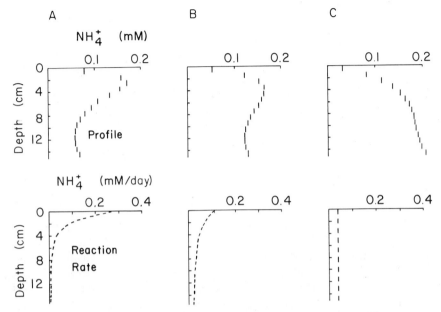

Figure 19. The vertical pore water NH_4^+ profiles (upper row) expected for different reaction rate distributions (lower row) when burrow size and abundance are fixed at $r_1 = 0.05$ cm and $r_2 = 2.1$ cm, respectively. The depth-integrated reaction rate is the same in each case.

by infauna than is the buildup of solutes controlled by first-order kinetics, that is, the kinetics of reaction determine the response of an element to animal activities. This comes about because, the greater the pore water depletion in the latter case, the faster the solid phase dissolves to replenish the dissolved phase, whereas in the former case production is relatively independent of pore water buildup. These assumed kinetics are of course idealized and representative of end-member behavior. For instance, production by bacteria of metabolites like NH_4^+ is unlikely to be entirely independent of pore water NH_4^+ concentrations.

The shape of vertical profiles depends strongly on the depth dependence of reaction rates as well as the abundance and size of infauna. Fig. 19 illustrates the pore water NH_4^+ profile shape expected for three NH_4^+ production rate distributions, each of which has a different depth dependence of production but each with the same integrated or total rate of NH_4^+ formation over the burrowed interval ($\int_0^L Rdx =$ constant). The number and size of burrows present in the sediment are the same in each case. Because the integrated production rate is constant, the total flux of NH_4^+ from the bioturbated zone at steady state is also the same. Maxima and minima in concentration distributions are produced by lateral diffusion when reaction rates are strongly attenuated with depth (Fig. 19A).

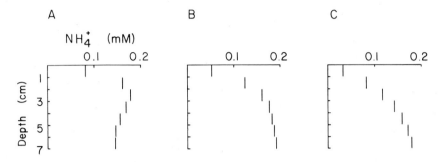

Figure 20. Vertical pore water profiles expected for the respective reaction rates A, B, and C of Fig. 19 but assuming that the burrowed zone is restricted to 7 cm.

Burrows tap the low-production-rate sediment horizons and prevent solute buildup at depth much as a water well on land would deplete a particular stratum of an aquifer having a low rate of groundwater replenishment. When reaction rates are less attenuated, as would occur in areas of rapid particle reworking, profiles have a parabolic shape that in itself might not suggest extensive irrigation unless reaction rates were independently known (Fig. 19C). The effect of varying burrowing depth for the same reaction rates is also plotted in Fig. 20 to show that the resulting solute profile shapes are comparable to the case of a lower attenuation of reaction rate over a larger burrowed interval, L. These examples illus-

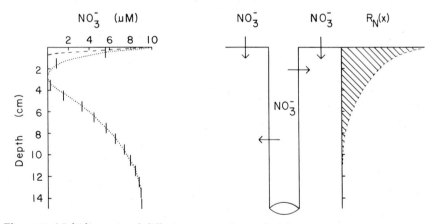

Figure 21. Multidimensional diffusion can produce subsurface maxima in pore water constituents that otherwise would be depleted in the simple one-dimensional case. Expected NO_3^- distributions for $R_N = -0.01\exp(-0.5x)$ and either no burrows (dashed line) or burrows of radius $r_1 = 0.05$ cm, abundance $r_2 = 2.5$ cm, and length $L = 15$ cm are plotted for illustration. Depletion of reactive organic substrate at depth implied by the attenuation of R_N allows resupply of pore water NO_3^- by lateral diffusion from burrows.

trate that profile shape alone is not indicative of any unique reaction rate or burrow size distribution. Independent determinations of both must be made in order to interpret solute distributions.

Inverted stratigraphies of pore water constituents whose sources are in overlying water can also be generated by multidimensional diffusion when reactive substrate depletion with depth occurs. To illustrate this, a hypothetical NO_3^- profile is shown in Fig. 21 for an NO_3^- reduction rate of $R = -0.01\exp(-0.5x)$ and microenvironment dimensions of $r_1 = 0.05$ cm, $r_2 = 2.5$ cm $(N \sim 510/m^2)$, and $L = 15$ cm. The concentration gradient, B, at the base of the irrigated zone is arbitrarily set here at $B = 0$ but presumably would be <0 in the natural situation where NO_3^- would be depleted again beneath the burrowed zone. The rate function is approximated by a zeroth order equation consistent with control of denitrification by the quality of decomposed organic matter present rather than simple first-order kinetics, as is appropriate when NO_3^- itself is also limiting (Vanderborght and Billen, 1975; Billen, 1978; Sørensen, 1978). R_0 and α were chosen to be in the range of reported NO_3^- fluxes into sediments (Billen, 1978). In environments where substrate depletion takes place very close to the sediment–water interface, such as in most areas of the deep sea or eroded areas of seafloor, the depth scale of solute minima such as shown in Fig. 21 would be greatly shortened and might not be detected because of coarseness in sampling intervals.

Of the three models considered, the cylinder microenvironment model is capable of most closely simulating solute distributions within the bioturbated zone. Although idealized, because it incorporates the fundamental changes in solute diffusion geometry caused by macrobenthos, it is relatively realistic and allows development of intuition in relating one-dimensional pore water data to a multidimensional transport–reaction regime. It is also capable of predicting the magnitude of effects on bottom chemistry by benthic populations having different sizes and abundances if reaction rates are known. Coupling between solid phase reworking and solute transport owing to reversible adsorption is not taken into account by the cylinder model, but this should not be significant unless the linear adsorption coefficient is greater than ~ 10 (Schink and Guinasso, 1978).

5. Macrofaunal Influence on Sediment–Water Exchange Rates

The exchange of solutes between sediments and overlying waters is a major component of material cycles in aquatic environments. Because macrobenthos control pore water solute concentration profiles, they influence methods of predicting fluxes and in some cases the actual magnitude of a flux. If sediment solute distributions were controlled by one-dimensional vertical transport and reaction processes then in principle the diffusive flux of a constituent across the sediment–water interface could be simply calculated from Fick's first law:

$$J = -\phi D_s \left(\frac{\partial C}{\partial x}\right)_{x=0} \tag{9}$$

where J is the solute diffusive flux at the interface, all other symbols are defined as before, and advective flow is ignored.

In this context two end-member examples will be considered. The first is pore water solutes controlled by zeroth-order or pseudo-zeroth-order reaction kinetics so that $R = R_0 \exp(-\alpha x) + R_1$ and the second is solutes subject to first-order reactions, or $R = k(C_{eq} - C)$, where R_0, α, R_1, k, and C_{eq} are constants. Dissolved NH_4^+ and $Si(OH)_4$ approximate these two cases, respectively (Berner, 1974; Schink et al., 1975; Hurd, 1973; Kamatani and Riley, 1979).

At steady state the diffusive flux of a solute produced or consumed by zeroth-order reactions is independent of transport. If $R = R_0 \exp(-\alpha x) + R_1$ over the interval $0 \leq x \leq L$, then:

$$J = \frac{\phi R_0}{\alpha}[1 - \exp(-\alpha L)] - \phi R_1 L \tag{10}$$

assuming no significant flux from the region below L. Usually the term $\exp(-\alpha L)$ is insignificant for organic matter decomposition reactions. The overall magnitude of diffusive exchange in this case is fixed by production or consumption rates and in principle is not affected by macrofaunal activities. The geometry of exchange can be greatly altered, however, and prediction of fluxes from vertical pore water gradients by use of equation (9) made invalid. The one-dimensional transport model is incorrect in this case because a portion of solute exchange takes place both at the sediment–water interface ($x = 0$) and across burrow walls at varying depths within a deposit.

The vertical solute concentration gradient at the interface is a function of the sizes and abundances of infauna, which determine the geometry of diffusion (Fig. 18), and the depth dependence of the reaction rate (Figs. 19 and 20). In the case of the three production rate distributions and corresponding solute profiles in Fig. 19, for example, the total flux of NH_4^+ from the upper 15 cm is identical at 0.56 μmol/cm² per day. In

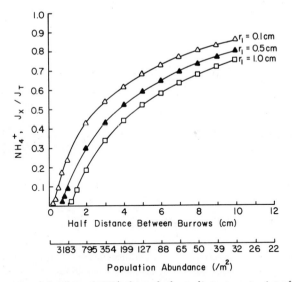

Figure 22. The ratio of the flux of NH_4^+ through the sediment–water interface at $x = 0$, J_x, to the total flux of NH_4^+ from both burrow walls and the sediment–water interface, J_T, as a function of burrow size and abundance. The reaction rate assumed is that of Fig. 12C. The ratio J_x/J_T is a measure of the ratio of the molecular diffusion coefficient and the apparent transport coefficient required to match the total flux with the vertical pore water concentration gradient. After Aller (1980a).

contrast, the fluxes predicted from the vertical gradient alone in each case are 0.26, 0.16, and 0.092 μmol/cm² per day from left to right, respectively (Fig. 19) or 46%, 29%, and 16% of the total expected value. The lower the attenuation of reaction rate with depth in a burrowed sediment the less accurate application of equation (9) is in predicting the flux from the vertical concentration gradient alone. The cylinder microenvironment model can be used to predict the variation in the proportion of NH_4^+ released by vertical diffusion across the sediment–water interface relative to total released. This is shown in Fig. 22 as a function of animal size and abundance for the reaction rate distribution of Fig. 19A.

The ratio of the solute flux predicted from equation (9) to that actually measured is sometimes used to determine a value of the effective transport coefficient D_s' necessary to match the vertical concentration gradient with the measured flux (Martens and Klump, 1980). This coefficient is identical to that described previously for modeling pore water distributions in the bioturbated zone. The flux ratio plotted in Fig. 22 is therefore a measure of the ratio of the bulk sediment molecular diffusion coefficient to the effective diffusion coefficient, D_s/D_s'.

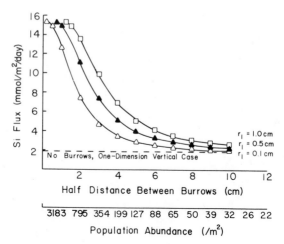

Figure 23. The total flux of dissolved silica from sediment as a function of burrow size and abundance. Because of concentration-dependent dissolution, the flux increases as flushing of the sediment by burrows increases. The assumed reaction rate is $k = 0.2/\text{day}$ with $C_{eq} = 577\ \mu M$. Solid-phase Si is taken as unlimited; in practice elevated fluxes could not be maintained indefinitely. After Aller (1980a).

For solutes subject only to a first-order reaction rate $R = k(C_{eq} - C)$ the sediment–water flux at steady state is:

$$J = \phi \sqrt{kD_s}\,(C_T - C_{eq}) \tag{11}$$

assuming that the concentration gradient across L is 0 and the value $(k/D)L$ is large. The presence of D_s in the expression demonstrates that the flux in this case is not independent of transport. The greater the effective transport rate the larger the flux. This is illustrated in Fig. 23 utilizing the cylinder model for a rate constant $k = 0.2/\text{day}$, a diffusion coefficient $D_s = 0.687$, an overlying water concentration $C_T = 0.074$ mM, a basal gradient $B = 0.060$ mM/cm, and a solubility $C_{eq} = 0.577$ mM. For simplicity, solid-phase silica is assumed to be in excess and not limiting to the reaction. In contrast to ammonium, the greater the flushing of sediment by burrows in this instance, the greater the total flux of silica. The variation, as a function of animal size and abundance, in D_s/D_s' necessary to explain a given silica flux from a vertical pore water gradient is similar to but less pronounced than for the zeroth-order case (Aller, 1980a). Although the total dissolved silica flux from the bottom may be greatly increased, a corresponding increase in solute supply by additional dissolution of the solid phase tends to balance the diffusive loss into burrows and lessens the effect of irrigation on the pore water concentration profile as compared with constituents like NH_4^+.

These examples show that irrigated tube structures and active burrowing influence pore water constituents differently depending on the kinetics of reactions controlling solute loss or supply and reaction rate distribution. The ratio D_s/D_s' determined from the discrepancy between measured fluxes and that predicted from a vertical concentration gradient for one constituent is therefore unlikely to be applicable to all solutes in a deposit.

6. Reaction Rates

Aside from changing the geometrical distribution of reaction rates by redistributing and repackaging particles, macrobenthos stimulate the rates of microbially mediated decomposition reactions (Fenchel, 1970; Hargrave, 1970, 1976; Harrison and Mann, 1975; Fenchel and Harrison, 1976; Aller, 1978; Yingst and Rhoads, 1980). This occurs in several ways:

1. Surface area of organic detritus is mechanically increased or existing surfaces reexposed during feeding.
2. Grazing maintains microbial populations in a high-productivity phase of rapid growth.
3. Metabolite buildup is decreased and electron acceptor supply increased by irrigation, particle reworking, and multidimensional diffusion.
4. Mucus secretions provide new reactive substrate.
5. Subduction or capture of resuspended reactive organic matter is increased during reworking (discussed in Section 3).

An example of such stimulation is shown by measurement of the oxygen uptake flux to various naturally occurring particles before and after ingestion by different invertebrates (Fig. 24) (Hargrave, 1976). Increased uptake is presumably caused by one or more of the factors listed above. Uptake decreases to preingestion rates after a few days.

Because at least one component of stimulation, reexposure of surfaces, is related directly to particle reworking activities, in the strictest sense the reaction rate, R, in previous models is a function of the particle mixing rate, D_B. The extent of dependence remains to be determined. Limits on the effect of mechanical disturbance can be set by measuring metabolic rates in sediments that have been vigorously stirred ($D_B \to \infty$) compared with physically undisturbed material ($D_B \to 0$). Westrich (1983) reports that SO_4^{2-} reduction rates in nearshore sediments are two times higher in mixed than unmixed samples. Similar but less dramatic results have been obtained in our own laboratory for a variety of anaerobic decomposition reactants or products indicating elevated rates for periods

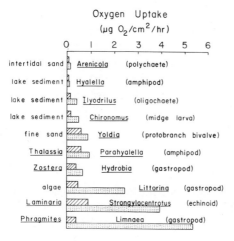

Figure 24. Oxygen consumption rates of different material before (cross hatched) and after (stippled) ingestion by benthic invertebrates. Elevated consumption rates are sustained for several days following defecation. After Hargrave (1976).

of 1–2 weeks after mixing. This implies that decomposition rates in sediments may vary between these two extremes of mixed and unmixed on a spatial scale comparable to the size of individual mobile macrofauna regardless of other stimulatory effects.

Greenwood (1968) suggests a simple model as to how such mechanical stimulation may occur. Physical segregation of bacteria from organic substrate and adsorption or physical hinderance of exoenzyme movement by clays are envisioned as major factors limiting decomposition at the microscale of sediment aggregates (Fig. 25). The mechanical disaggregation of such clusters by macro- and meiofauna results in a new arrangement in which relatively unutilized substrate is exposed and pore water exchanged. A short-term increase of microbial activity results.

In surface-oxidized material the interior of such aggregates is often

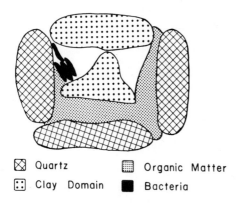

Figure 25. Model sedimentary aggregate illustrating role of physical and chemical (reactive clay surfaces) barriers to utilization by microorganisms of organic matter in a deposit. After Greenwood (1968).

anoxic despite abundant oxygen in surrounding pore waters. These aggregates, which are usually fecal pellets, can be modeled as distinct sedimentary microenvironments themselves just as can be done for burrows (Greenwood, 1968; Jørgenson, 1977b; Anderson and Meadows, 1978; Aller and Yingst, 1978). In order to maintain a chemical microenvironment distinct from that imposed by gradients from the sediment–water interface or adjacent burrows, it is necessary that D_s (the bulk diffusion coefficient) within the aggregate be less than D_s for the surrounding sediment or that reaction rates be locally stimulated. The compacted, low-porosity nature of many fecal pellets should indeed restrict diffusion within pellets compared with ambient sediment, and data like those shown in Fig. 24 imply that metabolic rates may be stimulated as well. Because fecal pellet size, shape, mucus content, and compactness vary between individuals and animal species, considerable microscale heterogeneity in biogeochemical properties and reaction rates is expected at the sediment–water interface where newly formed pellets are deposited.

Creation of chemical microenvironments by macrofauna may also determine local distribution of meiofauna around biogenic structures (Bell et al., 1978; Aller and Yingst, 1978; Thistle, 1979). Because meiofaunal grazing of bacteria is known to influence decomposition (Fenchel and Harrison, 1976) this effect will further accentuate spatial variation in reaction rate.

7. Chemistry of the Burrow Habitat

The fact that burrow walls are permeable to solute diffusion means that sedimentary reactions can modify the burrow habitat. A continual flux of solutes from surrounding sediment into burrows takes place. The magnitude of this flux is determined in any given case by concentration gradients around a burrow and sediment permeability. In order to hold the composition of burrow water constant or within a restricted range of variation different from pore water, continual irrigation of the burrow must balance the flux of solutes from the surrounding sediment. The steady-state balance for a constituent not subject to precipitation in the burrow is given by:

$$F = \frac{AJ}{V_0} = \frac{v}{V_0}(C_B - C_T) \tag{12}$$

where F is the rate of change of burrow water solute concentration, J is the solute flux (constant) from the tube walls, A is the inner surface area of the burrow, V_0 is the burrow volume, v is the irrigation rate (volume/

time), C_B is the steady-state solute concentration in burrow water, and C_T is the solute concentration in overlying seawater. If the solute of interest is also excreted by the burrow inhabitant, an additional source term must be added to F.

This simple mass balance model has been shown to work well for an intertidal population of *Amphitrite ornata*, a terebellid polychaete (Aller and Yingst, 1978). The rate of change, F, in that case was estimated from changes in the burrow water composition during tidal recession when irrigation could not take place (Fig. 26). Mass balance models of this type have also been used to calculate irrigation rates necessary to supply oxygen to burrow inhabitants (Mangum and Burnett, 1975).

The flux of solutes from surrounding sediment determines in part the excursions in burrow water composition during intertidal periods. Metabolic activity of burrow inhabitants (including bacteria) also alters burrow waters during the intertidal interval; oxygen, for example, is known to be depleted in exposed burrows (Jones, 1955; Thompson and Pritchard,

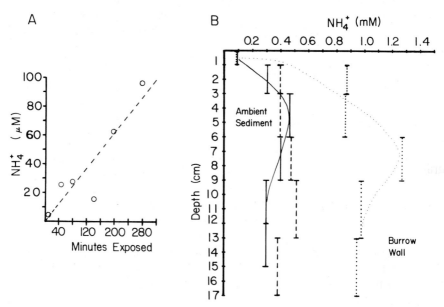

Figure 26. (A) NH_4^+ concentration in *Amphitrite ornata* burrows exposed on a tidal flat versus time of exposure. Except for the first sample (t = 10 min), which represents a single burrow, each point represents the average concentration in two or more burrows. (B) NH_4^+ concentration in pore water from concentric sample intervals around one branch of an *A. ornata* burrow and from surrounding sediment. ······, 0.5- to 1.5-cm radial interval from burrow axis; ----, 1.5- to 3.75-cm radial interval; ———, sediment 30 cm from axis. Each bar represents the vertical interval over which the sample was taken. After Aller and Yingst (1978).

1969). Assuming that C_B must on the average be within a range consistent with the physiological tolerances of a species, then the solute flux determines the minimum irrigation effort an animal must expend to control the burrow environment when connection with overlying water is possible. Because the flux to an individual burrow is controlled by reaction rates and solute buildup in surrounding sediment, it is necessarily influenced by the presence or absence of other animals. This can be illustrated using the cylinder model developed previously and equation (8) (Aller, 1980b). As an example, F for NH_4^+ has been calculated over a range of burrow sizes and interburrow distances from the relation

$$J_r = -\phi D_s \left(\frac{\partial C}{\partial r}\right)_{r=r_1} \quad (13)$$

using the reaction rate distribution of Fig. 12 and the remaining model values as in Fig. 17 (see page 79), and integrating over burrow length (15 cm). If C_B and C_T are assumed constant then the factor $F \cdot V_0$ is proportional to the irrigation rate necessary to balance the solute influx. This value is also plotted in Fig. 27 as a function of burrow size and density.

Excursions in burrow water compositions during intertidal periods, the irrigation effort required to keep burrow waters at steady state, or the standing concentration of sediment-derived solutes in the burrow at a particular irrigation rate can all be lowered in two ways: (1) by decreasing interburrow distances or (2) by increasing burrow size. Of the two strategies, crowding together in patches of high burrow abundance is the most effective at reducing irrigation requirements. The construction of impermeable burrow walls would accomplish the same result.

These simple consequences of diffusion geometry are apparently utilized by animals inhabiting chemically undesirable sediments. It is known that early colonizers of defaunated substrates such as waste dumps or eroded areas of the bottom are often small tube dwellers in high abundance (Grassle and Grassle, 1974; McCall, 1977; Rhoads et al., 1977; Pearson and Rosenberg, 1978). These sediments commonly contain elevated levels of metabolites or toxins. The small size of such animals is consistent with the often relatively low oxygen contents of these environments (Pearson and Rosenberg, 1978), while crowding or high population density can be interpreted as an extremely effective adaptation to a chemically hostile substratum (Fig. 27) (Aller, 1980b). Crowding may also have the beneficial effect of physically stabilizing the seafloor against erosion (Schäfer, 1972; Rhoads et al., 1978; Rhoads and Boyer, Chapter 1).

Consideration of animal adaptations to sediment chemistry therefore leads to the inference that high population density and rapid reproduction

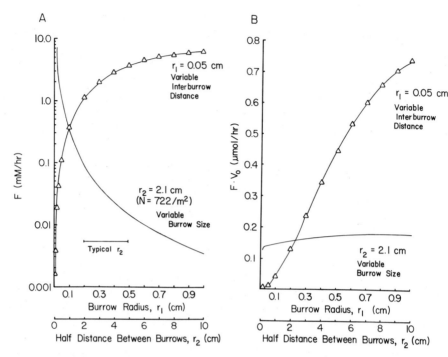

Figure 27. (A) The expected initial rate of change, F, of burrow water NH_4^+ concentrations in exposed burrows as a function of burrow size and abundance of surrounding burrows. The NH_4^+ production rate in surrounding sediment is that of Fig. 12C. (B) The irrigation rate expressed as the factor $F \cdot V_0$ (see text) required to "stat" a burrow at a constant NH_4^+ concentration as a function of burrow size and interburrow distances. Crowding together dramatically decreases the irrigation work required of individuals to detoxify the burrow habitat or, correspondingly, decreases the concentration of sediment-derived solutes in the burrow at any particular irrigation rate. After Aller (1980b).

in this context may represent specific adaptations to the chemical environment rather than to the usual assumption that high fecundity is characteristic of poor competitors. Overall, very little is known about the chemical characteristics of the infaunal habit. This example, which is speculative, illustrates that future research in this area will probably provide much insight into the ecology of soft-bottom benthos.

8. Spatial and Temporal Patterns in Sediment Chemistry

Because the effects of macrofauna on sediment chemistry are determined by the number, kinds, and sizes of animals, sediment composition varies in space and time in accordance with the particular benthic community present (Rhoads et al., 1977; Aller, 1980c). This variation should occur regardless of otherwise uniform substrate characteristics such as grain size or quantity of organic matter.

Temporal patterns in the kinds of animals inhabiting a deposit are known to occur on ecological time scales (Pearson and Rosenberg, 1978; Rhoads and Boyer, Chapter 1). Successional patterns in sediment chemistry occur in conjunction with these changes in community structure (Rhoads et al., 1977). With respect to sediment chemistry, many of the most important changes in animal type involve a simple increase in burrowing depth and size of individuals with time. Progressively deeper irrigation and particle reworking cause increased pore water exchange, the subduction of fresh organic matter, reoxidation of reduced compounds, and increased biogeochemical heterogeneity. The role of sediment chemistry in actually causing or allowing successive replacement of fauna is at present speculative, as described previously.

On longer time scales, bioturbation as a process has changed progressively through geologic time with both the depth and intensity of reworking increasing since the Precambrian (Thayer, 1979; Bambach and Sepkoski, 1979). The corresponding pattern of change in chemical diagenetic processes over geologic time must have been similar to spatial or temporal patterns in sediment chemistry associated with differences in faunal distributions in recent sediments. In the absence of significant reworking, pore water solute profiles were presumably similar to those predicted by one-dimensional transport reaction models. Saturation of pore water with respect to a variety of reduced phases must have occurred on the average closer to the sediment–water interface. Nutrients like PO_4^{3-}, which may become incorporated into reduced phases, or like Si, which are subject to concentration-dependent dissolution, were almost certainly less efficiently recycled to the water column than in present-day bioturbated sediment. Sulfide oxidation was probably less important because of less extensive particle reworking, and, partly as a result, the dissolution of carbonate was probably less favored (Aller, 1982). Microenvironmental heterogeneity in diagenetic reactions must have also been minimal in the absence of macrofaunal construction activity. Some systematic differences in the authigenic mineral components of sediments through geologic time from otherwise equivalent depositional environments can therefore be expected to occur along with the differences in sedimentary structures already documented (Bambach and Sepkoski, 1979; Sepkoski and Bambach, 1979).

9. Summary

In the preceding discussion I have emphasized some of the general principles involved in understanding and quantifying the effects of macrofauna on sediment chemistry. Specific effects depend on the functional groups of animals present in the benthic community, characteristics of the depositional environment such as resuspension and net deposition rates, and the kinetics of reaction of the constituents of interest. A lack of detailed consideration of these factors in construction of quantitative models can lead to unwarranted or at least misleading conclusions, such as the often-repeated statement that the net flux of solutes from or to bottom sediments must be increased by bioturbation. The interactions between different groups of sediment-dwelling organisms as determined by, or reflected in, sediment chemistry remain one of the most intriguing and understudied aspects of marine biogeochemistry.

Appendix: Solutions to Model Equations

Equation (5) with conditions 5.1–4 has the following solutions:

Zone 1

$$A(x) = \beta_1 \exp(\alpha_1 x) + \beta_2 \exp(\alpha_2 x)$$

Zone 2

$$A(x) = \beta_3 \exp(-\lambda x/\omega)$$

where:

$$\alpha_1 = \frac{\omega - \sqrt{\omega^2 + 4D_B\lambda}}{2D_B}$$

$$\alpha_2 = \frac{\omega + \sqrt{\omega^2 + 4D_B\lambda}}{2D_B}$$

$$\beta_1 = \frac{J_0\alpha_2 \exp(\alpha_2 L_1)}{\alpha_1(D_B\alpha_2 - \omega)\exp(\alpha_1 L_1) - \alpha_2(D_B\alpha_1 - \omega)\exp(\alpha_2 L_1)}$$

$$\beta_2 = \frac{-J_0\alpha_1 \exp(\alpha_1 L_1)}{\alpha_1(D_B\alpha_2 - \omega)\exp(\alpha_1 L_1) - \alpha_2(D_B\alpha_1 - \omega)\exp(\alpha_1 L_1)}$$

$$\beta_3 = B_1 \exp(\alpha_1 L_1) + \beta_2 \exp(\alpha_2 L_1)\exp(\lambda L/\omega)$$

Equation (7) with conditions 7.1–4 has the following solutions:

Zone 1

$$C_1 = C_T - \frac{R_1 x^2}{2D_1} + \frac{R_1 L_2 x}{D_1} - \frac{R_0 x}{D_1 \alpha}[\exp(-\alpha L_2)] + \frac{R_0}{D_1 \alpha^2}[1 - \exp(-\alpha x)]$$

Zone 2

$$\begin{aligned} C_2 = C_1 &- \frac{R_1(x - L_2)^2}{2D_2} + \frac{R_1 L_1(L_2 - L_1)}{D_1} \\ &+ \frac{R_1(L_2 - L_1)^2}{2D_2} + \frac{R_1 L^2}{2D_1} - \frac{R_0 L_1}{D_1 \alpha}[\exp(-\alpha L_2)] \\ &+ \frac{R_0}{D_1 \alpha^2}[1 - \exp(-\alpha L_1)] + \frac{R_0}{D_2 \alpha}(L_1 - x)[\exp(-\alpha L_2)] \\ &+ \frac{R_0}{D_2 \alpha^2}[\exp(-\alpha L_1) - \exp(-\alpha x)] \end{aligned}$$

Equation (8) with conditions 8.1–3 and $R = R_0 \exp(-\alpha x) + R_1 + k(C_{eq} - C)$ has the following solution:

$$C(x,r) = C_T + Bx + \frac{2}{LD_s} \sum_{n=0}^{\infty} \left[\frac{G_n}{\mu_n^2}\right] \left[\frac{U_0(\mu_n r)}{U_0(\mu_n r_1)} - 1\right] \sin(\lambda_n x)$$

$$- \frac{2B}{L} \sum_{n=0}^{\infty} \left[\frac{(-1)^n U_0(\mu_n r)}{\lambda_n^2 U_0(\mu_n r_1)}\right] \sin(\lambda_n x)$$

where:

$$n = 0, 1, 2 \ldots$$
$$\lambda_n = (n + \tfrac{1}{2})(\pi/L)$$
$$\mu_n = (k/D_s + \lambda_n^2)^{1/2}$$
$$G_n = \frac{k(C_T - C_{eq})}{\lambda_n} - \frac{R_1}{\lambda_n} + \frac{(-1)^n kB}{\lambda_n^2} + R_0 \frac{[\alpha e^{-\alpha L}(-1)^n - \lambda_n]}{\alpha^2 + \lambda_n^2}$$

$$U_0(\mu_n r) = K_1(\mu_n r_2) I_0(\mu_n r) + I_1(\mu_n r_2) K_0(\mu_n r)$$

The functions $I_v(z)$ and $K_v(z)$ are the modified Bessel functions of the first and second kind, respectively, of order v.

ACKNOWLEDGMENTS. This work has been supported by NSF Grant OCE-7900971 and successor OCE-811415. Fellowship support from the Sloan Research Foundation is also acknowledged. J. Pasdeloup deciphered and typed the manuscript. I thank J. Y. Yingst, J. E. Mackin, and P. W. McCall for critical comments or editorial suggestions.

References

Aller, R. C., 1978, Experimental studies of changes produced by deposit feeders on pore water, sediment, and overlying water chemistry, *Am. J. Sci.* **278**:1185–1234.

Aller, R. C., 1980a, Quantifying solute distributions in the bioturbated zone of marine sediments by defining an average microenvironment, *Geochim. Cosmochim. Acta* **44**:1955–1965.

Aller, R. C., 1980b, Relationships of tube-dwelling benthos with sediment and overlying water chemistry, in: *Marine Benthic Dynamics* (K. R. Tenore and B. C. Coull, eds.), pp. 285–308, University of South Carolina Press, Columbia.

Aller, R. C., 1980c, Diagenetic processes near the sediment–water interface of Long Island Sound. I. Decomposition and nutrient element geochemistry (S, N, P), *Adv. Geophys.* **22**:237–350.

Aller, R. C., 1982, Carbonate dissolution in shallow water marine sediments: Role of physical and biological reworking, *J. Geol.* **90**:79–95.

Aller, R. C., and Cochran, J. K., 1976, ^{234}Th/^{238}U disequilibrium in near-shore sediment: Particle reworking and diagenetic time scales, *Earth Planet. Sci. Lett.* **29**:37–50.

Aller, R. C., and Dodge, R. E., 1974, Animal–sediment relations in a tropical lagoon, Discovery Bay, Jamaica, *J. Mar. Res.* **32**:209–232.

Aller, R. C., and Yingst, J. Y., 1978, Biogeochemistry of tube-dwellings: A study of the sedentrary polychaete *Amphitrite ornata* (Leidy), *J. Mar. Res.* **36**:201–254.

Aller, R. C., Benninger, L. K., and Cochran, J. K., 1980, Tracking particle associated processes in nearshore environments by use of ^{234}Th/^{238}U disequilibrium, *Earth Planet. Sci. Lett.* **47**:161–175.

Anderson, J. G., and Meadows, P. S., 1978, Microenvironments in marine sediments, *Proc. R. Soc. Edinburgh Sect. B* **76**:1–16.

Amiard-Triquet, C., 1974, Etude experimentale de la contamination par le cerium 144 et le fer 59 d'un sédiment à *Arenicola marina* L. (Annelida Polychete), *Cah. Biol. Mar.* **15**:483–494.

Aston, S. R., and Chester, R., 1973, The influence of suspended particles on the precipitation of iron in natural waters, *Estuarine Coastal Mar. Sci.* **1**:225–231.

Baas-Beckling, L. G. M., Kaplan, I. R., and Moore, D., 1960, Limits of the natural environment in terms of pH and oxidation–reduction potentials, *J. Geol.* **68**:243–284.

Bambach, R. K., and Sepkoski, J. J., Jr., 1979, The increasing influence of biologic activity on sedimentary stratification through the Phanerozoic, *Geol. Soc. Am.* **11**:383.

Baumfalk, Y. A., 1979, Heterogeneous grain size distribution in tidal flat sediment caused by bioturbation activity of *Arenicola marina* (Polychaeta), *Neth. J. Sea Res.* **13**:428–440.

Bell, S. S., Watzin, M. C., and Coull, B. C., 1978, Biogenic structure and its effect on the spatial heterogeneity of meiofauna in a salt marsh, *J. Exp. Mar. Biol. Ecol.* **35**:99–107.

Benninger, L. K., Aller, R. C., Cochran, J. K., and Turekian, K. K., 1979, Effects of biological sediment mixing on the ^{210}Pb chronology and trace metal distribution in a Long Island Sound sediment core, *Earth Planet. Sci. Lett.* **43**:241–259.

Ben-Yaakov, S., 1973, pH buffering of pore water of recent anoxic marine sediments, Limnol. Oceanogr. **18**:86–94.

Berner, R. A., 1974, Kinetic models for the early diagenesis of nitrogen, sulfur, phosphorus and silicon in anoxic marine sediments, in: *The Sea* (E. D. Goldberg, ed.), Volume 5, pp. 427–450, John Wiley and Sons, New York.

Berner, R. A., 1976a, The benthic boundary layer from the viewpoint of a geochemist, in: *The Benthic Boundary Layer* (I. N. McCave, ed.), pp. 33–55, Plenum Press, New York.

Berner, R. A., 1976b, Inclusion of adsorption in the modelling of early diagenesis, Earth Planet. Sci. Lett. **29**:333–340.

Berner, R. A., 1980, *Early Diagenesis—A Theoretical Approach*, Princeton University Press, Princeton, New Jersey.

Berner, R. A., 1981, A rate model for organic matter decomposition during bacterial sulfate reduction in marine sediments, Colloq. Int. C.N.R.S. **293**:35–44.

Billen, G., 1978, A budget of nitrogen recycling in North Sea sediments off the Belgian coast, Estuarine Coastal Mar. Sci. **7**:127–146.

Brenchley, G. A., 1978, On the regulation of marine infaunal assemblages at the morphological level: A study of the interactions between sediment stabilizers, destabilizers, and their sedimentary environment, Ph.D. dissertation, Johns Hopkins University, Baltimore, Maryland.

Cadée, G. C., 1976, Sediment reworking by *Arenicola marina* on tidal flats in the Dutch Wadden Sea, Neth. J. Sea Res. **10**:440–460.

Carpenter, R., Peterson, M. L., and Bennett, J. T., 1982, ^{210}Pb-derived sediment accumulation and mixing rates for the Washington continental slope, Mar. Geol. **48**: (in press).

Claypool, G. E., and Kaplan, I. R., 1974, The origin and distribution of methane in marine sediments, in: *Natural Gases in Marine Sediments* (I. R. Kaplan, ed.), pp. 99–139, Plenum Press, New York.

Cochran, J. K., and Aller, R. C., 1979, Particle reworking in sediments from the New York Right Apex: Evidence from ^{234}Th/^{238}U disequilibrium, Estuarine Coastal Mar. Sci. **9**:739–747.

Coyer, P. E., and Mangum, C. P., 1973, Effect of temperature on active and resting metabolism in polychaetes, in: *Effects of Temperature on Ectothermic Organisms* (W. Wieser, ed.), pp. 173–180, Springer-Verlag, New York.

Cullen, D. J., 1973, Bioturbation of superficial marine sediments by interstitial meiobenthos, Nature **242**:323–324.

Dapples, E. C., 1942, The effect of macro-organisms upon near-shore marine sediments, J. Sediment. Petrol. **12**:118–126.

DeMaster, D. J., Nittrouer, C. A., Cutshall, N. H., Larsen, I. L., and Dion, E. P., 1980, Short lived radionuclide profiles and inventories from Amazon continental shelf sediments, Eos **61**:1004.

Fager, E. W., 1964, Marine sediments: Effects of a tube-building polychaete, Science **143**:356–359.

Fenchel, T., 1970, Studies on the decomposition of organic detritus derived from the turtle grass *Thalassia testudinum*, Limnol. Oceanogr. **15**:14–20.

Fenchel, T., and Harrison, P., 1976, The significance of bacterial grazing and mineral cycling for the decomposition of particulate detritus, in: *The Role of Terrestrial and Aquatic Organisms in Decomposition Processes* (J. M. Anderson and A. Macfadyen, eds.), Proc. Symp. Br. Ecol. Soc. **17**:285–299.

Fenchel, T., Kofoed, L. H., and Lappalainen, A., 1975, Particle size-selection of two deposit feeders: The amphipod *Corophium volutator* and the prosobranch *Hydrobia ulval*, Mar. Biol. **30**:119–128.

Filipek, L. H., and Owen, R. M., 1980, Early diagenesis of organic carbon and sulfur in outer shelf sediments from the Gulf of Mexico, *Am. J. Sci.* **280**:1097–1112.

Fisher, J. B., Lick, W. J., McCall, P. L., and Robbins, J. A., 1980, Vertical mixing of lake sediments by tubificid oligochaetes, *J. Geophys. Res.* **85**:3997–4006.

Foster-Smith, R. L., 1978, An analysis of water flow in tube-living animals, *J. Exp. Mar. Biol. Ecol.* **34**:73–95.

Froelich, P. N., Klinkhammer, G. P., Bender, M. L., Luedtke, N. A., Heath, G. R., Cullen, D., Dauphin, P., Hammond, D., Hartman, B., and Maynard, V., 1979, Early oxidation of organic matter in pelagic sediments of the eastern equatorial Atlantic: Soboxic diagenesis, *Geochim. Cosmochim. Acta* **43**:1075–1091.

Goldberg, E. D., and Koide, M., 1962, Geochronological studies of deep sea sediments by the ionium/thorium method, *Geochim. Cosmochim. Acta* **26**:417–450.

Goldhaber, M. B., and Kaplan, I. R., 1974, The sulfur cycle, in: *The Sea* (E. D. Goldberg, ed.), Volume 5, pp. 569–655, John Wiley and Sons, New York.

Goldhaber, M. B., Aller, R. C., Cochran, J. K., Rosenfeld, J. K., Martens, C. S., and Berner, R. A., 1977, Sulfate reduction diffusion and bioturbation in Long Island Sound sediments: Report of the FOAM group, *Am. J. Sci.* **277**:193–237.

Grassle, J. F., and Grassle, J. P., 1974, Opportunistic life histories and genetic systems in marine benthic polychaetes, *J. Mar. Res.* **32**:253–284.

Greenwood, D. J., 1968, Measurement of microbial metabolism in soil, in: *The Ecology of Soil Bacteria* (T. R. G. Gray and D. Parkinson, eds.), pp. 138–157, University of Toronto Press, Toronto.

Grill, E. V., and Richards, R. A., 1964, Nutrient regeneration from phytoplankton decomposing in sea water, *J. Mar. Res.* **22**:51–69.

Grundmanis, V., and Murray, J. W., 1977, Nitrification and denitrification in marine sediments from Puget Sound, *Limnol. Oceanogr.* **22**:804–813.

Guinasso, N. L., Jr., and Schink, D. R., 1975, Quantitative estimates of biological mixing rates in abyssal sediments, *J. Geophys. Res.* **80**:3032–3043.

Hammond, D. E., and Fuller, C., 1979, The use of radon-222 as a tracer in San Francisco Bay, in: *San Francisco Bay: The Urbanized Estuary* (T. J. Conomos, ed.), pp. 213–230, American Association for the Advancement of Science, San Francisco, California.

Hammond, D. E., Simpson, H. J., and Mathieu, G., 1975, Methane and radon-222 as tracers for mechanisms of exchange across the sediment–water interface in the Hudson River Estuary, in: *Marine Chemistry in the Coastal Environment* (T. M. Church, ed.), ACS Symposium Series 18, pp. 119–132, American Chemical Society, Washington, D.C.

Hammond, L. S., 1981, An analysis of grain size modification in biogenic carbonate sediments by deposit-feeding holothurians and echinoids (Echinodermata), *Limnol. Oceanogr.* **26**:898–906.

Hanor, J. S., and Marshall, N. T., 1971, Mixing of sediment by organisms, in: *Trace Fossils* (B. F. Perkins, ed.), Louisiana State University Miscellaneous Publication 71-1, pp. 127–136, Louisiana State University Press, Baton Rouge.

Hargrave, B. T., 1970, The effect of a deposit-feeding amphipod on the metabolism of benthic microflora, *Limnol. Oceanogr.* **15**:21–30.

Hargrave, B. T., 1972, Aerobic decomposition of sediment and detritus as a function of particle surface area and organic content, *Limnol. Oceanogr.* **17**:583–596.

Hargrave, B. T., 1976, The central role of invertebrate faeces in sediment decomposition, in: *The Role of Terrestrial and Aquatic Organisms in Decomposition Processes* (J. M. Anderson and A. Macfadyen, eds.), *Proc. Symp. Br. Ecol. Soc.* **17**:301–321.

Hargrave, B. T., and Phillips, G. A., 1977, Oxygen uptake of microbial communities on solid surfaces, in: *Aquatic Microbial Communities* (J. Cairns, Jr., ed.), pp. 545–587, Garland, New York.

Harrison, P. G., and Mann, K. H., 1975, Detritus formation from eelgrass (*Zostera marina* L.): The relative effects of fragmentation, leaching, and decay, *Limnol. Oceanogr.* **20**:924–934.

Hoffman, R. J., and Mangum, C. P., 1970, The function of coelomic cell hemoglobin in the polychaete *Glycera dibranchiata*, *Comp. Biochem. Physiol.* **36**:211–228.

Hurd, D. C., 1973, Interactions of biogenic opal, sediment and seawater in the central equatorial Pacific, *Geochim. Cosmochim. Acta* **37**:2257–2282.

Hylleberg, J., 1975, Selective feeding by *Abarenicola pacifica* with notes on *Abarenicola vagabunda* and a concept of gardening in lugworms, *Ophelia* **14**:113–137.

Jones, J. D., 1955, Observations on the respiratory physiology and on the haemoglobin of the polychaete genus *Nephthys*, with special reference to *N. hombergii* (Aud. et M. Edw.), *J. Exp. Biol.* **32**:110–125.

Jørgensen, B. B., 1977a, The sulfur cycle of a coastal marine sediment (Linfjorden, Denmark), *Limnol. Oceanogr.* **22**:814–831.

Jørgensen, B. B., 1977b, Bacterial sulfate reduction within reduced microniches of oxidized marine sediments, *Mar. Biol.* **41**:7–17.

Jørgensen, B. B., 1978, A comparison of methods for the quantification of bacterial sulfate reduction in coastal marine sediments. II. Calculation from mathematical models, *Geomicrobiol. J.* **1**:29–48.

Jumars, P. A., Nowell, A. R. M., and Self, R. F. L., 1981, A Markov model of flow–sediment–organism interactions: Model formulation and sensitivity analysis, in: *Sedimentary Dynamics of Continental Shelves* (C. A. Nittrouer, ed.), pp. 155–172, Elsevier, Amsterdam.

Kamatani, A., and Riley, J. P., 1979, Rate of dissolution of diatom silica walls in seawater, *Mar. Biol.* **55**:29–35.

Khripounoff, A., and Sibuet, M., 1980, La nutrition d'echinodermes abyssaux. I. Alimentation des holothuries, *Mar. Biol.* **60**:17–26.

Korosec, M. A., 1979, The effects of biological activity on transport of dissolved species across the sediment–water interface of San Francisco Bay, M.S. thesis, University of Southern California, Los Angeles.

Krishnaswami, S., Benninger, L. K., Aller, R. C., and Van Damm, K. L., 1980, Atmospherically-derived radionuclides as tracers of sediment mixing and accumulation in nearshore marine and lake sediments: Evidence from ^{7}Be, ^{210}Pb, and ^{239,240}Pu, *Earth Planet. Sci. Lett.* **47**:307–318.

Krüger, F., 1959, Zur Ernährungsphysiologie von *Arenicola marina* L., *Zool. Anz. Suppl.* **22**:115–120.

Lerman, A., 1978, Chemical exchange across sediment–water interface, *Annu. Rev. Earth Planet. Sci.* **6**:281–303.

Lerman, A., 1979, *Geochemical Processes: Water and Sediment Environments*, John Wiley and Sons, New York.

Levinton, J. S., 1980, Particle feeding by deposit feeders: Models, data and a prospectus, in: *Marine Benthic Dynamics* (K. R. Tenore and B. C. Coull, eds.), pp. 423–439, University of South Carolina Press, Columbia.

Li, Y.-H., and Gregory, S., 1974, Diffusion of ions in sea water and in deep sea sediments, *Geochim. Cosmochim. Acta* **38**:703–714.

Livingston, H. D., and Bowen, V. T., 1979, Pu and ^{137}Cs in coastal sediments, *Earth Planet. Sci. Lett.* **43**:29–45.

Luedtke, N. A., and Bender, M. L., 1979, Tracer study of sediment–water interactions in estuaries, *Estuarine Coastal Mar. Sci.* **9**:643–651.

McCaffrey, R. J., Myers, A. C., Davey, E., Morrison, G., Bender, M., Luedtke, N., Cullen, D., Froelich, P., and Klinkhammer, G., 1980, The relation between pore water chemistry and benthic fluxes of nutrients and manganese in Narragansett Bay, Rhode Island, *Limnol. Oceanogr.* **25**:31–44.

McCall, P. L., 1977, Community patterns and adaptive strategies of the infaunal benthos of Long Island Sound, *J. Mar. Res.* **35**:221–266.

Mangum, C. P., 1964, Activity patterns in metabolism and ecology of polychaetes, *Comp. Biochem. Physiol.* **11**:239–256.

Mangum, C. P., and Burnett, L. E., 1975, The extraction of oxygen by estuarine invertebrates, in: *Physiological Ecology of Estuarine Organisms* (F. J. Vernberg, ed.), pp. 147–163, University of South Carolina Press, Columbia.

Mangum, C. P., Santos, S. L., and Rhodes, W. R., Jr., 1968, Distribution and feeding in the onuphid polychaete, *Diopatra cuprea* (Bosc), *Marine Biol.* **2**:33–40.

Mare, M. F., 1942, A study of a marine benthic community with special reference to the microorganisms, *J. Mar. Biol. Assoc. U. K.* **25**:517–574.

Martens, C. S., and Klump, J. V., 1980, Biogeochemical cycling in an organic-rich coastal marine basin. I. Methane sediment–water exchange processes, *Geochim. Cosmochim. Acta* **44**:471–490.

Myers, A. C., 1977, Sediment processing in a marine subtidal sandy bottom community. I. Physical aspects, *J. Mar. Res.* **35**:609–632.

Noshkin, V. E., and Bowen, V. T., 1973, Concentrations and distributions of long-lived fallout radionuclides in open ocean sediments, in: *Radioactive Contamination of the Marine Environment*, pp. 671–686, International Atomic Energy Agency, Vienna, Austria.

Nittrouer, C. A., Sternberg, R. W., Carpenter, R., and Bennett, J. T., 1979, The use of Pb-210 geochronology as a sedimentological tool: Application to the Washington continental shelf, *Mar. Geol.* **31**:297–316.

Nixon, S. W., Kelly, J. R., Furnas, B. N., Oviatt, C. A., and Hale, S. S., 1980, Phosphorus regeneration and the metabolism of coastal marine bottom communities, in: *Marine Benthic Dynamics* (K. R. Tenore and B. C. Coull, eds.), pp. 219–242, University of South Carolina Press, Columbia.

Okubo, A., 1971, Oceanic diffusion diagrams, *Deep Sea Res.* **18**:789–802.

Olsen, C. R., Simpson, H. J., Ping, T.-H., Bopp, R. F., and Trier, R. M., 1981, Sediment mixing and accumulation rate effects on radionuclide depth profiles in Hudson estuary sediments, *J. Geophys. Res.* **86**:11020–11028.

Pearson, T. H., and Rosenberg, R., 1978, Macrobenthic succession in relation to organic enrichment and pollution of the marine environment, *Oceanogr. Mar. Biol. Annu. Rev.* **16**:229–311.

Peng, T. H., Broecker, W. S., and Berger, W. H., 1979, Rates of benthic mixing in deep-sea sediment as determined by radioactive tracers, *Quat. Res.* (N.Y.) **11**:141–149.

Powell, E. N., 1977, Particle size selection and sediment reworking in a funnel feeder, *Leptosynapta tenuis* (Holothuroedea, Synaptidae), *Int. Rev. Ges. Hydrobiol.* **62**:385–408.

Pryor, W. A., 1975, Biogenic sedimentation and alteration of argillaceous sediments in shallow marine environments, *Geol. Soc. Am. Bull.* **86**:1244–1254.

Redfield, A. C., 1934, On the proportion of organic derivatives in sea water and their relation to the composition of plankton, in: *James Johnstone Memorial Volume*, pp. 176–192, University Press, Liverpool.

Rhoads, D. C., 1974, Organism–sediment relations on the muddy sea floor, *Oceanogr. Mar. Biol. Annu. Rev.* **12**:263–300.

Rhoads, D. C., and Stanley, D. J., 1965, Biogenic graded bedding, *J. Sediment. Petrol.* **35**:956–963.

Rhoads, D. C., Aller, R. C., and Goldhaber, M. B., 1977, The influence of colonizing benthos on physical properties and chemical diagenesis of the estuarine seafloor, in: *Ecology of Marine Benthos* (B. C. Coull, ed.), pp. 113–138, University of South Carolina Press, Columbia.

Rhoads, D. C., Yingst, J. Y., and Ullman, W. J., 1978, Seafloor stability in central Long Island Sound. Part I. Temporal changes in erodibility of fine-grained sediment, in: *Estuarine Interactions* (M. L. Wiley, ed.), pp. 221–244, Academic Press, New York.

Richards, F. A., 1965, Anoxic basins and fjords, in: *Chemical Oceanography* (J. P. Riley and G. Skirrow, eds.), Volume 1, pp. 611–645, Academic Press, New York.

Riedl, R. J., Huang, N., and Machan, R., 1972, The subtidal pump: A mechanism of interstitial water exchange by wave action, *Mar. Biol.* **13**:210–221.

Rijken, M., 1979, Food and food uptake in *Arenicola marina*, *Neth. J. Sea Res.* **13**:406–421.

Santschi, P. H., Li, Y.-H., Bell, J. J., Trier, R. M., and Kawtaluk, K., 1980, Pu in coastal marine environments, *Earth Planet. Sci. Lett.* **51**:248–265.

Schäfer, W., 1972, *Ecology and Paleoecology of Marine Environments* (G. Y. Craig, ed.; I. Oertel, translator), University of Chicago Press, Chicago.

Schink, D. R., and Guinasso, N. L., Jr., 1977, Effects of bioturbation on sediment–seawater interaction, *Mar. Geol.* **23**:133–154.

Schink, D. R., and Guinasso, N. L., Jr., 1978, Redistribution of dissolved and adsorbed materials in abyssal marine sediments undergoing biological stirring, *Am. J. Sci.* **278**:687–702.

Schink, D. R., Guinasso, N. L., Jr., and Fanning, K. A., 1975, Processes affecting the concentration of silica at the sediment–water interface of the Atlantic Ocean, *J. Geophys. Res.* **80**:3013–3031.

Self, R. F. L., and Jumars, P. A., 1978, New resource axes for deposit feeders? *J. Mar. Res.* **36**:627–641.

Sepkoski, J. J., Jr., and Bambach, R. K., 1979, The temporal restriction of flat-pebble conglomerates: An example of co-evolution of organisms and sediments, *Geol. Soc. Am. Abst.* **11**:256.

Sholkovitz, E., 1973, Interstitial water chemistry of the Santa Barbara Basin sediments, *Geochim. Cosmochim. Acta* **37**:2043–2073.

Smethie, W. M., Jr., Nittrouer, C. A., and Self, R. F. L., 1981, The use of radon-222 as a tracer of sediment irrigation and mixing on the Washington continental shelf, in: *Sedimentary Dynamics of Continental Shelves* (C. A. Nittrouer, ed.), pp. 173–200, Elsevier, Amsterdam.

Smith, K. L., Jr., 1978, Benthic community respiration in the N.W. Atlantic Ocean: In situ measurements from 40 to 5200 m., *Mar. Biol.* **47**:337–347.

Sørensen, J., 1978, Capacity for denitrification and reduction of nitrate to ammonia in a coastal marine sediment, *Appl. Environ. Microbiol.* **35**:301–305.

Stumm, W., and Morgan, J. J., 1970, *Aquatic Chemistry*, John Wiley and Sons, New York.

Thayer, C. W., 1979, Biological bulldozers and the evolution of marine benthic communities, *Science* **203**:458–461.

Thistle, D., 1979, Harpacticoid copepods and biogenic structures: Implications for deep-sea diversity maintenance, in: *Ecological Processes in Coastal and Marine Systems* (R. J. Livingston, ed.), pp. 217–231, Plenum Press, New York.

Thompson, R. K., and Pritchard, A. W., 1969, Respiratory adaptations of two burrowing crustaceans, *Callianassa californiensis* and *Upogebia pugettensis* (Decapoda, Thalassinidea), *Biol. Bull.* **136**:274–287.

Thorstenson, D. C., 1970, Equilibrium distribution of small organic molecules in natural waters, *Geochim. Cosmochim. Acta* **34**:745–770.

Turekian, K. K., Cochran, J. K., and DeMaster, D. J., 1978, Bioturbation in deep-sea deposits: Rates and consequences, *Oceanus* **21**:34–41.

Turekian, K. K., Cochran, J. K., Benninger, L. K., and Aller, R. C., 1980, The sources and sinks of nuclides in Long Island Sound, *Adv. Geophys.* **22**:129–164.

Ullman, W. J., and Aller, R. C., 1982, Diffusion coefficients in nearshore marine sediments, *Limnol. Oceanogr.* **27**:552–556.

Vanderborght, J. P., and Billen, G., 1975, Vertical distribution of nitrate in interstitial water of marine sediments with nitrification and denitrification, *Limnol. Oceanogr.* **20**:953–961.

Vanderborght, J. P., Wollast, R., and Billen, G., 1977, Kinetic models of diagenesis in disturbed sediments. I. Mass transfer properties and silica diagenesis, *Limnol. Oceanogr.* **22**:787–793.

van Straaten, L. M. J. U., 1952, Biogene textures and the formation of shell beds in the Dutch Wadden Sea. I–II, *Koninkl. Nederl. Akad. Wet. Proc. Ser. B* **55**:500–516.

Westrich, J., 1983, Ph.D. dissertation, Yale University, New Haven, Connecticut (in preparation).

Whitlatch, R. B., 1974, Food-resource partitioning in the deposit-feeding polychaete *Pectinaria gouldii*, *Biol. Bull.* **147**:227–235.

Yingst, J. Y., and Rhoads, D. C., 1980, The role of bioturbation in the enhancement of microbial turnover rates in marine sediments, in: *Marine Benthic Dynamics* (K. R. Tenore and B. C. Coull, eds.), pp. 407–422, University of South Carolina Press, Columbia.

Zeitzschel, B., 1980, Sediment–water interactions in nutrient dynamics, in: *Marine Benthic Dynamics* (K. R. Tenore and B. C. Coull, eds.), pp. 195–218, University of South Carolina Press, Columbia.

II

Recent Freshwater Environments

Chapter 3

The Effects of Benthos on Physical Properties of Freshwater Sediments

PETER L. McCALL and MICHAEL J. S. TEVESZ

1. Introduction	105
2. Freshwater Sediments and Macrobenthos	106
3. Macrobenthos Life-Styles	113
3.1. Chironomids	113
3.2. Amphipods	116
3.3. Oligochaetes	118
3.4. Bivalves	120
4. Mixing of Sediments	124
4.1. Effects on Sediment Stratigraphy	124
4.2. Effects on Particle Size Distribution	139
4.3. Effects on Mass Properties of Sediments	146
5. Biogenic Modification of Sediment Transport	150
5.1. Field Observations	150
5.2. Laboratory Experiments	157
5.3. Some Comparisons	160
6. Ecological Interactions and Sediment Properties	161
6.1. Spatial–Temporal Variations of Fauna	161
6.2. Interactions among Macrofauna	163
7. Future Work	166
References	168

1. Introduction

In this chapter we review the effects of freshwater benthos on the physical (as opposed to chemical) properties of the bottom. Specifically, we will focus our discussion on the effects of macrobenthos (adult length >1 mm) on fine-grained bottoms (sediments that contain approximately 50% by

---

PETER L. McCALL • Department of Geological Sciences, Case Western Reserve University, Cleveland, Ohio 44106.   MICHAEL J. S. TEVESZ • Department of Geological Sciences, Cleveland State University, Cleveland, Ohio 44115.

weight silt–clay-sized particles) of lakes and slow-flowing rivers. There are three reasons for this approach. The first is that there is simply too little known about freshwater meio- and microbenthos to merit a review of their effects on sediment properties. More importantly, the macrobenthos are probably the most potent modifiers of sediment properties by virtue of their size relative to sediment grains, their population density, their ability to move through a relatively large volume of sediment, and their feeding and respiratory habits. Finally, we have restricted ourselves to fine-grained bottoms because these comprise the bulk of freshwater lake sediments and because the structure of this sediment is more easily altered than is that of sediments with larger grain size. Some information from slow-flowing river bottoms is included with lakes because their sediments and faunas are very much alike (Hynes, 1970).

We will first describe some typical macrobenthic communities of lakes and isolate some important organisms for study. Next we will look at the biogenic modification of sediment grain size, mass properties, organic matter, stratigraphy, and erodibility by these organisms. Finally, we will look briefly at the temporal effects on sediment properties produced by succession and seasonality and suggest some directions for future research.

2. Freshwater Sediments and Macrobenthos

In a recent review of sedimentary processes in lakes, Sly (1978) delineated three major differences between lacustrine and marine environments. These are:

1. The smaller size of most lakes compared to oceanic environments means that long-period waves are less frequent and that bottom energy levels are below those of oceanic systems; for this reason, well-sorted coarse sands and gravels are confined to the shallowest portions of larger lakes and the average grain size of lake bottoms is smaller than that in shelf environments (Fig. 1).
2. Because lakes are closed systems with respect to sediment transport, and because of the usually larger ratio of drainage basin to lake area, sedimentation rates in lakes are higher (by about a factor of ten) than in many marine environments.
3. Lakes are almost tideless, so that intertidal areas are nearly absent.

In many respects, except for major differences in salinity fluctuations and tidal motions, the shallower regions of large lakes resemble nearshore marine estuarine environments, particularly in their often mutually high

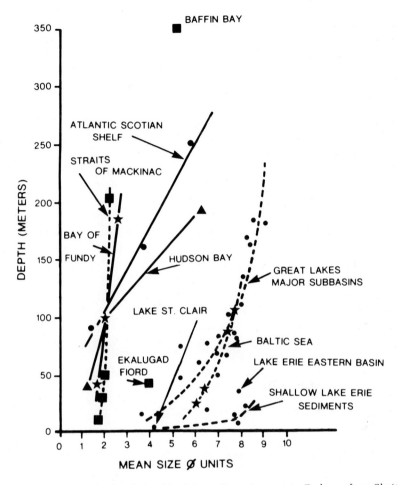

Figure 1. Particle size–depth relationships in aquatic environments. Redrawn from Sly (1978).

sedimentation rates and dominance of silt–clay-sized sediment. The number of important mineral phases in lakes is not large. Lacustrine sediments are mostly quartz, feldspar, calcite, dolomite, the clay minerals illite and smectite, apatite, and iron sulfides and oxides (Jones and Bouser, 1978). There is a wider range of organic matter in muddy lakes (*gyttja*, with mostly autocthonously produced organic matter, as opposed to *dy*, or peaty allocthonous sediment)—from close to zero up to about 50% organic matter by weight (Wetzel, 1975)—than in most marine environments, where organic carbon content is typically less than 10% by weight.

Soft-bottom marine communities are dominated in terms of numbers and biomass by polychaete annelids, bivalve molluscs, and amphipods (Sanders, 1956, 1968; Sanders and Hessler, 1969). The soft bottoms of freshwater lakes and slow-flowing rivers are dominated by chironomid insect larvae and oligochaete annelids that together appear to fill a niche similar to that of the marine polychaetes, bivalve molluscs of two types, and amphipods, not necessarily all together or in that order of importance. Although lakes are variable bodies of water, examples will show some typical patterns of community composition and distribution.

Lake Esrom is a well-studied moderately eutrophic lake located in Denmark. It has an area of 17.3 km^2, is about 8 km long and 2–3 km wide, and has a maximum depth of 22 m. Rooted vegetation occurs in a relatively flat narrow belt along the shore to a depth of about 6 m (28% of total lake area). Steep slopes are present from 6–10 m with less steep slopes below 10 m. The bottom of the lake below 10 m (52% of total lake area) is covered with silt–clay sediment, and at the greatest depth little oxygen is present in the water from July to November (Jonasson, 1972). As in many other lakes, sedimentation rate and organic matter content of the sediment increase with depth below the epilimnion (Fig. 2).

In most lakes at least two macrofaunal communities can be distinguished: a shallower *littoral* community whose boundaries are deter-

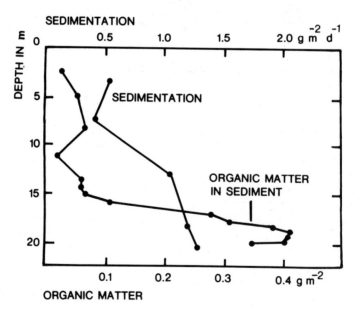

Figure 2. Sedimentation rates and organic matter concentrations in Lake Esrom. Redrawn from Jonasson (1978).

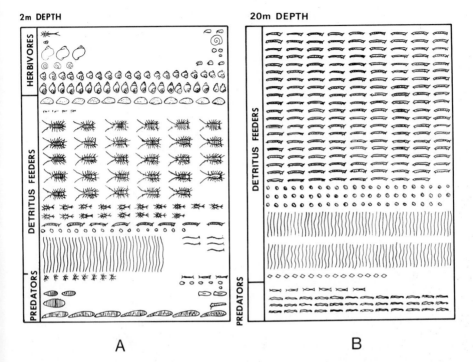

Figure 3. (A) Littoral (2 m) community of Lake Esrom. Numbers per 0.02 m². Herbivores: *Centroptilum luteolum, Micronecta minutissima, Lymnaea pereger, Gyraulus albus, Valvata piscinalis, Bithynia tentaculata, Oxyethira flavicornis, Planorbis planorbis, Armiger crista,* and *Eurycercuc lamellatus*. Detritus feeders: *Dreissena polymorpha, Canthocamptus staphylinus, Asellus aquaticus, Caenis moesta,* Chironomidae, *Pisidium* sp., and Oligochaeta and Ceratopogonidae. Predators: Hydracarina, *Helobdella stagnalis, Glossophonia complanata, Herpobdella octoculata* and *H. testacea,* Tanypodinae, *Candona candida* and *C. neglecta, Polycelis tenuis,* and Leptoceridae. (B) Profundal community (20 m) of Lake Esrom. Numbers per 0.02 m². Detritus feeders: *Chironomus anthracinus, Pisidium casertanum* and *P. subtruncatum,* and *Potamothrix hammoniensis* and coccoons. Predators: *Procladius pectinatus* and *Chaorborus flavicans*. Redrawn from Jonasson (1978).

mined by the shore of the lake and the depth to which rooted vegetation extends, and a deeper *profundal* community that encompasses the lake bottom below the zone of rooted vegetation. About 300 benthic metazoan species are found in the littoral macrophyte community of Lake Esrom (Jonasson, 1978). The relative abundance of macrofaunal dominants is shown in Fig. 3A. The abundance of macrophytes in this shallow environment gives rise to a multitude of microenvironments and thus high species diversity and a wide range of trophic types, notably grazing herbivores and carnivores in addition to deposit and suspension feeders. In the muddy profundal of Esrom (>10 m), diversity drops to 20 species,

Table Ia. Relative Abundances and Biomass of Different Animal Groups in Zoobenthos Samples Taken from Lake Võrtsjärv in July–August, 1965[a]

	Relative abundance (percent)	
	Littoral	Sublittoral and profundal
Chironomidae	35.8	64.5
Oligochaeta	9.4	25.1
Mollusca		
Small		
Gastropoda	8.6	2.8
Pisidiidae	8.0	1.2
Big clams		
Unionidae	0.3	0.1
Dreissena	2.3	
Varia		
Turbellaria	<0.1	0.1
Nematoda	1.6	1.0
Hirudinea	4.6	0.3
Asellus	16.9	
Gammarus	0.5	
Araneida	0.1	
Hydracarina	2.0	4.4
Ceratopogonidae	4.1	0.3
Diptera (others)	0.3	
Ephemeroptera	0.8	0.1
Trichoptera	2.9	
Lepidoptera	<0.1	
Coleoptera	1.4	
Heteroptera	<0.1	
Sialis	<0.1	
	Biomass	
Total abundance m^{-2}	2270	907
Total weight (g · m^2) with large Mollusca	20.08	9.42
Total weight (g · m^2)	41.69	14.50

[a] From Timm (1975).

and the dominant species are all deposit and/or suspension feeders (Fig. 3B). Chironomid insect larvae, tubificid oligochaetes, and sphaerid bivalves are the most abundant organisms here. Jonasson (1972, 1978) defines two other communities in Lake Esrom that are not found in all (or even most) lakes: a muddy sublittoral community (6–10 m) distinguished primarily by the overwhelming dominance in terms of numbers and bio-

Table Ib. Relative Abundance of the Bottom Fauna at 30 and 50 m Depth in the Central Part of Lake Mälaren, 1969–1976[a]

	Relative abundance (percent)	
	30 m	50 m
Crustacea	4.0	52.2
Pontoporeia affinis		
Diptera, Chironomidae	27.5	6.5
Procladius spp.		
Micropsectra spp.		
Tanytarsus spp.		
Phaenopsectra coracina		
Monodiamesa sp.		
Mollusca	6.3	4.8
Pisidium personatum		
P. casertanum		
P. henslowanum		
P. subtruncatum		
Oligochaeta	57.4	35.4
Potamothrix hammoniensis		
Limnodrilus hoffmeisteri		
Tubifex tubifex		
T. ignotus		
Aulodrilus pluriseta		
Peloscolex ferox		
Nemertini	2.5	0.7
Prostoma sp.		
Turbellaria	2.1	0.4
Other taxa	0.1	0.1

[a] From Wiederholm (1978), 0.6-mm sieve mesh.

mass of the large freshwater bivalve *Dreissena polymorpha*, and a gravel surf zone community. *Dreissena* is restricted to Eurasian lakes; in North America we find corbiculid and the larger unionid clams taking its place, although they are not necessarily present as a distinct depth-defined community dominant.

Timm (1975) described the benthic macrofauna of a larger (270 km^2) but shallower (average depth 2.7 m) eutrophic lake in Estonia, Lake Vôrtsjärv. The littoral region, 0–1.5 m, covers 15% of the lake area, while the profundal, 1.5–6 m, makes up 62% of the lake area. Once again we see the dominance of chironomids, oligochaetes, and molluscs (Table Ia) and the increasing importance of chironomids and oligochaetes in the profundal zone (Fig. 4). Species lists from the profundal of the larger oligotrophic Great Slave Lake in Canada and Lake Mälaren in Sweden

Table Ic. Comparisons of Relative Abundances and Biomass of Organisms at Depths of 0–100 m in the Delta Region and in the Main Part of Great Slave Lake[a]

Organism	Relative abundance (percent)	
	Main lake	Delta
Amphipoda	73.6	60.8
Sphaeriidae	4.7	2.0
Oligochaeta	10.5	31.2
Chironomidae	4.8	4.0
Ostracoda	4.2	0.9
Gastropoda	0.7	0
Nematoda	0.3	0.8
Miscellaneous	1.1	0.3
	Biomass	
Total organisms m^{-2}	1539	2328
Total weight (g · m^2)	2.5	3.29

[a] From Rawson (1953).

(Table Ib,c) illustrate the importance of these groups and also the occasional dominance of crustacean amphipods.

The lessons we should take away from these descriptions are that most of the bottom of most lakes is covered with fine-grained sediments (silts and clays) with varying amounts of organic matter and that the

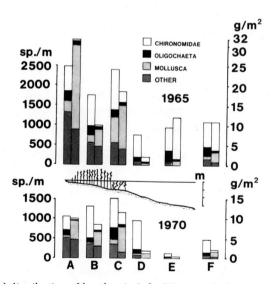

Figure 4. Vertical distribution of benthos in Lake Vôrtsjärv. Redrawn from Timm (1975).

Effects of Benthos on Freshwater Sediment Physical Properties

dominant organisms on these bottoms in terms of numbers and biomass are typically chironomids, oligochaetes, amphipods, and bivalves—a small variety (pisidiids) and a large variety (unionids).

3. Macrobenthos Life-Styles

We will examine the activities of macrobenthos that affect sediment properties in greater detail later, but it will be useful here to describe briefly the life positions, feeding, mobility, and life histories of important macrobenthos (see also Chapter 4).

3.1. Chironomids

The Chironomidae (common name "midges") are a family of dipteran insects whose closest familiar relatives are the mosquitoes. The family is distributed worldwide, mostly in aquatic environments. Members of the subfamily Chironominae and genus *Chironomus* are particularly abundant in the profundal zone of lakes. The life history of the group has been reviewed by Oliver (1971). Most of the life cycle is spent in the larval stage (Fig. 5), which lasts from several weeks in some tropical forms to 2 years in a few more northerly species. The adult flying stage lasts from a few days to a few weeks at most and is used for dispersal, mating, and oviposition. All the energy required to complete the life cycle is obtained in the larval stage, because the adults do not feed. Most temperate species

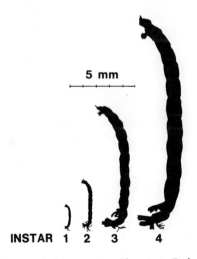

Figure 5. Relative size of instars of *Chironomus anthracinus*. Redrawn from Jonasson (1972).

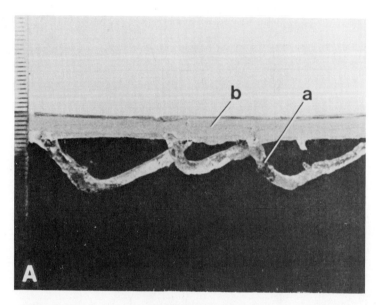

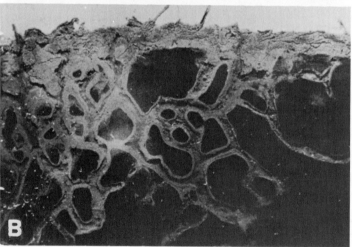

Figure 6. (A) Resin case of U-shaped tubes of Chironomus plumosus. Scale in mm. Reproduced from McLachlan and Cantrell (1976) with permission of the publisher. (B) Chironomid burrows from Lake Erie sediments. Depth of sediment column: 10 cm. (C) Conical burrow of C. anthracinus from Lake Esrom. Reproduced from Jonasson (1972) with permission of Oikos.

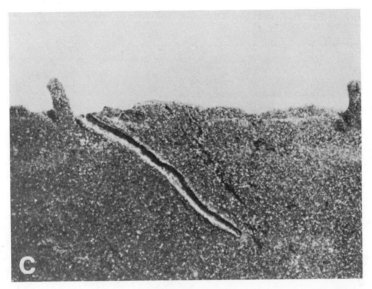

Figure 6. (cont'd)

have one or two generations per year, and a few have one generation every 2 years. Eggs are deposited in water and larvae develop in the lake bottom, where they undergo four molts (instars) before pupating and emerging from the lake as flying adults. The size and weight of larval chironomids depends on a variety of factors such as temperature and food supply, but fourth-instar chironomids are likely to be 10–22 mm in length and weigh 1–2 mg dry (Jonasson, 1972; Mackey, 1977; Hilsenoff, 1967; M. S. Johnson and Munger, 1930).

Larvae burrow into bottom sediments; most larvae live in the upper 8–10 cm of the substratum, with occasional occurrences at 40- to 50-cm depths (Cole, 1953; Ford, 1967; Hilsenoff, 1966). The burrowing and feeding behavior of a few chironomid species has been studied. Most chironomids live in tubes lined with a transparent, fibrous salivary secretion. It is thought that the very abundant and widespread species *C. plumosus* typically constructs a U-shaped tube in the bottom (Fig. 6A) and filter feeds by drawing water by means of anterioposterior undulations of its body through a net spun in the lumen of the tube (Walshe, 1947, 1951). But *C. plumosus* feeds in another very different way as well. Larvae may perform as surface deposit feeders, scraping the top 1–2 mm of sediment around their burrows for detritus and bacteria attached to fine-grained sediment particles; they may also occasionally feed on pieces of higher vegetation (Konstantinov, 1969). And not all burrows are U-shaped. Some are straight tubes; some are of more complex designs (Fig. 6B) (McLachlan and Cantrell, 1976).

Table II. Faunal Groups Captured in Bottom Emergence Trap, Western Basin, Lake Erie, 1979[a]

Faunal group	No. m^{-2} captured 28 August–26 September (29 days)	No. m^{-2} captured 26 September–17 October (21 days)
Oligochaeta (Naididae)	6	2
Chironomidae	50 (+ 112 pupae)	13
Nematoda	0	0
Bivalvia (Sphaeridae)	9	4
Amphipoda	246	110

[a] Area = 0.46 m².

The amount of time and effort allocated to one feeding method or the other is not well known, but it is fairly well established that the bulk of the food of mud-dwelling *C. plumosus* comes from bottom sediment (Konstantinov, 1969; McLachlan, 1977). Burrowing behavior and feeding method may be under environmental control. For instance, we have observed simpler tube structure and evidence of more surface deposit feeding (greater abundance of sticky, salivary threads extending radially from tube openings) by *C. plumosus* in Lake Erie in sediments with a very high water content (>85% water by weight) and during low-oxygen conditions under the summer hypolimnion. Jonasson (1972) described the burrowing and feeding behavior of a related profundal species, *C. anthracinus*. This species normally inhabits a conically shaped tube (Fig. 6C) that extends above the interface like a chimney. During the summer stagnation period in Lake Esrom, *C. anthracinus* feeds on surface deposits in a circular area around its tube openings; when oxygen disappears from the water, feeding stops and the "chimney" decays; when overlying water is rich in oxygen and phytoplankton, the larvae apparently filter feed and construct small fanlike extensions of their tubes that may close the tube under storm conditions and prevent the entry of sediment into the burrow.

Mud-dwelling chironomids are very mobile, leaving the bottom (usually at night) and undergoing migration into the water column (Mundie, 1959). *C. plumosus* is one of the most mobile benthos in Lake Erie, as measured by its ability to leave the bottom and swim to a height of 1 m above the bottom (Table II). The individual frequency of reburrowing, however, is unknown.

3.2. Amphipods

There are several hundred species of these freshwater crustaceans (see Fig. 7 for morphology), but only a few—*Hyalella azteca*, several

Effects of Benthos on Freshwater Sediment Physical Properties   **117**

species of *Gammarus* (e.g., *fasciatus, pseudolimnaeus, lacustris*), and *Pontoporeia hoyi* (= *affinis*)—are widespread and abundant constituents of soft-bottom lake communities. *Hyalella* and *Gammarus* are usually found in the warmer and shallower regions of lakes (Bousfield, 1958; Hargrave, 1970a), while *P. hoyi* is more abundant in the deeper and colder parts of lakes in northern Europe, Asia, and North America, where the average temperature is less than 20°C (Bousfield, 1958; Smith, 1972). *P. hoyi* is the most abundant and productive component of the macrobenthos in the deeper regions of some northern lakes, including the upper Great Lakes (Juday and Birge, 1927; Eggleton, 1937, 1939; Adams and Kregar, 1969; Henson, 1970; Mozley and Alley, 1973; Freitag *et al.*, 1976). *Pontoporeia* is also found in shallow regions, especially in arctic and subarctic lakes and rivers (J. W. Moore, 1979).

Adult amphipods are typically 5–10 mm long and weigh 1–2 mg dry (Juday and Birge, 1927; Hargrave, 1970a; Alley and Chin, 1978). Their population density in lakes ranges from a few hundred to over 14,000 per square meter (Marzolf, 1965; Kraft, 1979). The life cycle is variable, depending on food, temperature, and oxygen conditions, as well as on the particular species under consideration. *Gammarus* species typically live for less than 1 year, mature rapidly, and produce up to six groups of eggs during their reproductive phase; gammarids are in reproductive condition in all months except October–January in north temperate regions (Macan and Mackereth, 1957; Hynes, 1955). *P. hoyi* is unusual, because its lifespan is about 2.5 years and reproduction does not begin until about 2 years.

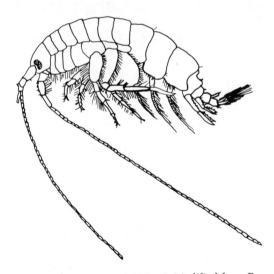

Figure 7. The amphipod *Pontoporeia* (×9). Modified from Pennack (1978).

The female produces one brood, usually but not always in winter, and then dies (Juday and Birge, 1927; Segerstrale, 1967).

While most freshwater amphipods are labeled as scavengers (Pennak, 1978), most profundal amphipods probably derive most of their nutrition from bacteria and algae obtained from surface deposit feeding (Marzolf, 1965; Hargrave, 1970b,c). In soft muds, amphipods live in simple burrows in the top 2 cm of sediment with the majority in the top 1 cm (Hargrave, 1970a; Krezoski et al., 1978; Heuschele, 1980). Like chironomids, they are quite mobile and undergo extensive and frequent migrations into the overlying water column (Table II) (Mundie, 1959).

3.3. Oligochaetes

Oligochaetes are typically bilaterally symmetrical, segmented, hermaphroditic annelids. Eleven families of oligochaetes are found in freshwater environments (Brinkhurst and Jamieson, 1971), but only three—the Tubificidae, Naididae, and Lumbriculidae—are widespread and abundant in lakes. Of these, the Tubificidae are easily the most important, if importance is measured by widespread distribution, population density and biomass, and ability to alter sediment properties. Along with chironomids, tubificid oligochaete worms are usually the dominant macrofauna in lake profundal regions. Population densities are variable and may range from a few hunmdred individuals per square meter to several million individuals per square meter (Palmer, 1968). Higher population densities are associated with fine-grained sediment (Fig. 8), higher inputs of organic matter to sediments, and high temperatures (Brinkhurst and Jamieson, 1971; Ladle, 1971; M. G. Johnson and Brinkhurst, 1971; Birtwell and

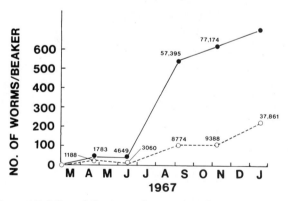

Figure 8. Numbers of *Tubifex tubifex* in sand (○) and mud (●) cultures. Numbers represent dry weight of worms in micrograms. Redrawn from Ladle (1971).

Figure 9. Oligochaete burrows in Lake Erie sediments. Depth of sediment column: 15 cm.

Arthur, 1980). Individual dry weights range from less than 0.2 mg per mature individual (Jonasson and Thorhauge, 1976b) to over 3.4 mg per individual (Ladle, 1971; McCall et al., 1979). Organism length is hard to measure, since oligochaetes are capable of wide variation of extension and contraction. Contracted lengths of preserved worms are typically 2–5 cm (e.g., Birtwell and Arthur, 1980), although lengths up to 15 cm have been reported (Caspers, 1980).

Tubificid life histories are poorly known because it is often impossible to identify field collections of immature tubificids to the species level (Brinkhurst and Jamieson, 1971; Bonomi and DiCola, 1980). The life history of even a single species is variable from locality to locality depending on temperature and productivity, among other things. Higher temperatures and high levels of food decrease the time required to reach maturity and lengthen the breeding period (Kennedy, 1966; Bonomi and DiCola, 1980).

Some tubificids take 2 years to reach maturity and breed for a short period during the summer months (Brinkhurst, 1964). Other species like the cosmopolitan species *Limnodrilus hoffmeisteri* and perhaps *Tubifex*

tubifex have variable life histories, but usually reach maturity in a year or less and often breed all year around (Brinkhurst and Kennedy, 1965; Kennedy, 1966). There is one report of *L. hoffmeisteri* having four or five generations a year in a eutrophic reservoir in South Wales (Potter and Learner, 1974). The majority of breeding worms appear to die after reproduction, but some worms are capable of resorbing sexual organs and repeated breeding (Brinkhurst and Jamieson, 1971; Poddubnaya, 1980). A few tubificids and most Naididae reproduce asexually.

Tubificids are infaunal burrowers (Fig. 9). They are found mostly in the upper 20 cm of sediment, but they feed primarily in the top 2–8 cm (Sorokin, 1966; Davis, 1974; McCall and Fisher, 1980). Tubificids typically live and feed head down in the sediment. Some portion of the posterior of the worms may project above the sediment–water interface. The posterior is sometimes waved about under low-oxygen conditions for the purpose of respiration. The worms selectively ingest silt and clay particles at depth and feed on attached microflora, primarily bacteria (Brinkhurst and Chua, 1969; Brinkhurst et al., 1972; McCall et al., 1979). Fecal pellets are deposited at the sediment–water interface. Tubificids move about in sediment and are able to move laterally for short (<1 m) distances in laboratory microcosms, but they do not swim actively above the interface in the manner of chironomids and amphipods (Table II) (Brinkhurst, 1974). Most lateral movement must be due to passive current transport of adults and coccoons, which are millimeter-sized, ovoid egg containers dropped into the sediment by breeding tubificids.

3.4. Bivalves

3.4.1. Unionacea

The Unionacea (particularly Unionidae and Margaritiferidae) are distributed worldwide, but they are most abundant and diverse (about 220 species) in eastern North America in the Mississippi River drainage basin (Pennak, 1978). Unionids are found on all substratum types from gravel to mud and in both rivers and lakes (R. I. Johnson, 1970; Tevesz and McCall, 1979). While it is true that unionids are typically found in quite shallow water only a few meters deep, large populations have also been recorded from deeper profundal muds. Wood (1953), for example, found abundant unionids in the deepest part of the western basin of Lake Erie at depths of 12 m.

Unionaceans are the largest members of the permanent freshwater infauna. Adult lengths range from 4 to more than 14.6 cm, with average lengths of about 7 cm (e.g., Matteson, 1948; Okland, 1963; Tudorancea,

1969; Crowley, 1957; Hendelberg, 1960). Wood (1953) collected 14 species of unionids from western Lake Erie and found that average individual wet weights of species ranged from 1 to 25 g; the average for all species was about 10 g. J. B. Fisher and Tevesz (1976) found the average dry weight of *Elliptio complanata* in Lake Pocotopaug, Connecticut, to be about 1 g per individual (average length 7.4 cm). Population densities range from a few tens of individuals (Wood, 1953; Okland, 1963; J. B. Fisher and Tevesz, 1976) to several hundred per square meter (Tudorancea, 1969). As a consequence, unionids typically comprise 90% or more of the standing crop biomass of the infaunal macrobenthic community in which they occur (Tudorancea and Gruia, 1969; Wood, 1953; Okland, 1963; Negus, 1963; J. B. Fisher and Tevesz, 1976).

Among unionaceans sexes are separate, and the eggs of the female are fertilized by sperm suspended in the water column and drawn into the mantle cavity by the suspension-feeding female. Fertilized eggs are carried in the gills, which function as a brood pouch (LeFevre and Curtis, 1912). Valved glochidia larvae are expelled from the brood pouch either 4–6 weeks after fertilization in short (summer) breeders or several months later in long (winter) breeders, in which eggs are fertilized in the latter part of the summer and carried through the winter to be discharged in the spring and summer (LeFevre and Curtis, 1912; Matteson, 1948; Yokley, 1972). A single female may release a few thousand to over two million glochidia larvae (Coker et al., 1921). Glochidia larvae settle to the bottom, where they die unless taken into the mouth of (hookless glochidia) or brushed by (hooked glochidia) the proper species of fish. Glochidia attach themselves to the gill filaments or sometimes the fins of fish (several hundred to several thousand glochidia per fish), where they develop for 2–3 and, exceptionally, 4 weeks, after which they drop from the fish as juveniles and take up their life on the bottom (LeFevre and Curtis, 1912; Matteson, 1948; Coker et al., 1921). Unionaceans grow rapidly in the first 2–5 years of life, after which growth slows markedly. Juveniles mature in 2–5 years, may reproduce once per year for many years, and live to be from 5 to 116 years old; most species live 15–40 years (Stansbery, 1967; Haranghy, 1971; Yokley, 1972; LeFevre and Curtis, 1912; Stober, 1972; Hendelberg, 1960; Crowley, 1957). Growth rates and longevity are of course dependent on many factors, but where temperatures are held approximately constant, growth rates of some species decline in deeper water and finer sediments, while variability of growth rate decreases and longevity increases (Fig. 10).

Unionaceans are filter feeders, so their feeding activity does not mix sediments in the way that the feeding activity of oligochaetes does. The typical life position of a unionid on soft sediments is shown in Fig. 11. Life position, however, is variable. Some individuals live with their an-

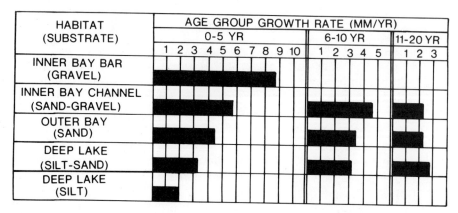

Figure 10. Unionid growth rates as a function of sediment type. Redrawn from Stansbery (1970).

terior margin and apertures projecting a centimeter above the bottom, while others have been reported living up to 20 cm below the sediment–water interface (McCall et al., 1979; Van Cleave, 1940; Baker, 1928). While the food of unionaceans has been reported to consist of zooplankton, phytoplankton, and organic detritus (Pennak, 1978; Coker et al., 1921), the proximity of the inhalent aperture to the sediment–water interface on soft, easily resuspended bottoms probably means that nutrition is also obtained from bacteria growing on ingested fine-grained sediments in a manner similar to that of the brackish-water filter-feeding bivalve

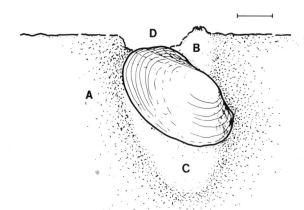

Figure 11. *Lampsilis radiata* in infaunal life position. A, sediment unaffected by burrowing activities; B,C, zone of high water content produced by burrowing activity; D, depression in sediment–water interface produced by burrowing activity. Scale bar: 2 cm. Reproduced from McCall et al. (1979) with permission of the *Journal of Great Lakes Research*.

Rangia cuneata, described by Tenore (1968). At any rate, copious amounts of mucus-laden sediment are deposited on the sediment surface as pseudofeces by the unionid *Lampsilis radiata* in Lake Erie (McCall et al., 1979). The clams are quite active at temperatures above 10°C (Stansbery, 1970) and frequently move about on top of and in the top 10–15 cm of surface sediment (McCall et al., 1979).

3.4.2. Pisidiidae

The Pisidiidae (formerly Sphaeriidae) are considerably different from the Unionacea. Individual lengths (2–20 mm) and dry weights (0.5–2 mg) of pisidiids are much smaller than those of unionaceans (Foster, 1932; Thomas, 1965; Mackie et al., 1976; Heard, 1977; Jonasson, 1972). Population densities of pisidiids in the profundal zone of lakes are higher than those of unionaceans, typically ranging from a few hundred to a few thousand (2600) individuals per square meter. Pisidiids occur over a wider range of depths and they reach their maximum abundance at greater depths than unionaceans (Jonasson, 1972; Thut, 1969; Nalepa and Thomas, 1976; Hamilton, 1971). In Lake Michigan, for example, unionids occur no deeper than 30 m and are typically found at depths of less than 10 m (Reigle, 1967), while pisidiids reach their maximum abundance at 40–50 m and occur to depths of 225 m (Eggleton, 1937).

The life history of pisidiids has been recently summarized by Heard (1977). Individuals are hermaphroditic. Fertilization is internal; young are incubated in gill structures and are released as miniature adults. There is no parasitic larval stage as there is in unionids. Relatively few young are brooded at one time—usually 2–4 young in the genus *Sphaerium*, more in *Sphaerium* species from temporary habitats, and up to 20 in the genus *Pisidium*—and young at any one time are typically of different sizes and at different stages of development. Most individuals live 1.5–2 years and produce only one brood; some species live 3–4 years and some produce two broods. Young are produced throughout the year, although a particular species may reach a peak in reproduction in the summer or winter months.

Many pisidiids are filter feeders on detritus, bacteria, algae, and protists (Monakov, 1972; Meier-Brook, 1969), but deposit feeding by common and abundant species using the ciliated part of the bivalve foot to collect fine-grained particles has also been reported (Mitropolsky, 1966; Monakov, 1972). Most pisidiids are infaunal. Some species live in "normal" bivalve position, with siphons just below the interface with the ventral portion of the shell and foot pointing down into the sediment (Meier-Brook, 1969; McCall, 1979). Other pisidiids of the genus *Pisidium* have the unique life position shown in Fig. 12. The clam burrows into the

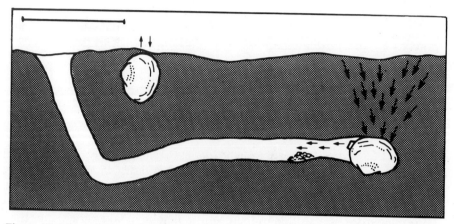

Figure 12. Life position of *Pisidium* (right) and *Sphaerium* (left). Arrows indicate feeding and respiratory currents. Scale bar: 5 mm. Adapted in part from Meier-Brook (1969).

bottom at a steep angle and then turns parallel to the interface and forms a canal several millimeters below the interface. This becomes the exhalent canal, water being drawn into the clam from surrounding interstitial spaces (Meier-Brook, 1969). Jonasson (1972) records a similar life position in a horizontal burrow for the common species *P. casertanum* and further adds that burrows appear to be concentrated at the boundary between oxidized and reduced sediments. Most pisidiids are indeed found in the upper 2 cm of sediment, but Gale (1976) found that young and newborn *S. transversum* are able to burrow to at least 16 cm below the sediment–water interface. Not all young burrow deeply, and those that do become metabolically inactive and can live for more than a month in anaerobic sediments. This deep-burrowing behavior is not unique to *S. transversum* and may be an adaptation to escape temporarily unfavorable circumstances at the interface (Gale, 1976).

4. Mixing of Sediments

4.1. Effects on Sediment Stratigraphy

4.1.1. Kinds of Mixing and Factors Affecting Mixing

Infaunal macrobenthos can mix the sediments in which they live by either their feeding activities or their burrowing activities or both. Feeding by freshwater deposit feeders can mix sediments by highly ordered or random movement of sediment particles. The deposit feeding tubificid oligochaetes provide the best example of ordered mixing of freshwater sediments.

Tubificids usually feed head and mouth down in the sediment with their posterior end at or above the sediment–water interface. Sediment particles and associated bacteria are ingested at depth. Bacteria are digested on their passage through the gut (Wavre and Brinkhurst, 1971; Brinkhurst et al., 1972), whereupon sediment particles are packed into mucus-bound fecal pellets and deposited at the sediment–water interface. Thus the worm receives a food supply constantly renewed by gravity. Moreover, the food supply is not diluted with recently digested sediment, and the worm can remain sedentary yet be fairly well protected by sediment from predators. Mixing by such feeding has been labeled "conveyor belt" mixing by Rhoads (1974) and *convective mixing* by J. B. Fisher et al. (1980). Both terms refer to the circulation of sediment particles vertically through the surface layers of the deposit under the feeding action of benthos. A laboratory experiment by Robbins et al. (1979) illustrates this type of mixing. They placed a 1-mm layer of illite coated with ^{137}Cs on top of a sediment column that contained 50,000 worms m^{-2}. ^{137}Cs is a gamma emitter that sorbs strongly to illite clay particles. It is thus an excellent material for tracing nondestructively the movement of sediment particles by deposit feeders. ^{137}Cs activity throughout the sediment column was monitored with an external gamma detector. ^{137}Cs dropped below the sediment–water interface as a fairly homogeneous layer as worms fed below the layer and deposited fecal material on top of it. This process continued until day 49 when a secondary peak was formed at the sediment–water interface after the ^{137}Cs reached the top of the worm feeding zone at ~3 cm depth (Fig. 13). This mixing process can be analyzed mathematically as a combination of convection (primarily) and diffusion (to account for gradual peak broadening) of sediment particles. Such a model has been constructed by J. B. Fisher et al. (1980); it is discussed in detail there and in Chapter 7.

The other type of biogenic sediment mixing is the random movement of particles. It is not the peculiar result of a deposit feeding strategy, but rather is created by suspension feeders, deposit feeders, as well as omnivores and is a result of burrowing as much as feeding. Thus it is probably the more common type of biogenic mixing. Another experiment by Robbins et al. (1979) illustrates this kind of mixing. In this experiment, ^{137}Cs-labeled clay was added to a sediment column containing the amphipod *Pontoporeia hoyi* (10,000 m^{-2}). The activity profile evolved from a nearly Gaussian shape to a uniform distribution in the top 1.5 cm in 17–20 days (Fig. 14). The net effect of amphipod feeding and burrowing activity is to mix sediments randomly. Particles can be treated as if they were diffusing down from the interface. Appropriate models of this kind of mixing

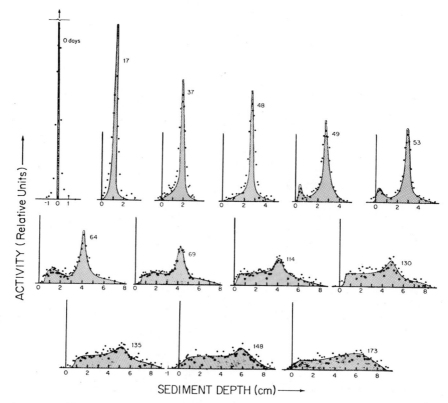

Figure 13. Effect of tubificids on the distribution of ^{137}Cs in laboratory sediments. From Robbins *et al.* (1979) with permission of Elsevier Scientific Publishing Co.

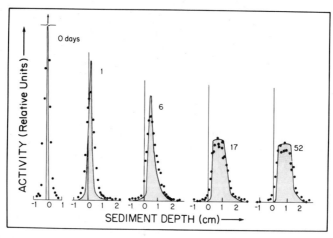

Figure 14. Effect of an amphipod (*Pontoporeia hoyi*) on the distribution of ^{137}Cs in laboratory sediments. Redrawn from Robbins *et al.* (1979) with permission of Elsevier Publishing Co.

are discussed in Robbins et al. (1979) and in Chapter 7. Unionids, pisidiids, and chironomids also probably mix sediment in this fashion.

The configuration of the volume of biogenic sediment mixing depends not only on the type of organism involved but also on population density and distribution (horizontal and vertical) and individual organism size. A single large active organism (e.g., a unionid) may have more important effects than many small individuals (e.g., pisidiids or tubificids). Because large infaunal species tend to be more rare and patchily distributed than small infauna, the amount and the depth of sediment mixed by large species will also be more patchily distributed than those of sediment mixed by small, more evenly distributed organisms. The patchy distribution of deep (>15 cm) mixing of sediment-associated trace metals (Kovacik and Walters, 1973; Walters et al., 1974) and pollen and ^{210}Pb (Robbins et al., 1978) in the western basin of Lake Erie coincident with the distribution of large unionid clams in the lake may be an example of such patchy mixing. Nevertheless, most infaunal benthos, large and small, are patchily distributed (e.g., Alley and Anderson, 1968; Brinkhurst, 1965; Mozley and Garcia, 1972; Henson, 1962; Freitag et al., 1976), and it may appear unlikely that geochemists and geochronologists can tell anything of value from the few cores that they are restricted to analyzing from any one lake habitat. The reason they can is probably that over time the various animal patches will have visited most spots in the habitat. If sediment-associated particles are inert or slowly reactive, they will experience time-averaged biogenic mixing. More rapidly reacting substances should more closely track the shifts of infaunal populations.

4.1.2. Depth and Rate of Sediment Mixing

4.1.2a. Oligochaetes. Oligochaetes are among the most potent movers of sediment in fresh water. Because of this and because they feed at depth in the sediment, the vertical distribution of tubificids has received more attention than that of any other freshwater invertebrates. Figure 15 shows depth distributions of tubificids in cores collected from seven different lakes. The depth distribution of these animals is extremely variable. While in most lakes more than 90% of the tubificids will be found in the top 10–12 cm of sediment, there are exceptions. In Lake George, for instance (Fig. 15E), 85% of the worm population is found below this depth. A better generalization is that depth distributions are not fixed to a certain small range worldwide, but are finely tuned to different lake environments. We will look at two factors influencing depth distribution, namely worm population density and oxygen content of overlying water.

When sediment cores collected by the same investigator within one lake (Figs. 15A,B) or from two nearby lakes (Figs. 15C,D) are compared, worms are found to live deeper in cores with higher worm population

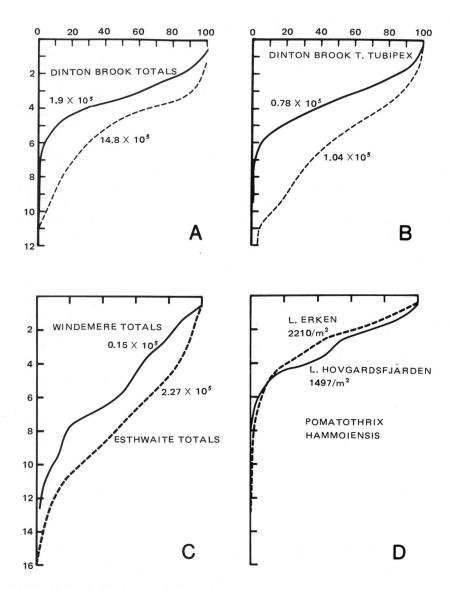

Figure 15. Depth distribution of tubificids in various lakes. The x axis represents percent of fauna found deeper than depth indicated on y axis (in cm). (A) Dinton Brook (total tubificids). Data from Brinkhurst and Kennedy (1965). (B) Dinton Brook (*Tubifex tubifex*). Data from Brinkhurst and Kennedy (1965). (C) Esthwaite Water and Lake Windemere. Data from Stockner and Lund (1970). (D) Lakes Erken and Hovgardsfjärden. Data from Milbrink (1973). (E) Lake George and Green Bay. Data from Burgis *et al.* (1973) and J. A. Fisher and Beeton (1975). In A–D, solid lines denote low worm abundances and dashed lines denote high worm abundances. In E, the solid lines indicate data from Green Bay, and the dashed line indicates data from Lake George.

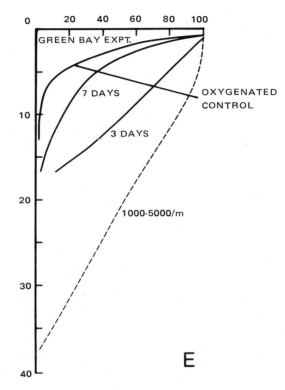

Figure 15. (cont'd)

densities. In comparable cores with similar population densities, depth distributions are also similar. For example, 50% of the worms in Dinton Brook (Fig. 15A) are found between 2.5 and 4.5 cm when the population density is ~200,000 m^{-2} and only 10% of the population is found below 5 cm. When abundances increase to ~1,480,000 m^{-2}, 50% of the population is found between 4 and 7 cm, and over 40% of the population is below 5 cm (Fig. 15A). The median depth of worms in cores is 1–2 cm deeper in cores with high worm abundances (Figs. 15A–C). In comparable cores with similar abundances, depth distributions are also similar, especially at low abundances (Fig. 15D). The correlation of abundance and depth of occurrence in comparable cores probably has little to do with the need for a greater volume of living space. There is no absolute relation between abundance and depth of occurrence if all cores are compared together. We think it is more likely that the worms adjust their *feeding* depth lower when abundances are higher. McCall and Fisher (1980) report that with densities of ~65,000 worms m^{-2} the zone of maximum feeding was at a depth of 7.7 cm; with ~48,000, at about 5.5 cm. In an experiment by Robbins et al. (1979) in different sediment (Fig. 13), the feeding depth of 3 cm was displaced downward to about 5–6 cm when more worms were added to the experimental microcosm.

There is some evidence that oligochaetes are selective feeders on sediment bacteria (Brinkhurst et al., 1972; Wavre and Brinkhurst, 1971) and that they feed in the reduced zone of the sediment. A greater abundance of worms will circulate the "sediment conveyor belt" faster (see Section 4.1.1), and so sediment will likely need to fall to a greater depth before the microbial food supply is replenished. In addition, the oxidized surface layer is somewhat thickened by their feeding activity (see Chapter 4), so there is a smaller reducing environment for the growth of anaerobes. If these hypotheses are correct, then depth distribution would be a function not only of worm abundance but also of sedimentological factors such as grain size (surface area for bacterial attachment), organic matter input (microbial food supply), and pore water geochemistry (microbial environment), and it should not be surprising that the depth distribution of oligochaetes varies widely among different lakes.

J. A. Fisher and Beeton (1975) described the effect of low dissolved oxygen in water overlying sediment on the depth distribution of the oligochaete Limnodrilus hoffmeisteri from Green Bay, Lake Michigan. Exposure to anoxic water drove the worms down into the sediment for a period of 3 days; the median depth of occurrence fell from 3 to 11 cm (Fig. 15E). There was a subsequent upward migration for the next 4 days, but the worms still lived deeper than they did in an oxygenated control aquarium (Fig. 15E). Perhaps this is a partial explanation for the anomalously great depths of occurrence that Burgis et al. (1973) report for Lake George oligochaetes. Tropical Lake George is subject both to frequent storms that disturb the surface layers of the sediment and to intervening periods of stratification and anoxic hypolimnetic waters (Ganf and Viner, 1973). Additional factors such as sediment water content (Berg, 1938; Stockner and Lund, 1970) and worm length and biomass (see McCall and Fisher, 1980) are probably also important determinants of depth distribution. Feeding depths may be different for different species. Krezoski and Robbins (1980), for example, used the same radiotracer methods as J. B. Fisher et al. (1980) but utilized different sediments (from Lake Huron instead of Lake Erie) and different oligochaetes (the lumbriculid Stylodrilus heringianus instead of the tubificids Tubifex tubifex and L. hoffmeisteri) and observed different faunal feeding depths (3–5 cm instead of 6–8 cm).

It is the depth of *feeding* that is primarily responsible for controlling the depth of sediment mixing by oligochaetes. There is typically a fairly narrow 2-cm zone of maximum feeding that is shallower than the deepest penetration of the worms (Fig. 16). While worms burrowed to depths of

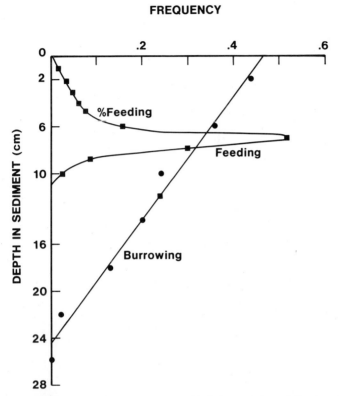

Figure 16. Feeding depth versus burrowing depth for Tubifex tubifex in a laboratory microcosm. Population density: 10^5 worms m^{-2}.

25 cm in this case, the depth of maximum feeding was 7–8 cm. Oligochaetes burrow to greater depths than those at which they commonly live, and they feed at even shallower depths. Burrowing itself contributes only slightly to sediment mixing. While correlations of mixing depth and worm distribution have been reported (Krezoski et al., 1978), this will not always or even often be the case. It is known, for instance, that most tubificids cease feeding at low oxygen levels (see McCall and Fisher, 1980), so that even if worms migrate deeper under low-oxygen conditions—as in the study of J. A. Fisher and Beeton (1975)—they do not necessarily mix the sediments to greater depths.

A number of workers have performed experiments on factors affecting the rate of mixing (feeding rate) by oligochaetes (see Table III). The reported mixing rates are quite variable, and no single value can be used to estimate mixing rate. At any one temperature, mixing rates vary by factors of from 3 to 160, indicating that mixing rate is strongly dependent

Table III. Comparison of Tubificid-Induced Subduction Velocities Reported by Various Authors

Source/organism(s)	Subduction velocity (cm · day^{-1} · m^{-2} per 100,000 individuals)							
	2–4°C	5–7°C	8–10°C	11–13°C	14–16°C	17–19°C	20–22°C	23–25°C
Alsterberg (1925)								
Limnodrilus and *Tubifex*					0.55			
Irlev (1939)								
Tubifex tubifex						0.02		
Ravera (1955)						0.05		
Peloscolex ferox		0.01	0.01					
Bythonomus lemani		0.01	0.01					
Psamnorytes barbatus		0.01	0.01					
Poddubnaya (1961)								
Temperature experiments								
Limnodrilus hoffmeisteri	0.04	0.09	0.20		0.30	0.60	0.68	
L. udekemianus	0.06	0.12	0.22		0.36	0.66	0.77	
Other experiments								
L. hoffmeisteri					0.47	0.16	0.23, 1.67	
L. udekemianus					0.41	0.19	0.27, 1.62	

Wachs (1967)								
T. tubifex	0.04	0.07	0.14	0.17	0.26	0.52	0.64	0.73
Appleby and Brinkhurst (1970)								
T. tubifex	0.03			0.12	0.27	0.27	0.34	
L. hoffmeisteri	0.02			0.08		0.13	0.38	
Peloscolex multisetosus	0.01			0.05	0.07	0.01	0.01	
Davis (1974)								
Limnodrilus spp.			0.47					
McCall and Fisher (1980)								
T. tubifex	0.03	0.02			0.11		0.20	
M. J. S. Tevesz and D. M. Lipinski								
Mixed tubificids						0.10		
Robbins et al. (1979)								
T. tubifex							0.10	
J. B. Fisher et al. (1980)								
T. tubifex							0.12	
Krezoski and Robbins (1980)								
Stylodrilus heringianus				0.18				

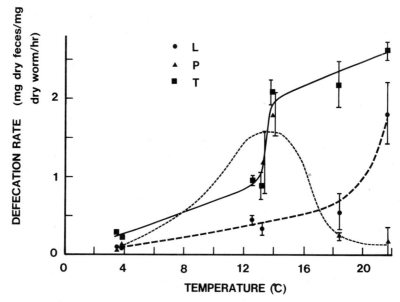

Figure 17. Feeding rate as function of temperature for *Limnodrilus hoffmeisteri* (L), *Tubifex tubifex* (T), and *Peloscolex multisetosus* (P). Redrawn from Appleby and Brinkhurst (1970).

on the particular species of worm studied. For any one species, water temperature appears to be the most important determinant of mixing rate. Maximum rates are typically 5–10 times higher than minimum rates over the seasonal range of temperatures experienced by one species; for any one species–temperature combination, mixing rate ranges only by about a factor of 3 (Table III). Mixing rates are not linearly dependent on temperature, and different species of worms have different feeding optima at different temperatures (Fig. 17) (Wachs, 1967; Appleby and Brinkhurst, 1970; McCall and Fisher, 1980). Smaller but statistically significant effects on mixing rates have also been found for worm abundance, presence of other worm species, and sediment type (Wachs, 1967; Zahner, 1967; Appleby and Brinkhurst, 1970; McCall and Fisher, 1980). McCall and Fisher (1980) also found significant interactions between the effects of temperature and substratum and temperature and abundance. This means that the relation of feeding rate to temperature is different on different substrata and at different worm abundances.

Given these variations in mixing depths and rates, few generalizations concerning mixing by worms are safe. Given the positive correlation of mixing depth and worm size and population density, the negative correlation of mixing depth and oxygen level, and the positive correlation of mixing rate and temperature, we expect that warm eutrophic lakes with a high input of organic matter that will support a greater number of larger worms will have the most thoroughly and deeply mixed sediments. Of course, if oxygen levels are so low that anoxia is frequent and of long duration then mixing will cease. Burgis et al. (1973) report mixing to 20 cm in shallow, eutrophic, tropical Lake George, but an undetermined portion of this mixing is due to storm disturbances of the bottom. J. B. Fisher et al. (1980) estimate that mixing rates by tubificids can exceed sedimentation rates by factors of ten or more in some parts of Lake Erie. Robbins et al. (1977, 1978) have found mixed profiles of radioisotopes 2–6 cm or more in thickness in other portions of the Great Lakes, which they attribute largely to biological processes.

Stockner and Lund (1970) report mixing of sediments primarily by tubificids but also by chironomids in Esthwaite Water and Lake Windemere (see Fig. 15C for tubificid depth distribution). They used the presence of live phytoplankton in sediments as a tracer of mixing. The difficulty with this tracer was that the longevity of live algal species in muds was not well known. Lund (1954) did laboratory experiments in which live algae were incubated in dark anoxic muds. None survived longer than 3 years. Three years would correspond to a sediment depth of only 0.45 cm (1.5 mm $\cdot$ year^{-1}) in these lakes and the presence of any live algae deeper than this would be due to mixing. Stockner and Lund (1970) found live algae deep in sediment cores but believed that biological mixing could not account for their presence and so concluded that algae could remain viable for 183–290 years. We think that their results are in fact compatible with biological mixing. Many viable species were found to 8–10 cm in Windemere (lower tubificid abundance) and about 15 cm in Esthwaite (higher tubificid abundance). The number of viable species changed by 100% seasonally in the upper 5 cm of sediment, which is compatible with rapid mixing or ventilation of sediments (probably by chironomids). The greatest depth to which viable species were found was 25 cm in Windemere and 35 cm in Esthwaite. (No measure of relative *abundance* of algae was available.) Both tubificids and chironomids were present at these depths, although at greatly reduced abundances. Still, it does not appear unreasonable to us to expect that a few algae could be brought down to this depth, even though wholesale mixing is quite unlikely.

4.1.2b. Other Macrobenthos. Mixing of sediment by benthos other than oligochaetes has received little attention. Ford (1967) studied the vertical distribution of Chironomidae in muddy stream bottoms. He found that over 95% of the chironomids in both the laboratory and the field occurred in the top 5 cm of sediment. Stockner and Lund (1970) found chironomids almost entirely in the top 6 cm of the sediment of the two

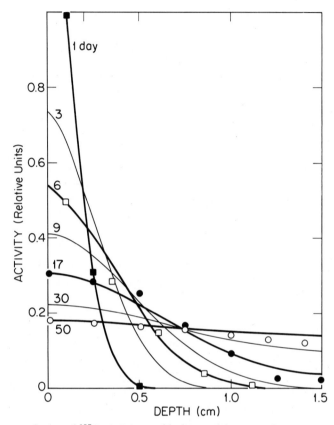

Figure 18. Time evolution of ^{137}Cs activity profile for particle mixing by *Pontoporeia hoyi*. Calculated activity is represented as a solid line; the points represent laboratory data. Reproduced from Robbins *et al.* (1978) with permission of Elsevier Scientific Publishing Co.

English lakes, Lake Windemere and Esthwaite Water, although they did find two specimens at 55 and 65 cm below the surface in Esthwaite sediment. Milbrink (1973) found that most chironomids in four Swedish lakes were restricted to the upper 3–4 cm of sediment. One larger species, *Chironomus plumosus*, was found to depths of 15 cm. In Lake Erie, we find that burrows of fourth instars of this same species reach 8–11 cm below the sediment surface. Timm (1975) found a similar vertical distribution of chironomids: Small chironomids were found primarily in the top 5 cm of the mud of Lake Vôrtsjärv, while large *C. plumosus* were most abundant 10–15 cm below the sediment–water interface. The actual mixing of sediments by chironomids has hardly been examined. Stockner and Lund (1970) report that Kleckner (1967) performed laboratory experiments in which chironomids and tubificids mixed glass microbeads into sediment: "Generally only the top 3 cm were disturbed to a significant degree." McLachlan (1977) found that C. plumosus thoroughly worked laboratory sediments by feeding and active burrowing while employing grazing as a feeding method. The same species left sediment "virtually undisturbed" while suspension feeding. The depth and rate of mixing were not reported.

Robbins et al. (1979) studied the mixing of sediment by the amphipod *Pontoporeia hoyi* from Lake Michigan in a laboratory experiment using ^{137}Cs-labeled clay as a tracer of mixing. *Pontoporeia* mixes by burrowing and surface deposit feeding rather than by conveyor belt deposit feeding. Sediments were only mixed to a depth of 1.5–2 cm. Particle mixing by *Pontoporeia* can be accurately described by a simple eddy diffusion model of particle movement (Fig. 18). The upper 1.5 cm were uniformly mixed at 7°C by a population of 16.0×10^3 amphipods m^{-2} in 17–30 days.

Pisidiid and unionid bivalves also mix sediment by their burrowing activities. Although pisidiids are occasionally found several centimeters below the sediment–water interface (Berg, 1938), most burrow in the top 2.5 cm, and it is unlikely that they account for any significant mixing below this depth. McCall et al. (1979) report that the unionid *Lampsilis radiata* from Lake Erie (average length 6.3 cm) plows muddy sediments to a depth of 2–3 cm during semiinfaunal crawling and mixes sediments to a depth of 10 cm during burrowing. Larger unionids could mix sediments incompletely to a maximum depth of about 20 cm.

4.1.3. Complicated Stratigraphic Effects

Mixing alters sediment stratigraphy. The magnitude of this mixing effect on sedimentary materials (and thus age determination) is dependent on the mixing rate, mixing depth, and sedimentation rate (Chapter 7), making any generalizations applicable to all lake environments suspect. An observation that bears closer examination than it has received to date is that not all sedimentary materials show the same response to biological mixing. That is, examination of some sedimentary parameters may indicate thorough biogenic mixing in the upper few centimeters of sediment, while others show no such evidence. For example, the laboratory work and field observations of Robbins et al. (1978) on ^{210}Pb and ^{137}Cs tell us that the top 5–6 cm of sediment in Lake Erie are biogenically mixed, but vertical profiles of some metals (Walters et al., 1974) and organic carbon and nitrogen (G. Matisoff, unpublished data) show no evidence of mixing. This may be a result of very high and time-dependent inputs of these materials. Pore water materials that react and establish equilibria faster than the sediment is mixed may also be transparent to biogenic mixing. Håkanson and Källström (1978) show a radiograph of sediment from Lake Ekoln in Sweden that they know to be mixed by tubificids, chironomids,

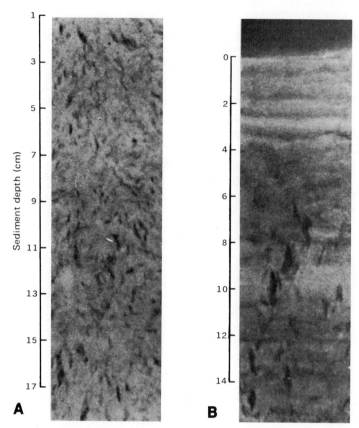

Figure 19. Radiographs of sediment cores. (A) From 29 m in central Lake Ekoln. (B) From 32 m in southern Lake Ekoln. (C) From 8 m in western Lake Erie; depth of sediment column: 16 cm. A and B from Håkanson and Källström (1978) with permission.

and pisidiid clams. The core shows laminae of light and dense materials in an otherwise apparently homogeneous core of silt–clay material. The laminae are most strongly developed in the top 4 cm of sediment. We have observed a similar phenomenon in Lake Erie, where laminae are most abundant in the top 7 cm of sediment (Fig. 19). Laminae are not normally present when biological mixing occurs and are usually good evidence of dominance by physical processes (D. G. Moore and Scruton, 1957; Rhoads, 1975). However, where mixing is of the ordered type that results from tubificid feeding, this need not be the case. Laminae produced near the sediment–water interface by waves and currents may remain in existence until they descend to the base of the feeding zone (3–8 cm). The base of the feeding zone itself may produce a lamina of large dense (quartz instead of clay) material that is left behind as a lag deposit of

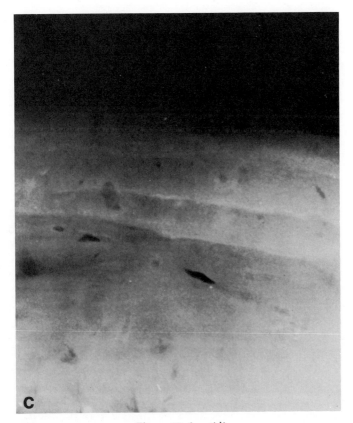

Figure 19. (cont'd)

rejected food material by these selective deposit feeders. Källström and Axelson (1978) also suggest that some laminae are biologically produced, with environmental factors such as oxygen in the overlying water determining the location of an ever-changing base of the biologically disturbed zone. Finally, Howard and Elders (1970) suggest that some laminae can survive intense mixing (at least in a marine sand with burrowing amphipods). Thus mixing is not always a simple process with simple effects, and where sizable benthic populations are present, some caution needs to be exercised before it is concluded that mixing does not occur in a particular environment.

4.2. Effects on Particle Size Distribution

Alterations in the size of particles comprising soft bottoms can result in alterations of the rate and mode of sediment transport from the bottom.

In addition, the volume and arrangement of interparticle spaces can be changed with a resulting change in the chemical flux of dissolved materials through the sediment. Macrobenthos alter particle size by making or destroying fecal pellets and pseudofeces.*

The net effect of sizable populations of deposit feeders on particle size distributions in mud bottoms is the addition of fine sand-size fecal pellets to a silt–clay matrix. There has been little work on the pelletization of freshwater muds. Iovino and Bradley (1969) studied the pelletization of sediment by chironomid larvae of Mud Lake, which they believed to be a modern analogue of Eocene oil shale environments. Mud Lake is unusual because it is very shallow (only about 45 cm deep) and alkaline (pH 7.7–10.2) and has no phytoplankton and no clastic sediments. The sediment is comprised almost entirely of pellets of blue-green algae egested by *Chironomus* inhabiting the lake bottom at densities of a few hundred per square meter. Laboratory populations of second instars fed blue-green algae produced pellets averaging 168 μm × 174 μm. Fourth instars produced pellets 342 μm × 168 μm. Pellets from the lake averaged 364 μm × 142 μm. Laboratory populations fed green algae produced no coherent pellets at all. The method of feeding on the two food sources was also different—grazing on the blue greens and filter feeding on the greens. McLachlan and McLachlan (1976) studied the formation of bottom sediment in a newly created temperate lake where the parent material for sediment formation was both inorganic clastic sediment and organic matter (7–10% by weight). They found that *Chironomus plumosus* increased the amount of both fine (<105 μm) and coarse (>250 μm) material in the sediment. Fines increased by about a factor of 4.5 and were the result of mechanical disturbance of the mud by the chironomids. The amount of coarse particles was over 200 times higher in sediment with chironomids than in sediment with none, although most of the coarse particles (93%) were associated with the tube silk rather than free fecal pellets (7%). This squares with our casual observations of chironomids in laboratory aquaria. When *C. plumosus* are introduced to sediments pelletized by tubificid oligochaetes, pellets disappear in a circular area around chironomid burrows, while massive aggregations of sediment, mucus, and silk become associated with the chironomid tubes.

Similar laboratory observations on the mayfly nymph *Hexagenia*, along with published accounts of its burrowing activities (Lyman, 1943;

---

* By *particle size* we mean the size of naturally produced aggregates of bottom material, which may be amalgams of organic and inorganic material, and not the size of mineral grains measured by standard sedimentological techniques. Pseudofeces are aggregates of mineral grains and organic material that are manipulated by an organism but not ingested. They are usually produced by suspension feeders and are less cohesive and more amorphous than fecal pellets.

B. P. Hunt, 1953), suggest that Hexagenia may be another important pellet disaggregator, but no experiments have been done to determine the magnitude of its effects. Other fecal pellet producers have been identified among the Gastropoda (Calow, 1975), Amphipoda (Hargrave, 1976; Lautenschlager et al., 1978), and Copepoda (Ferrante and Parker, 1977), but their quantitative significance in altering physical properties of sediment remains to be measured.

Tubificid oligochaetes are the pelletizers that have received the most attention. McCall (1979) found that the top 0.5–1.0 cm of silt–clay mud in western Lake Erie consisted of fecal pellets in various stages of decay. Constituent mineral particles that made up the pellets had a median grain diameter of 1.5 µm. Fresh pellets were cylindrical mineral–organic aggregates averaging 280 µm × 70 µm. Even when pellets were subjected to vigorous mechanical disturbance, mineral grains were still bound in an organic matrix and some pellet pieces (80 µm long) remained (Fig. 20). Laboratory experiments on pellet production and decay rates indicated that in situ tubificid abundances of 10^4 m^{-2} should be able to pelletize the top 1–2 cm of sediment in this part of the lake and field observations showed the presence of a pellet layer for most of the year. Settling velocity measurements (Table IV) confirmed the existence of mineral grain aggregates in lake deposits and showed a two-orders-of-magnitude increase in settling velocity owing to pelletization. The net effect of pelletization of these fine muds is to create a surface with larger grains that are more difficult to erode and larger interparticle voids that are more permeable to the flow of water, as experiments described later will show.

But pelletization can have an opposite effect, depending on the initial particle size distribution of the sediment. Tevesz et al. (1980) examined the effects of tubificid feeding on a sandy river sediment (mean diameter 100 µm). They added worms (4×10^4 m^{-2}) to homogenized river sediment in the laboratory. A pronounced vertical sorting of sediment took place owing to the size-selective feeding of the worms at depth (Figs. 21 and 22). The worms fed in layer C (Fig. 21) on the <62.5-µm size fraction of particles, leaving behind a lag deposit with a higher sand concentration than deeper unreworked layer D. The top layer, A, consisted of sand-size fecal pellets composed primarily of <62.5-µm particles, and beneath this was a decayed pellet layer of similar composition (Fig. 22). Coulter counter analyses of the <62.5-µm fraction in the four layers further revealed that the worms fed selectively but not exclusively on clay-size (<4-µm) particles. The net effect of feeding on sandy sediments is to place lower-density sand-size pellets at the sediment surface; both layers A and B had higher water contents (by a factor of two) and more organic matter than unreworked layer D and were more easily transported by river currents than unreworked sediment. Such vertical sorting in sandy deposits has

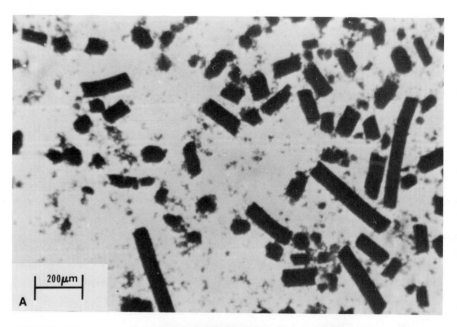

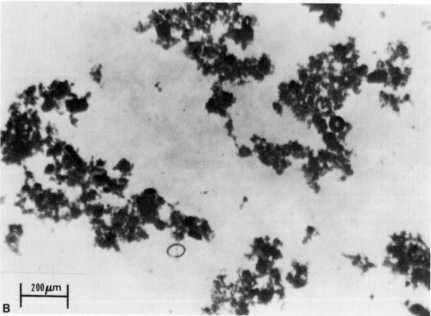

Figure 20. (A) Transmission light micrograph of the top 1 mm of sediment showing oligochaete fecal pellets (×12). (B) Transmission light micrograph of pellets disaggregated by stirring. Most sediment is bound in an organic matrix; few discrete pellets are visible (×12). (C) Transmission light micrograph of the top 1 mm of sediment from site 1 after treatment with blender and dispersant (×400). Note scale change.

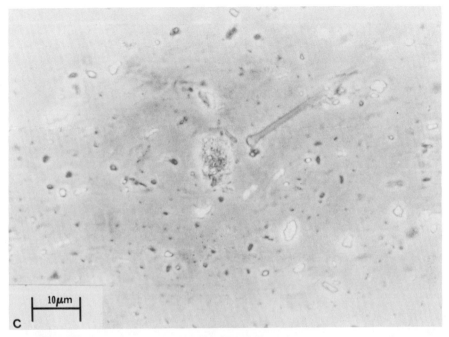

Figure 20. (cont'd)

also been reported by Kikuchi and Kurihara (1977) in submerged ricefield soils where the 0- to 5-cm layer of soil was enriched in 100-μm sand-size particles, while the deeper soil at 6–8 cm was depleted and soil deeper than 8 cm was unaffected.

Filter-feeding clams also aggregate sediment particles by the produc-

Table IV. Settling Velocities of Pelletized and Nonpelletized Sediment from Western Lake Erie[a]

Treatment	Median settling velocity (cm · sec^{-1})	Settling medium
Individual tubificid pellets	1.03	Lake water
Top 1 mm of deposit	0.13	Lake water
Top 1 cm of deposit	0.06	Lake water
Highly degraded pellets	0.03	Lake water
Top 1 cm of deposit	0.02	Distilled water and dispersant
Top 1 cm of deposit by standard sizing methods	0.0002	Distilled water and dispersant

[a] From McCall (1979).

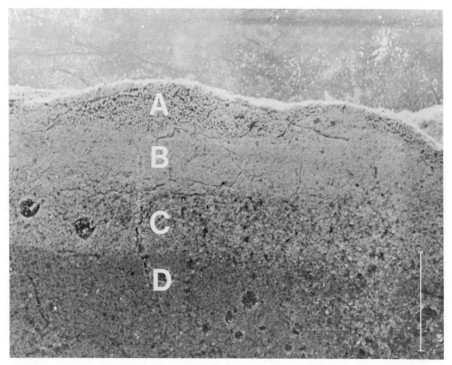

Figure 21. Sediment layering produced by tubificids in Vermilion (Ohio) River sediments. Scale bar: 1 cm. Reproduced from Tevesz et al. (1980) with permission of the Society of Economic Paleontologists and Mineralogists.

tion of mucally bound pseudofeces from silt- and clay-size material suspended in the water column. This biodeposition phenomenon has been quantified for only one freshwater species, *Corbicula fluminea*. Prokopovich (1969) found that the silt- and clay-size sediment (10–20% of sediment by weight) contained in a 113-mile-long concrete-lined canal consisted almost entirely of the excreta of this clam. The 30-m-wide, 5-m-deep canal contained $1-2 \times 10^4$ *Corbicula* in its upper portions. The average clam (1 cm long; 1 g wet weight) was able to filter about 0.5 liters of water per day and abstract completely silt–clay particles from the water column (average concentration 30 ppm). Prokopovich calculated that 620 tons of clams in the upper portions of the canal produced 3350 tons of excreta per year. Moreover, the amount of fine sediment was greater in the upper, faster-flowing (1.2 m · sec^{-1}) portions of the canal than in the slower-moving lower reaches, just the opposite of what would be expected from hydrodynamic considerations. The greater abundances of clams in the upper portion of the canal and the binding properties of mucus probably account for the discrepancy. Thus biodeposition by clams can dramatically alter both the size distribution and location of sediment deposits. Stanczykowska (1978) noted that the very common European bivalve, *Dreissena polymorpha*, produced large amounts of excreta that were saturated with mucus and enriched in bacteria that enabled large

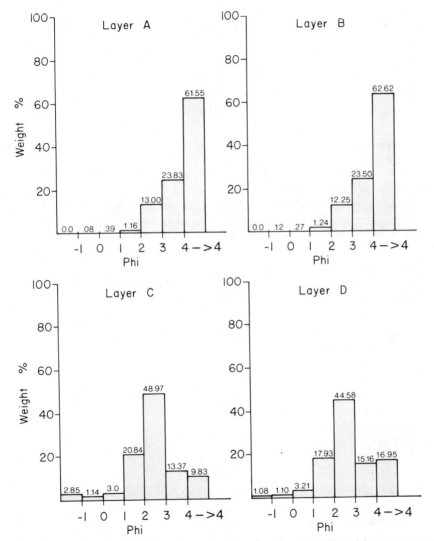

Figure 22. Mean grain size distribution ($N = 3$) of wet sieved sediment from laboratory aquaria containing worms. Letters denote layers pictured in Fig. 21. Reproduced from Tevesz et al. (1980) with permission of the Society of Economic Paleontologists and Mineralogists.

numbers of chironomids to live among the clams. McCall et al. (1979) also noted that the large unionids accounting for over half the standing crop biomass of macrobenthos in western Lake Erie produced copious amounts of pseudofeces, although they did not quantify their observation. The biodeposition phenomena are probably widespread and of great local importance.

4.3. Effects on Mass Properties of Sediments

Mass properties refers not to the properties of individual sediment particles such as size and density, but to those properties that arise from the arrangement or fabric of the particles in a deposit. We will discuss the effect of macrobenthos on water content and permeability.

4.3.1. Water Content

Water content is a measure of the compactness of a sediment: The fewer interparticle connections, the more open the fabric and the higher the water content. Water content is reported by benthic ecologists as a percent by weight of water of the total weight of water plus sediment, and by civil engineers and geotechnologists as the ratio of water weight to sediment weight. We use the former measure here. Water content is usually taken as one measure of fine-grained bottom stability. Other things being equal, bottoms with higher water contents are more easily eroded than bottoms with lower water contents (Fig. 23). This property is further discussed in Section 5. Water content is also related to sediment porosity, which in turn is one factor controlling the diffusivity of chemical species (see Chapter 4, Section 3.3).

The effect of tubificid oligochaetes on water content arises not so much from their burrowing activity as from their selective feeding and pelletizing activities. The size of the effect is dependent on the sand content of the deposit they inhabit. McCall and Fisher (1980) report that tubificids at densities of $50–100 \times 10^3$ m^{-2} have no effect on the water content of fine-grained Lake Erie muds (top 1 cm water content 70–80%, <3% sand), despite the creation of many vertical burrows and a pellet layer in the top 1 cm of deposit. Apparently a fabric that is opened in one part of the deposit (burrows, interpellet voids) is compacted in an adjacent portion of the sediment (pellets, deposit adjacent to burrows). This takes place over such a short lateral distance that no net change in water content is measured. However, where there is a significant admixture of sand, selective feeding can greatly alter water content. Tevesz et al. (1980) found that selective feeding of tubificids on a sandy river deposit (mean grain

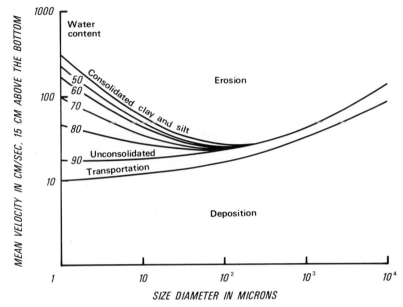

Figure 23. The influence of water content on the critical erosion velocity of fine-grained (<100-μm) sediments. Redrawn from Postma (1967).

size 110 μm, 83% sand) resulted in a vertical sorting of particles and increased water content in the top 1.5 cm of deposit (48% with worms, 27% with no worms).

Burrowing by mobile infauna can also increase water content of the top few centimeters of fine-grained sediment. This phenomenon can be seen in both fine muds and sands, but it is likely that the effect of burrowing is greater where a vertical sorting by grain density can occur in addition to the injection of water into the deposit during burrowing. McCall et al. (1979) have produced quantitative information on the effect of mobile burrowers in freshwater sediments on water content. They found that the unionid clam *Lampsilis radiata*, a mobile infaunal filter feeder (mean length 6.3 cm), increased water contents from 62% to 72% in the top centimeter of western Lake Erie sediment and from 42% to 63% in the 3- to 4-cm layer. As an alternative to measuring water content we have also used a rotational viscometer to measure the torque required to turn a T-bar spindle at 0.5 rpm as another measure of sediment compactness (see Myers, 1977). Vertical profiles of sediment compactness close to (≤2 cm) and far away from (≥2 cm) four individual *L. radiata* have been measured (Fig. 24). Close to the clams, compactness was reduced by 16–70% to a depth of 7 cm. At greater distances, however,

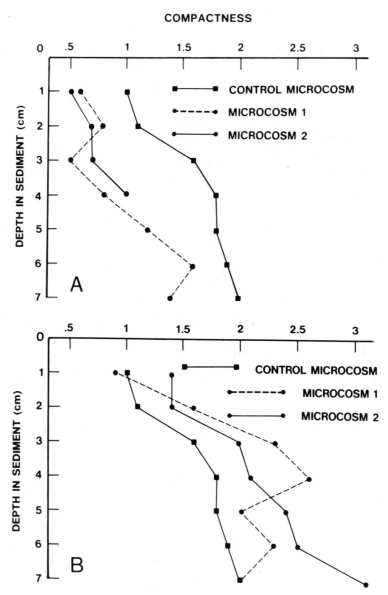

Figure 24. Mean compactness measured as torque (1 unit ≃ 575 ergs) as a function of depth in sediment. (A) Measurements made ≤ 2 cm from bivalve; (B) measurements made ≥ 2 cm from bivalve. Microcosms 1 and 2 contain *Lampsilis radiata*.

compactness was *increased* by from 0% to 58%. This is another of what Rhoads and Boyer (Chapter 1, Section 3.3) call "far-field effects," and is probably the result of foot probing during burrowing and valves adducting during cleansing activities.

4.3.2. Permeability

Burrows shorten the path length through which water flows within sediments. Fecal pellet layers also produce this effect by creating a smaller number of larger interparticle voids. McCall and Fisher (1980) found no evidence for biological advection of pore water through oligochaete burrows, but found that enhanced diffusion in the pelletal layer of sediment was sufficient to explain the flux of a chloride tracer through the sediment (see Chapter 4, Section 3.3). These experiments were done in laboratory aquaria without vigorous water circulation. Nevertheless, waves and currents may create pressure gradients sufficient to ventilate passively muddy sediments made permeable by biological activity. McCall and Fisher (1980) measured the permeability (ability of a column of water to pass through a thickness of sediment) of western Lake Erie sediment with and without tubificids. Permeability increased upon addition of 10^5 worms m^{-2} by a factor of two to four to a steady state value in a period of 3 days. This increase was due to the burrows made by the worms, since a thick pelletal layer could not form over this time interval. We have performed one experiment to test the effect of a current flow over pel-

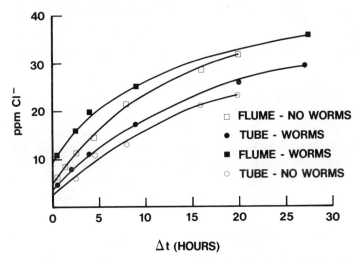

Figure 25. Effects of tubificid worms and flow on chloride diffusion across the sediment–water interface.

letized sediment. Chloride was added to western Lake Erie sediment to achieve a pore water concentration of 280 ppm. Deionized water was then placed over the sediment and the chloride concentration in the overlying water measured over time. Chloride flux of sediment with and without worms was measured in an annular flume with a current flow beneath that required to move sediment particles and in a tube with aeration but no currents (Fig. 25). Sediments with worms had a pellet layer of 0.5-cm thickness when the experiment began. Pelletized sediment with a current flow had higher initial flux rates of chloride than sediment without a pelletized layer or without a flow. Although more extensive experimentation is required, this experiment provides some support for the idea that macrobenthos can increase passive ventilation of the top layer of muddy sediments.

5. Biogenic Modification of Sediment Transport

The effects of grain size distribution, water content or porosity, and sediment shear strength on erosion depend on the particular sediment examined and on differing techniques of analysis (Sundborg, 1956; Postma, 1967; Southard, 1974). For some cohesive sediments, there is no discernible relation between measured physical properties and erosion of surface sediment (Parthenaides and Paaswell, 1968). Examination of the literature concerning the effects of marine biota on sediment transport (see Chapter 1) reveals at least four biological activities that might also be important in modifying sediment transport in freshwater environments:

1. Burrowing macrofauna and meiofauna can increase sediment water content and increase erodability (Rhoads and Young, 1970; Cullen, 1973; Rhoads et al., 1978).
2. Deposit feeders in high-water-content muds can produce sand-size pellets that are more difficult to move than fluid muds.
3. Dense patches of tube dwellers can increase sediment stability and decrease erosion (Mills, 1969; Yingst and Rhoads, 1978).
4. Bacterial and algal films can stabilize sediments and make them difficult to erode; surface deposit feeders can break up the film and make sediment less stable (Scoffin, 1970; Holland et al., 1974; Rhoads et al., 1978; Yingst and Rhoads, 1978; Boyer, 1980).

5.1. Field Observations

McCall and Fisher (1980) appear to be the only workers to have examined the effects of biota on the transport of freshwater muds in any detail. We will review their results and add some previously unpublished data from that study.

McCall and Fisher (1980) collected 38 box cores from nine locations in Lake Erie (Fig. 26, Table V). The box cores were inserted into an annular flume where the bottom stress needed to initiate rolling, τ_c, and observable entrainment of particles into the flow, τ_e, were measured. The nine location averages of sediment median grain size (measured by standard geological techniques), percent clay, percent organic matter, and percent water content were simultaneously regressed on location averages of critical entrainment stress obtained under aerobic bottom water conditions.

Average physical properties of sediment such as grain size, clay content, and water content are not good predictors of critical entrainment stress of the fine silt and clay sediments in Lake Erie. More detailed measurements of these properties at each sampling site might clarify the relation to τ_e by unmasking small-scale variability hidden by average measurements. The observations are worth collecting, but the small amount of data currently available suggests that little would be gained. Eight measurements of median grain size were made over 2 years at both location 2 and location 6; the standard deviations of these measurements are 11% and 14% of the mean, respectively. Since the median grain size of sediment from locations 1–9 changed by a factor of 20, we might expect

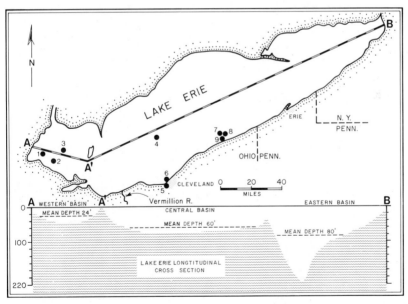

Figure 26. Lake Erie sampling localities for sediment entrainment studies. Redrawn from McCall and Fisher (1980).

Table V. Description of Box Cores Collected from Lake Erie

Box number	Location	Date collected	τ_c (dynes · cm^{-2})	τ_e (dynes · cm^{-2})	H$_2$O content (percent)	Comment[a]
1	1	6/16/76	0.21	0.82	71	
2	1	10/1/76	0.75	1.67	75	
3	1	10/1/76	0.56	1.52	78	
4	2	7/5/76	0.18	0.61	70	
5	2	10/1/76	0.61	1.44	71	
6	2	10/1/76	0.58	1.26	74	
7	3	7/6/76	0.35	0.74	72	
8	3	7/6/76	0.43	1.21	82	
9	3	5/23/77	0.35	1.21	78	
10	3	5/23/77	0.32	1.61	73	
11	3	9/1/77	0.32	1.51	86	A
12	3	9/1/77	0.38	1.14	84	
13	4	7/5/76	0.21	0.74	88	T
14	4	7/5/76	0.33	0.66	84	
15	4	9/29/76	0.46	1.07	87	
16	4	9/29/76	0.38	0.99	88	
17	4	9/29/76	0.38	1.05	84	
18	5	10/3/76	0.52	1.08	66	
19	5	10/3/76	0.48	1.05	60	
20	6	10/3/76	0.62	1.17	61	
21	6	10/3/76	0.86	1.39	59	T
22	7	5/19/77	0.33	0.84	68	
23	7	5/19/77	0.53	1.46	66	T
24	7	9/4/77	0.20	2.08	65	
25	7	9/4/77	0.68	2.15	71	A
26	7	9/4/77	1.05	2.68	71	A
27	8	5/6/77	0.45	1.20	64	
27	8	5/6/77	1.20	4.33	64	Second run
28	8	5/6/77	2.47	4.27	74	A,T
29	8	5/6/77	0.56	1.93	71	T
30	8	5/19/77	0.45	1.23	59	
31	8	5/19/77	0.73	2.20	61	
32	8	5/19/77	0.59	2.15	55	
33	8	9/4/77	0.67	2.08	—	A
34	8	9/4/77	0.45	2.06	64	A
35	8	9/4/77	0.41	1.55	68	
36	8	9/4/77	0.65	1.99	57	
37	9	5/19/77	1.44	3.27	70	
38	9	5/19/77	0.51	3.02	67	

[a] A, anoxic overlying water in lab; T, high tube density (>3000 tubes m^{-2}).

that any difference in τ_e owing to grain size differences would not be entirely masked by within-site variability. In fact, sediment water content is probably the most variable of the physical properties studied, and water content measurements were collected with each critical stress measurement. Power law and linear regressions of individual box core water content and τ_e explain only 17% and 11% of the data. The slope of the regression is not significantly different from zero (0.023 ± 0.025, p = 0.05). Neither is there any correlation of τ_e and water content within the nine collection sites.

These results appear to be contrary to those of other workers who showed grain size and water content to be major controls of sediment erosion (Postma, 1976; Southard, 1974; Fukuda, 1978; Fukuda and Lick, 1980). The results of Fukuda (1978) are particularly appropriate for comparison, because he studied the erosion of fine-grained sediments in Lake Erie that were "azoic" with respect to macrofauna. Fukuda studied two types of sediment from Lake Erie. One was an illitic clay derived from heavily weathered Chagrin Shale, which crops out along the lake and is a parent material for lake sediment. The other was lake bottom sediment collected with grab samplers and subjected to sieving and storage in the laboratory for several weeks. Some of his results are shown in Fig. 27. Data on illitic clay were obtained by adding fluid clay–water mixtures to box cores and then measuring τ_e as the sediment dewatered through self-compaction. Data obtained in the same way with the single difference that the whole flume bottom was covered with sediment gave the same results as those obtained with the box cores, indicating that the box core method is an acceptable way to determine τ_e.

The results show also that, without the action of other processes acting to modify lake sediment properties, entrainment is indeed related to water content: Higher water content results in lower τ_e (r = −0.92, N = 26). In the natural situation, however, other processes, many of them biogenic, act to change sediment erodibility in such a way that the effect of water content is masked. Furthermore, for a given water content all lake sediments are more difficult to erode than parent illitic clay sediments. In addition, average τ_es for sediments collected from sites 2 and 9 were 100% and 75% greater, respectively, in box cores collected with minimal disturbance of biogenically modified sediment surfaces by McCall and Fisher (1980) than in the grab-sampled and sieved sediment used by Fukuda (1978). Some factor or factors act to aggregate or bind particles together, making lake cores more difficult to erode. Finally, some other processes must be acting in more or less random fashion, some increasing and some decreasing erodibility, in such a way that the result is a collection of data points through which it is not possible to draw one smooth curve containing all the observations.

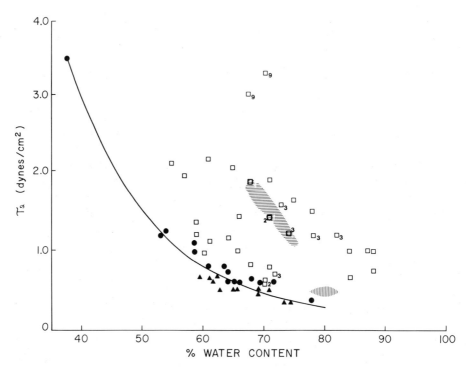

Figure 27. Critical entrainment stress, τ_e, of oxidized box cores as a function of sediment water content. ●, Box cores of shale-based sediment run by Fukuda (1978); ▲, runs made with the entire flume covered with shale-based sediment (also by Fukuda, 1978); □, box cores collected from locations 1–9. Area shaded with horizontal bars is to be compared with box cores from location 9; area shaded with vertical bars is to be compared with box cores from locations 2 and 3. Redrawn from McCall and Fisher (1980).

Some of the observed variations of τ_e may be related to biological activity. Critical entrainment stress appears to be higher (1) in box cores containing high densities of oligochaete worm tubes projecting out of the bottom to a height of 3 mm or more; (2) in box cores collected in the fall (mean temperature 1 m off bottom 11°C) compared to box cores collected in summer (mean temperature 22°C); and (3) in box cores accidentally exposed to water with low dissolved oxygen concentrations. The contrasts of otherwise similar box cores are shown in Table VI. Given the small sample sizes and the unknown underlying distributions of the variables, a nonparametric test of the differences in τ_e, either the Wilcoxon signed rank or sign test, is most desirable. Both substitute the sign of the difference between comparable box cores for the absolute difference, and both give the same result here. In all three contrasts all the differences between comparable cores are negative. The null hypothesis is that there is no difference between, say, summer and fall box cores. If that is true, then there is a 50% chance of a positive difference and a 50% chance of a negative difference for any pair of box cores. Since there are four pairs of box cores in each contrast, the probability of observing four negative differences in four comparisons is given by the binomial distribution as $(1/4)^4 = 0.063$. For a two-tail test of significance the probability of equally extreme fluctuations in the opposite direction (all positive differences) must be added and the probability rises to 0.125. This means that if the null hypothesis were true deviations as large as those observed in these contrasts would occur by chance 12.5% of the time.

We might expect dense fields of worm tubes to increase the critical entrainment stress, since they will act to promote sediment binding and prevent the transfer of stress to the sediment–water interface (but see Chapter 1, Section 4.1). Chironomid tubes are probably even more effective at binding sediment than tubificid oligochaete tubes because they have a larger diameter, greater length, and thicker walls constructed of a sticky tube silk. Edwards (1958) provides testimony to the binding power of chironomid tubes. He found masses of chironomid tubes floating in the effluent channel of a sewage works; the tube mats were associated with the spring emergence of *Chironomus riparius* (10^5 m^{-2}). The tube mats

Table VI. Contrasts of Box Cores Collected from Lake Erie

Contrast	Comparable box cores	Stress difference (dynes · cm^{-2})	Percent increase
Low worm tube density–high worm tube density	14–13	−0.08	12
	27–29	−0.73	61
	22–23	−0.62	74
	20–21	−0.22	19
			Mean = $\overline{41.5}$
High dissolved O$_2$–low dissolved O$_2$	24–$\overline{25,26}$	−0.34	16
	36–$\overline{33,34}$	−0.08	4
	28–27	−3.07	256
	12–11	−3.27	32
			Mean = $\overline{77}$
Summer–fall	1–$\overline{2,3}$	−0.78	95
	4–$\overline{5,6}$	−0.74	55
	$\overline{13,14}$–$\overline{15,16,17}$	−0.34	33
	$\overline{22,23}$–24	−0.93	45
			Mean = $\overline{57}$

were also produced in laboratory experiments in which it was found that methane bubbles generated at depth could lift a tube mass a few centimeters in thickness off the bottom of aquaria after emergence of the chironomids. Edwards hypothesized that irrigation currents of the insect larvae removed gas bubbles from the sediment under normal conditions so that the "vernal sloughing" of the tube layer, as he called it, occurred only following spring emergence and elevated temperatures (higher gas production). The effect of chironomids on sediment entrainment has not yet been the subject of any flume studies.

The oxygen content of the water overlying the box cores occasionally dropped to low levels (2 ppm or less), either because water circulation in the aquarium holding the box core ceased for a period of time or because a box core remained sealed for several days prior to leveling and insertion into the flume. When this happened, increased mucus binding of the surface was usually evident. In most cases, strings of mucus-bound sediment were visible on the sediment surface; in a few cases, the entire sediment surface was bound into a 1- to 2-mm-thick mat that could be peeled off the box core surface intact with forceps. In only one case (box core 8, location 3) was a similar mat found under fully oxidized conditions; here also critical entrainment stress was increased over the comparable box core 7. It is hypothesized that mucus binding of surface sediment increased τ_e in this case.

The summer–fall contrast is more difficult to account for. The only obvious difference in comparable cores collected at these times is bottom water temperature. It is possible, of course, that the seasonal change in sediment erodibility may not be directly related to temperature. For instance, microbial production of exopolymer decreases when nitrogen availability is low (Unz and Farrah, 1976) and the availability of nitrogen may change seasonally. However, many biological processes that affect sediment erodibility are temperature-related, so the increase in critical entrainment stress from summer to fall may also be related to biological activity in the sediments. Yingst and Rhoads (1978) documented the existence of a seasonal pattern of erodibility of marine (Long Island Sound) sediments in more detail. They proposed that differing temperature responses of sediment-stabilizing microbes, principally mucus-secreting bacteria, and sediment destabilizing fauna, principally burrowing macrofauna, together with phase lags in the response activity to temperature could produce seasonal changes in sediment properties. Such a seasonal cycle can have large effects not only on bottom processes but also on the overlying water column; McCall (1977), for instance, found that in warmer months about twice as much bottom material was resuspended into the water column in central Long Island Sound as in winter months, despite a trend in wind and wave disturbances in the opposite direction.

5.2. Laboratory Experiments

McCall and Fisher (1980) reported the results of an experiment in which Lake Erie sediment was added to box cores with worms added, with no worms added, and with formalin added. After incubating for 20 days at 7°C and 24°C, τ_e for each box was determined in the annular flume. At 7°C, τ_e of pelletized sediment was 39% higher than τ_e of nonpelletized sediment. There was no significant difference in τ_e between boxes with and without formalin treatment, although the formalin treatment did make sediment more difficult to erode. At this temperature the most important biogenic modification is pelletization, which decreased erodibility by increasing natural particle size (no tubes were built above the sediment surface); microbial binding was evidently not effective in decreasing erodibility. The surface of the formalin-treated sediment may have been relatively more compact than that of the untreated sediment and thus more difficult to erode. Untreated sediment was subjected to surface mixing by meiofaunal ostracods and copepods that may have made sediment easier to erode. Indeed, the sediment treated at 7°C with no worms had the lowest τ_e of any sediment, perhaps because there was no pelletization, little microbial binding, and some loosening of surface sediment by meiofauna.

Sediment in boxes incubated at 24°C with no worms was 28% more difficult to erode than sediment incubated at 7°C with no worms. This was likely due to greater microbial activity at the higher temperature. There was no difference between sediments with and without worms, indicating that microbial mucus production is evidently as important as pelletization at this temperature. McCall and Fisher (1980) used parametric methods to discern these significant differences in average τ_e between treatments. Since the sediment in these laboratory box cores was reasonably homogeneous within treatments, we would really not be surprised if the errors were normally distributed, even though we would need more samples to actually test this idea. If we can assume a Gaussian distribution, then their method for making multiple comparisons (Tukey's T test), which is more efficient at detecting differences that actually exist than analogous nonparametric tests, is appropriate. Our confidence in the comparisons made should be tempered by the finding that only two contrasts are significant using nonparametric tests (Wilcoxon critical range test; Colquhoun, 1971). These are: (1) that the τ_e of box cores at 7°C with worms is greater than that of comparable cores without worms, and (2) that sediment incubated at 24°C with no worms had a larger τ_e than sediment incubated at 7°C with no worms.

Another experiment was performed to see if the increased mucus binding effect that was seen in field box cores subjected to low dissolved oxygen concentrations could be reproduced. Box cores receiving the same treatments as in the previous experiment were further incubated for 7 days in water low in dissolved oxygen (≈ 1 ppm), after which their τ_e was measured in the flume. As a further check on the dissolved oxygen effect, τ_e was measured a final time after incubation for 7 more days in fully oxidized water (≈ 10 ppm).

Critical entrainment stresses of box cores incubated under different dissolved oxygen concentrations are shown in Table VII. The increase in τ_e in box cores moved from high to low dissolved oxygen concentrations is also shown. In the comparisons that follow, both Gaussian and nonparametric methods give similar results. The increase in τ_e with decreased dissolved oxygen levels in box cores with active biota is significant; the average increase is about 30% and may be as much as 55–65%. There is no significant difference in τ_e between high and low oxygen levels in formalin-treated box cores. Finally, the effect of anoxia on τ_e in sediments with biological activity is temperature-dependent. There is a greater than twofold increase in the anoxic effect between 7°C and 24°C. It is not known with certainty which organisms produce all the mucal material observed in box cores subjected to low oxygen levels. Most macroinvertebrates, molluscs and tubificids included, secrete mucopolysaccharides (S. Hunt, 1970), but smaller organisms—in particular protistans, algae, fungi, and bacteria—are probably quantitatively more important in particle aggregation [see Harris and Mitchell (1973) for a review]. Bacterial polysaccharides are particularly effective in enhancing aggregation in soils (Martin, 1946; Geoghegan and Brian, 1948; Fehrmann and Weaver, 1978), sewage wastes (Pavoni et al., 1972), and marine sediments (Rhoads et al., 1978; Yingst and Rhoads, 1978). Mucus-secreting algae are able to stabilize shallow marine sediments (Frankel and Mead, 1973; Holland et al., 1974). Protists may also cause particle aggregation (Pillai, 1941; Curds, 1963). But since an amount sufficient to coat the sediment surface was produced in a short amount of time (7 days) under dark as well as light conditions, it is thought that this binding material is bacterial in origin. Why this material is produced when dissolved oxygen concentrations are low is also unknown. It is known that exopolymer production is highest in the death phases of bacterial cultures, perhaps because of cellular lysis (Pavoni et al., 1972; Harris and Mitchell, 1973). Low-oxygen conditions may induce this phenomenon in aerobic bacteria. Alternatively, low oxygen may favor the growth of some bacteria that secrete greater amounts of binding substance [J. Y. Yingst (personal communication) suggests *Thiovulum* as an example]. The decrease in critical entrainment stress upon reaeration of box cores might be due to a decline in exopolymer production and/or an increase in near-surface burrowing and deposit feeding activity, which break up the mucus layer. Surface burrowing and deposit

Table VII. The Effect of Anoxia on Critical Entrainment Stress

I. Critical entrainment stress (dynes · cm^{-2}) of sediment incubated at different levels of dissolved oxygen

Temperature (°C)	Treatment	Dissolved oxygen		
		High oxygen (1)	Low oxygen (2)	Reaerated (3)
7°	No worms (nw)	0.72	0.94	1.08
		0.62	0.77	0.83
	Worms (w)	0.81	0.85	0.75
		1.08	1.13	1.00
	Formalin (f)	0.80	0.81	—
		0.85	0.83	—
24°	No worms	0.93	1.55	0.76
		0.81	1.09	0.68
	Worms	0.87	1.09	0.86
		0.80	1.33	0.85
	Formalin	0.81	0.83	—
		0.84	0.85	—

II. Column (2) − column (1) differences

7°C			24°C		
nw	w	f	nw	w	f
0.05	0.22	0.01	0.23	0.62	0.02
0.05	0.16	−0.02	0.53	0.27	0.01
x̄ = 0.045	0.19	−0.005	0.375	0.445	0.015

III. Statistics

A. Hypothesis: Mean column difference, d̄, for treatments with biological activity is zero.
  1. Student's $t_{(n-1)} = \dfrac{\bar{d}}{(S/\sqrt{n})} = 3.568$, $p < 0.005$ (one-tailed), $N = 8$, $\bar{d} = 0.264 \pm 0.175$ (95% confidence).
  2. The sum of ranks for paired samples, Wilcoxon's T, equals zero, $p = 0.01$, $N = 8$. The median difference, $\bar{d} = 0.225$. At the 0.95 level, $0.05 < \bar{d} < 0.53$.
B. Hypothesis: Formalin-treated samples show no increase in τ_e from high to low oxygen levels.
  1. $t = 0.588$, $p < 0.75$, $N = 4$.
  2. Wilcoxon $T = 3.5$, $p = 0.375$, $N = 4$.
C. Hypothesis: The increase in τ_e from high to low dissolved oxygen levels is greater at 24°C than at 7°C in those cores with active biota.
  1. $t = 3.074$, $0.025 < p < 0.01$, $N = 4$. At 0.95 level $\bar{d} = 0.292 \pm 0.232$
  2. Unpaired sum of ranks, T', equals 10, $p = 0.05$, $n_1 = n_2 = 4$. At the 0.95 level $\bar{d} = 0.29 \pm 0.19$.

feeding are perhaps the most active processes in the destruction of the mucus layer because exopolymers are resistant to microbial decay (Harris and Mitchell, 1973; Martin and Richards, 1963), but the existence of easily degradable "glues" cannot be ruled out.

5.3. Some Comparisons

These laboratory experiments are useful to demonstrate the existence and potential importance of various biogenic modifications of sediment erodibility (i.e., pelletization, microbial binding, locomotory movements). However, they are not especially good predictors of field results. For example, the laboratory experiment might lead us to believe that the natural bottom should be most difficult to erode during periods of high temperature, but this is not apparently the case. The laboratory experiment lasted only a few weeks and employed fixed incubation temperatures whereas the lake bottom "incubates" for a longer period of time at high temperatures, and bottom temperature changes more slowly over a longer period of time. In addition, having been sieved and fluidized, the initial condition of laboratory sediment must have been different than that of natural bottom sediment at the beginning of summer. Natural bottom sediments record a history of past physical–chemical and biological modifications, and it may sometimes require a period of weeks or months for new processes to erase this history.

Comparing the results of different workers is difficult at this time. We all use flumes of different design with unknown effects on our results and we measure erosion in different ways. Few data are available that are reliable in the sense that the natural sediment surface is not severely altered. But results of McCall and Fisher (1980), Rhoads *et al.* (1978), and Young and Southard (1978) are comparable, if only because these workers preserved the biogenically created surface properties of the fine-grained sediment that they examined. However, because each investigator used a different criterion for measuring a "critical" erosion point, only internal comparisons can be made. Young and Southard (1978) compared the critical boundary shear velocity between (1) marine muds collected, mixed, and settled into core boxes from suspension in the laboratory and (2) *in situ* bottoms and bioturbated laboratory cores. They found a twofold reduction of the critical erosion point in the latter case. McCall and Fisher (1980) found a similar difference between disturbed and undisturbed bottoms in freshwater Lake Erie muds, but the sign of the contrast was reversed. Undisturbed cores collected *in situ* were 75–100% more difficult to erode than biota-poor laboratory sediment at the same water content. While it is clear that disturbing biogenically produced surfaces will drastically alter experimental results, the direction in which they will be altered cannot be predicted in the general case. Further specification of sediment physical properties and infauna is needed to make fair comparisons. Rhoads et al. (1978) found a 60% increase in critical mean suspension velocity from summer to fall (July–November) in marine muds from Long Island Sound. We found a 57% increase. Rhoads et al. (1978) found a 56% increase in critical rolling velocity of fine silt beads upon incubation in seawater containing bacteria. We found a 40–80% increase in τ_e where evidence of increased binding was present. Rhoads et al. (1978) found an 80–100% increase in critical erosion velocity when $\sim 10^4$ tube dwellers m^{-2} were added to Long Island Sound muds. We found a 41% incease in τ_e in Lake Erie box cores with >3000 tubes m^{-2}. Significant biogenic modifications of fine-grained bottoms are a widespread phenomenon. Despite great sedimentologic and faunal differences in marine and freshwater bottoms, functionally similar organisms will likely produce about the same degree of change in erodibility of fine-grained surface sediments.

6. Ecological Interactions and Sediment Properties

Organisms influence substratum properties. Sediment properties influence the distribution and abundance of benthos. Although most of the literature concerns marine environments [see reviews by Gray (1974) and Rhoads (1974), for example], Cummins and Lauff (1969), Oliver (1971), and Brinkhurst (1974), among others, have demonstrated the correlation between organisms and physical and chemical properties of freshwater bottoms. It is important to realize that benthic fauna and sediments constitute a complex feedback system in which numerous influences on benthos distribution and abundance can indirectly affect sediment properties.

6.1. Spatial–Temporal Variations of Fauna

In Lake Erie there appears to be a seasonal cycle of bottom erodibility. We have hypothesized that this cycle is due in part to the different response of groups of benthos to changing temperature (Section 5). There may be other explanations of changing faunal abundances that can enable us to predict changes in sediment properties. Rhoads and Boyer (Chapter 1) have described how sediment properties develop during ecological succession on marine fine-grained bottoms. Following some disturbance of the bottom that causes local mortality of the macrobenthos, a repeatable pattern of faunal change occurs. Dense aggregations of small tube dwellers and other near-surface dwelling infauna are followed in time by larger, deeper-dwelling infauna, which may be either deposit feeders or filter feeders (McCall, 1977, 1978). Sediment shear strength, water content, biogenic texture, microtopography, and erodibility are differentially altered in early and late successional stages (Chapter 1, Table III).

The field experimental work required to establish the existence of an orderly pattern of recovery from disturbance in freshwater lakes has yet to be completed. However, for the past few years one of us (PLM) and his students have been conducting such experiments in Lake Erie. The picture that has been emerging is that, following a disturbance of the bottom, large numbers of near-surface-dwelling ostracods and naid oligochaetes are early colonizers. They are soon followed by abundant tube-dwelling chironomids; later come deeper-dwelling deposit-feeding tubificids. Large burrowing infauna (unionids) are absent from the early stages of recovery. Since these organisms affect the bottom in different ways, it is likely that the "successional paradigm" can be of use in predicting changing sediment properties on disturbed bottoms in freshwater as well as marine environments. Whether or not such a model will be widely applicable is unknown at present. There is some suggestion from the literature that it might be. For example, Herdendorf and Reutter (1976) investigated the effect of dredging on benthic populations near a nuclear power station on Lake Erie. The benthic fauna was eradicated during construction of water intake and discharge pipelines. Less than a year was required for recolonization with populations in the dredged area exceeding the original numbers. Sweeney et al. (1975) sampled a dredge spoil disposal site in Lake Erie 5 years after dumping had ceased. We find no statistical difference in mean abundance of major macroinvertebrate groups at dump and reference stations. Paterson and Fernando (1969), Lellak (1969), and McLachlan (1974) documented macroinvertebrate colonization of newly created reservoirs, fishponds, and small lakes. Faunas of chironomids followed by oligochaetes were established in less than 2 years, and in some cases less than 3 months.

Patchy distributions of benthos unrelated to measured substratum properties could also be due to temporally varying local disturbances of the bottom. In a correlation analysis of Lake Superior macrobenthos, Freitag et al. (1976) found no strong or consistent correlation of physical parameters with benthos distributions. Partial correlation and multiple regression analyses of Lake Ontario macrobenthos and physical–chemical sediment properties were made by Nalepa and Thomas (1976), who found that the distribution of the oligochaetes *Limnodrilus hoffmeisteri* and *Tubifex tubifex* were related to chemical sediment properties only at deepwater stations. The abundance of the amphipod *Pontoporeia hoyi* was related to sediment grain size only at deep-water stations. At shallower (more disturbed?) stations there was no significant relationship.

6.2. Interactions among Macrofauna

Whether or not "successions" occur on lake bottoms, it is known that freshwater benthos interact to alter their own abundances. A change in community composition will also result in altered sediment properties. Amensalistic, competitive, and mutualistic interactions have been described. Jonasson and Thorhauge (1972) report that when *Chironomus anthracinus* is abundant in Lake Esrom, the abundance of the tubificid *Pomatothrix hammoniensis* decreases. In years when *C. anthracinus* is absent, the number of tubificids increases. Brinkhurst and Kennedy (1965) found a similar negative correlation between *Chironomus thummi* and tubificid oligochaetes in different parts of a polluted stream. We have also observed a negative interaction between two common species: *Chironomus plumosus*, a surface deposit feeder, and the tubificid *Limnodrilus hoffmeisteri*, a subsurface deposit feeder. In field colonization experiments, the presence of large numbers of *C. plumosus* (>3000 m^{-2}) was associated with reduced abundances of oligochaetes. In laboratory experiments, *L. hoffmeisteri* avoid sediments inhabited by *C. plumosus*, and migrate from sediment into which chironomids are introduced within 1 day. The reason for this avoidance behavior is presently unknown. Chironomids burrow using their prolegs and mouthparts, and this motion can sever tubificids if they are encountered. Other possibilities are that chironomids alter tubificid food resources by altering sediment chemistry or that they excrete some toxic metabolite. Tubificids feed in the reducing zone and chironomids oxidize sediment by pumping water through their burrows. Lastly, surface deposit-feeding activities of chironomids may inhibit tubificid respiration. Within the deposit feeding Tubificidae, both intraspecific and interspecific competition have been found to influence population abundance and distribution (Jonasson and Thorhauge, 1976a; Bonomi and DiCola, 1980; Milbrink, 1980; Lang and Lang-Dobler, 1980). Brinkhurst et al. (1972) suggest a mutualistic interaction among three species of tubificids from Toronto Harbor, *Tubifex tubifex, L. hoffmeisteri,* and *Peloscolex multisetous*. Feeding (pelletization) and growth of these species is greater in mixed culture than in pure culture, each species stimulating growth of the others in an undetermined manner. Sephton and Paterson (1980) found that the bivalve molluscs of Movice Lake increased the numerical abundance of nonbivalve benthos (primarily chironomids) living close to them. Stanczykowsa (1978) found a similar result near patches of the clam *Dreissena*. We have observed the presence of a thickened and more well-defined black iron sulfide zone and greater abundance and deeper penetration into the substratum by tubificid oligochaetes in laboratory microcosms with unionids (*Lampsilis radiata*, 20 m^{-2}) than in microcosms without clams (Fig. 28).

Many of these mutualistic interactions are probably mediated by changes in microbial food supply brought about by the activities of macrobenthos. Hargrave (1970b,c) for example, found that the burrowing and feeding activities of the freshwater amphipod *Hyalella azteca* stimulated bacterial respiration in bottom sediments from Marion Lake. Fukuhara *et al.* (1980) found that ammonia production in the 0- to 5-cm layer of submerged ricefield soils was doubled under the influence of feeding by *L. hoffmeisteri* and *Branchyura sowerbyi*; the number of aerobic bacteria decreased, while the number of sulfate reducers increased. We have found that the sediment-mixing activities of tubificids routinely double the total number of bacteria (measured by epifluorescent microscopy) in the top 5 cm of Lake Erie sediment (Fig. 29). Pringle (1980) reports an increase in diversity and biomass of diatoms in stream substrata containing chironomid tubes. Aller (Chapter 2, Section 6) discusses the ways in which stimulation of microbial production occurs. The effect of this activity is to change sediment properties either directly by changing the binding properties of the bottom or indirectly by altering the abundance of other benthos. Many deposit-feeding benthos feed selectively on bacteria (Marzolf, 1965; Hargrave, 1970b; Wavre and Brinkhurst, 1971).

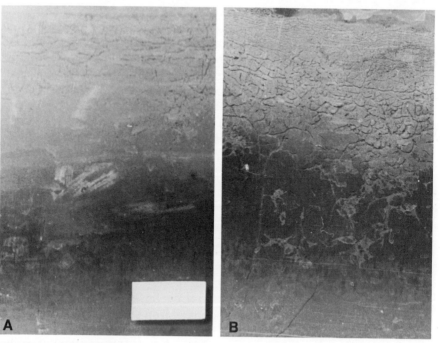

Figure 28. Tubificid burrows and black sulfide layer in laboratory aquaria (A) with no unionids and (B) with unionids (20 individuals m^{-2}). Scale bar: 1 cm.

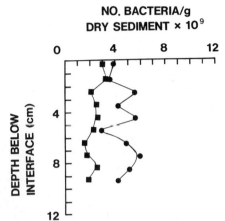

Figure 29. Effects of tubificid oligochaetes on bacterial abundances in Lake Erie sediments. ■, Microcosm with no worms; ●, microcosm with 5×10^4 worms m^{-2}.

Deposit-feeding tubificids are quite sensitive to changes in bacterial populations, as the following experiment shows. *L. hoffmeisteri* were placed in a 37-cm^2 microcosm tube containing two sediments. One half contained unaltered central basin sediment irradiated with a ^{60}Co source to stop bacterial activity. The tubificids avoided the irradiated sediment until

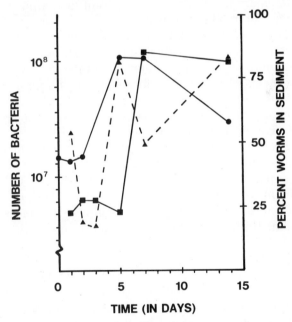

Figure 30. Oligochaete abundance in presterilized sediment as a function of time. ■, Abundance of *Limnodrilus hoffmeisteri*; ▲, bacterial abundance in experimental cell; ●, bacterial abundance in sterilized central basin mud to which no worms were added.

bacteria entered the log growth phase. Then the worms moved en masse to the irradiated sediment (Fig. 30). Thus changes in bacterial numbers can cause localization of macrofauna with resultant heterogeneities in sediment properties.

7. Future Work

On the basis of either organism size, abundance, or casual observation, we suggest that mayfly nymphs, amphipods, unionid clams, and meiofaunal ostracods, copepods, and nematodes are candidates for further study. Also, although it is clear that the cosmopolitan chironomids are very important in altering sediment chemistry (Chapter 4), probably too little has been made of their effects on physical sedimentary properties.

Workers in the field have been concerned primarily with the organismal effects on sediment stratigraphy, physical properties that bear on sediment transport, and physical–chemical properties that bear on diagenesis and flux of dissolved materials across the interface. (This last topic is the subject of Chapters 2 and 4). Models of biogenic sediment mixing are fairly well developed (Chapter 7). What is not well understood is which sedimentary materials are transparent to mixing and which are not. In addition, the mode of formation and preservation of laminae in lake sediments is poorly understood. Such sedimentary structures are a potentially useful tool in reconstructing the history of sediment transport (Nittrouer and Sternberg, 1981; Chapter 6) and deserve further study.

There are no standard methods for examining biogenic modification of sediment transport. Each worker uses a flume and bottom sampling device of different construction, and each measures critical erosion points in a different manner. Consequently, precise intercomparison of results is at present nearly impossible. We need some standard methods. Especially desirable would be an inexpensive quantitative way to measure the critical erosion point. It is, however, quite likely that we will eventually realize that the measurement of the point at which erosion is initiated is not enough. Then the problem will become how best to measure the flux rate of sediment from the bottom as a function of applied stress. One interesting substantive problem for future work is the determination of the rate of biogenic change of surface sediment properties that control sediment transport—how long, for instance, does it take for biota to return the bottom to an equilibrium condition following a large storm, or is there no equilibrium? Another is the measurement of the vertical gradient of erodibility of natural sediments. The few observations we have made (Table VI) suggest that the gradient is quite steep over the top few millimeters of sediment. Accurate calculation of the amount of sediment eroded during a storm, for instance, requires that we know this gradient. Finally, we echo the sentiment of Rhoads and Boyer (Chapter 1) that the quantitative assessment of the significance of microbial films is probably the single most important problem in understanding biologically controlled sediment transport.

We ought to begin also to examine the broader significance of our results. Over short time scales (perhaps a year or less) biogenic modifications of the bottom can have important effects on bottom stability and water column turbidity that can result in changes in sediment transport and benthic and pelagic community composition. But what about longer time scales? It is not clear, for example, how important biogenic control of erosion is compared to sediment transport in the 5-year, 50-year, or 100-year storm.

It may turn out that biogenic modification of soft bottoms has greater evolutionary significance than geological import, for it may be a powerful selective force in benthic communities (Wilson, 1980). Wilson's idea is that effects of organisms that are beneficial to their community can be selected for; they are not mere epiphenomena. The idea that marine communities function as "superorganisms" in which constituent species contribute to the maintenance of the whole community is not new (Dunbar, 1960; Margalef, 1968), but Wilson (1980) is the first to put the idea on what appears to be a firm evolutionary basis. Van Valen (1980) regards this idea of selection for benefactor organisms as one of major significance for evolutionary ecology. According to Wilson, species populations are subdivided into groups that are homogeneous with respect to genetic mixing (demes) and smaller groups that are homogeneous with respect to the manifestation of individual traits (trait groups). Trait groups can vary in their genetic composition. Variation among local groups of organisms that causes differential productivity of the groups can result in an increase in the number of the organisms that increase the fitness of their group. For example, earthworms improve the soil for plant growth and plants provide food for worms. Some parts of a field will have worms that aid plant growth better than other worms. More plants means more earthworms and the patches with more beneficient worms will expand. The temptation to explain the evolution of benefactors among marine and freshwater soft bottom benthos by analogous reasoning is great, since soft-bottom infauna so easily and greatly modify their surrounding environment and yet are sensitive to substratum properties. In addition, benthos are relatively sedentary organisms for much of their life cycle, concentrating movement on a brief dispersal stage. Thus, subdivided populations are probably common. It is probably not an accident that 20 of 42 examples of nontrophic interactions (not predation and not competition) that affect fitness (Wilson, 1980) are taken from benthic and soil communities.

ACKNOWLEDGMENTS. The original research published in this chapter was funded by grants from the National Science Foundation (OCFC-8005103) and the National Oceanic and Atmospheric Administration (NA80RAD0036), Office of Marine Pollution Assessment. Karen Toil typed and edited several drafts of the manuscript. Fred Soster worked up most of the data on colonization in Lake Erie. Walter Borowski did the oligochaete sediment preference experiment. Sharon Matis provided the chloride diffusion data. In developing ideas for this paper, we benefited greatly from discussions with D. C. Rhoads, J. B. Fisher, and W. J. Lick.

References

Adams, C. E., and Kregar, R. D., 1969, Sedimentary and faunal environment of eastern Lake Superior, in: *Proceedings of the 12th Conference on Great Lakes Research*, pp. 1–20, International Association for Great Lakes Research, Ann Arbor, Michigan.

Alley, W. P., and Anderson, R. F., 1968, Small scale patterns of spatial distribution of the Lake Michigan Macrobenthos, in: *Proceedings of the 11th Conference on Great Lakes Research*, pp. 1–10, International Association for Great Lakes Research, Ann Arbor, Michigan.

Alley, W. P., and Chin, J. J., 1978, Length–biomass relationships of the amphipod, *Pontoporeia affinis* of Lake Michigan, in: *Abstracts of the 20th Conference on Great Lakes Research*, p. 2, International Association for Great Lakes Research, Ann Arbor, Michigan.

Alsterberg, G., 1925, Die Nahrungszirkulation einiger Binnensetypen, *Arch. Hydrobiol.* **15**:291–338.

Appleby, A. G., and Brinkhurst, R. O., 1970, Defecation rate of three tubificid oligochaetes found in the sediment of Toronto Harbour, Ontario, *J. Fish. Res. Board Can.* **27**:1971–1982.

Baker, F. C., 1928, The freshwater mollusca of Wisconsin. Part II. Pelecypods, *Wisc. Geol. Nat. Hist. Surv. Bull.* **70**.

Berg, K., 1938, Studies on the bottom animals of Esrom Lake, *K. Dan. Vidensk. Selsk. Skr.* **8**:1–255.

Birtwell, I. R., and Arthur, D. R., 1980, The ecology of tubificids in the Thames estuary with particular reference to *Tubifex costatus* (Clapanedi), in: *Aquatic Oligochaete Biology* (R. O. Brinkhurst and D. G. Cook, eds.), pp. 331–382, Plenum Press, New York.

Bonomi, G., and DiCola, G., 1980, Population dynamics of *Tubifex tubifex*, studied by means of a new model, in: *Aquatic Oligochaete Biology* (R. O. Brinkhurst and D. G. Cook, eds.), pp. 185–204, Plenum Press, New York.

Bousfield, E. L., 1958, Freshwater amphipod Crustacea of glaciated North America, *Can. Field Nat.* **72**:55–113.

Boyer, L. F., 1980, Production and preservation of surface traces in the intertidal zone, Ph.D. dissertation, Department of the Geophysical Sciences, University of Chicago, Chicago, Illinois.

Brinkhurst, R. O., 1964, Observations on the biology of lake-dwelling Tubificidae, *Arch. Hydrobiol.* **60**:385–418.

Brinkhurst, R. O., 1965, The use of sludge worms (Tubificidae) in the detection and assessment of pollution, in: *Proceedings of the 8th Conference on Great Lakes Research*, pp. R09–10, International Association for Great Lakes Research, Ann Arbor, Michigan.

Brinkhurst, R. O., 1974, *The Benthos of Lakes*, St. Martin's Press, New York.

Brinkhurst, R. O., and Chua, K. E., 1969, A preliminary investigation of some potential

nutritional resources by three sympatric tubificid oligochaetes, *J. Fish. Res. Board Can.* **26:**2659–2668.
Brinkhurst, R. O., and Jamieson, B. G. M., 1971, *Aquatic Oligochaeta of the World*, University of Toronto Press, Toronto.
Brinkhurst, R. O., and Kennedy, C. R., 1965, Studies on the biology of the Tubificidae (Annelida, Oligochaeta) in a polluted stream, *J. Anim. Ecol.* **34:**429–443.
Brinkhurst, R. O., Chua, K. E., and Kaushik, N. K., 1972, Interspecific interactions and selective feeding of tubificid oligochaetes, *Limnol. Oceanogr.* **17:**122–133.
Burgis, M. J., Darlington, J. P. E. C., Dunn, I. G., Ganf, G. G., Gwahba, J. J., and McGowan, L. M., 1973, The biomass and distribution of organisms in Lake George, Uganda, *Proc. R. Soc. London Ser. B* **184:**271–298.
Calow, P., 1975, Defecation strategies of two freshwater gastropods, *Ancylus fluviatilis* Mull and *Planorbis contortus* Linn. (Pulmonata) with a comparison of field and laboratory estimates of food absorption rates, *Oecologia (Berlin)* **20:**51–63.
Caspers, H., 1980, The relationship of saprobial conditions to massive populations of tubificids, in: *Aquatic Oligochaete Biology* (R. O. Brinkhurst and D. G. Cook, eds.), pp. 503–506, Plenum Press, New York.
Coker, R. E., Clark, H. W., Shira, A. F., and Howard, A. D., 1921, Natural history and propagation of freshwater mussels, *Bull. U. S. Bur. Fish.* **37:**77–181.
Cole, G. A., 1953, Notes on the vertical distribution of organisms in the profundal sediments of Douglas Lake, Michigan, *Am. Midl. Nat.* **49:**252–56.
Colquhoun, O., 1971, *Lectures on Biostatistics*, Clarendon Press, Oxford.
Crowley, T. E., 1957, Age determination in *Anodonta*, *J. Conchol.* **24:**201–207.
Cullen, D. J., 1973, Bioturbation of superficial marine sediments by interstitial meiobenthos, *Nature* **242:**323–324.
Cummins, K. W., and Lauff, G. H., 1969, The influence of substrate particle size on the microdistribution of stream macrobenthos, *Hydrobiologia* **34:**145–181.
Curds, C. R., 1963, The flocculation of suspended matter by *Paramecium caudatum*, *J. Gen. Microbiol.* **33:**363–375.
Davis, R. B., 1974, Stratigraphic effects of tubificids in profundal lake sediments, *Limnol. Oceanogr.* **19:**466–488.
Dunbar, M. J., 1960, The evolution of stability in marine environments; natural selection at the level of the ecosystem, *Am. Nat.* **94:**129–136.
Edwards, R. W., 1958, Vernal sloughing of sludge deposits in a sewage effluent channel, *Nature* **180:**100.
Eggleton, F. E., 1937, Productivity of the profundal benthic zone in Lake Michigan, *Pap. Mich. Acad. Sci.* **22:**593–611.
Eggleton, F. E., 1939, The deep water bottom fauna of Lake Michigan, *Pap. Mich. Acad. Sci.* **21:**599–612.
Fehrmann, R. C., and Weaver, R. W., 1978, Scanning electron microscopy of *Rhizobium* spp. adhering to fine silt particles, *Soil. Sci. Am. J.* **42:**279–281.
Ferrante, J. G., and Parker, J. I., 1977, Transport of diatom frustules by copepod fecal pellets to the sediments of Lake Michigan, *Limnol. Oceanogr.* **22:**92–98.
Fisher, J. A., and Beeton, A. M., 1975, The effect of dissolved oxygen on the burrowing behavior of *Limnodrilus hoffmeisteri* (Oligochaeta), *Hydrobiologia* **47:**273–290.
Fisher, J. B., and Tevesz, M. J. S., 1976, Distribution and population density of *Elliptio complanata* (Mollusca) in Lake Pocotopaug, Connecticut, *Veliger* **18:**332–338.
Fisher, J. B., Lick, W. J., McCall, P. L., and Robbins, J. A., 1980, Vertical mixing of lake sediments by tubificid oligochaetes, *J. Geophys. Res.* **85:**3997–4006.
Ford, J. B., 1967, The vertical distribution of larval Chironomidae in the mud of a stream, *Hydrobiologia* **19:**262–272.

Foster, T. D., 1932, Observations on the life history of a fingernail shell of the genus *Sphaerium*, *J. Morphol.* **53**:473–497.

Frankel, L., and Mead, D. J., 1973, Mucilaginous matrix of some estuarine sands in Connecticut, *J. Sediment. Petrol.* **43**:1090–1095.

Freitag, R., Fung, P., Mothersill, J. S., and Prouty, G. K., 1976, Distribution of benthic macroinvertebrates in Canadian waters of Northern Lake Superior, *J. Great Lakes Res.* **2**:177–192.

Fukuda, M. K., 1978, The entrainment of cohesive sediments in freshwater, Ph.D. dissertation, Department of Mechanical and Aerospace Engineering, Case Western Reserve University, 210 pp.

Fukuda, M. K., and Lick, W., 1980, The entrainment of cohesive sediments in fresh water, *J. Geophys. Res.* **85**:2813–2824.

Fukuhara, H., Kikuchi, E., and Kurichara, Y., 1980, The effect of *Branchiura sowerbyi* on bacterial populations in submerged ricefield soil, *Oikos* **34**:88–93.

Gale, W. F., 1976, Vertical distribution and burrowing behavior of the fingernail clam, *Sphaerium transversum*, *Malacologia* **15**:401–409.

Ganf, G. G., and Viner, S. B., 1973, Ecological stability in a shallow equatorial lake (Lake George, Uganda), *Proc. R. Soc. London Ser. B* **184**:321–346.

Geoghegan, M. J., and Brian, R. C., 1948, Aggregate formation in soil. 1. Influence of some bacterial polysaccharides on the binding of soil particles, *Biochem. J.* **43**:5–13.

Gray, J. S., 1974, Animal–sediment relationships, *Oceanogr. Mar. Biol. Annu. Rev.* **12**:223–261.

Håkanson, L., and Källström, A., 1978, An equation of state for biologically active lake sediments and its implications for interpretations of sediment data, *Sedimentology* **25**:205–226.

Hamilton, A. L., 1971, Zoobenthos of fifteen lakes in the experimental lakes area, Northwestern Ontario, *J. Fish. Res. Board Canada* **28**:257–263.

Haranghy, L., 1971, Investigations on the life-span limitation of mussels as example of telometric growing animals, *Acta. Biol. Acad. Sci. Hung.* **22**:3–7.

Hargrave, B. T., 1970a, Distribution, growth, and seasonal abundance of *Hyalella azteca* (Amphipoda) in relation to sediment microflora, *J. Fish. Res. Board Can.* **27**:685–699.

Hargrave, B. T., 1970b, The utilization of benthic microflora by *Hyalella azteca* (Amphipoda), *J. Anim. Ecol.* **34**:424–437.

Hargrave, B. T., 1970c, The effect of a deposit-feeding amphipod on the metabolism of benthic microflora, *Limnol. Oceanogr.* **15**:21–30.

Hargrave, B. T., 1976, The central role of invertebrate feces in sediment decomposition, in: *The Role of Terrestrial and Aquatic Organisms in Decomposition Processes* (J. M. Anderson and A. MacFayden, eds.), pp. 301–321, Blackwell Scientific Publications, Oxford.

Harris, R. H., and Mitchell, R., 1973, The role of polymers in microbial aggregation, *Annu. Rev. Microbiol.* **27**:27–50.

Heard, W. H., 1977, Reproduction of fingernail clams (Sphaeriidae: *Sphaerium* and *Musculium*), *Malacologia* **16**:421–455.

Hendelberg, J., 1960, The fresh-water pearl mussel, *Margaritifera margaritifera* (L.), *Rep. Inst. Freshwater Res. Drottningholm* **41**:149–171.

Henson, E. B., 1962, Notes on the distribution of the benthos in the Straits of Mackinac region, in: *Proceedings of the 5th Conference on Great Lakes Research*, p. 174, International Association for Great Lakes Research, Ann Arbor, Michigan.

Henson, E. B., 1970, *Pontoporeia affinis* (Crustacea, Amphipoda) in the Straits of Mackinac region, in: *Proceedings of the 13th Conference on Great Lakes Research*, pp. 601–610, International Association for Great Lakes Research, Ann Arbor, Michigan.

Herdendorf, C. E., and Reutter, J. M., 1976, Pre-operational aquatic ecology monitoring program for the Davis–Besse nuclear power station, Unit 1. Progress Report July 1–December 31, 1975, CLEAR, Ohio State University, Columbus, Ohio.

Heuschele, S., 1980, Vertical distribution of benthos in profundal Lake Superior sediments, in: Titles and Abstracts, North American Benthological Society, 28th Annual Meeting, Savannah, Georgia, p. 22.

Hilsenoff, W. L., 1966, The biology of *Chironomus plumosus* in Lake Winnebago, Wisconsin, *Ann. Entomol. Soc. Am.* **59**:465–473.

Hilsenoff, W. L., 1967, Ecology and population dynamics of *Chironomus plumosus* (Diptera, Chironomidae) in Lake Winnebago, Wisconsin, *Ecology* **60**:1183–1194.

Holland, E. F., Zingmark, R. G., and Dean, J. M., 1974, Quantitative evidence concerning the stabilization of sediments by marine benthic diatoms, *Mar. Biol.* **27**:191–196.

Howard, J. D., and Elders, C. A., 1970, Burrowing patterns of haustoriid amphipods from Sapelo Island, Georgia, in: *Trace Fossils* (T. P. Crimes and J. C. Harper, eds.), pp. 243–262, Seel House Press, Liverpool.

Hung, B. P., 1953, The life history and economic importance of a burrowing mayfly, *Hexagenia limbata*, in southern Michigan lakes, Bulletin of the Institute of Fisheries Research, Michigan Conservation Department, No. 4, p. 151, Ann Arbor.

Hunt, S., 1970, *Polysaccharide–Protein Complexes in Invertebrates*, Academic Press, London.

Hynes, H. B. N., 1955, The reproduction cycle of some British freshwater Gammaridae, *J. Anim. Ecol.* **24**:352–360.

Hynes, H. B. N., 1970, *The Ecology of Running Waters*, University of Toronto Press, Toronto.

Iovino, S. J., and Bradley, W. N., 1969, The role of larval chironomidae in the production of lacustrine copropel in Mud Lake, Marion County, Florida, *Limnol. Oceanogr.* **14**:898–905.

Ivlev, V. S., 1939, Transformation of energy by aquatic animals, coefficient of energy consumption by *Tubifex tubifex* (Oligochaeta), *Int. Rev. Ges. Hydrobiol.* **38**:449–458.

Johnson, M. G., and Brinkhurst, R. O., 1971, Benthic community metabolism in Bay of Quinte and Lake Ontario, *J. Fish. Res. Board Can.* **28**:1715–1726.

Johnson, M. S., and Munger, F., 1930, Observations on excessive abundance of the midge *Chironomus plumosus* at Lake Pepin, *Ecology* **11**:110–126.

Johnson, R. I., 1970, The systematics and zoogeography of the Unionidae (Mollusca: Bivalvia) of the southern Atlantic slope region, *Harv. Univ. Mus. Comp. Zool. Bull.* **140**:263–450.

Jonasson, P. M., 1972, Ecology and production of the profundal benthos in relation to phytoplankton in Lake Esrom, *Oikos Suppl.* **14**:1–148.

Jonasson, P. M., 1978, Zoobenthos of lakes, *Verh. Int. Verein. Limnol.* **20**:13–37.

Jonasson, P. M., and Thorhauge, F., 1972, Life cycle of *Pomatothrix hammoniensis* (Tubificidae) in the profundal of a eutrophic lake, *Oikos* **23**:151–158.

Jonasson, P. M., and Thorhauge, F., 1976a, Population dynamics of *Pomatothrix hammoniensis* in the profundal of Lake Esrom with special reference to environmental and competitive factors, *Oikos* **27**:193–203.

Jonasson, P. M., and Thorhauge, F., 1976b, Production of *Pomatothrix hammoniensis* in the profundal of eutrophic Lake Esrom, *Oikos* **27**:204–209.

Jones, B. F., and Bouser, C. J., 1978, The mineralogy and related chemistry of lake sediments, in: *Lakes: Chemistry, Geology, Physics* (A. Lerman, ed.), pp. 179–236, Springer-Verlag, New York.

Juday, C., and Birge, E. A., 1927, *Pontoporeia* and *Mysis* in Wisconsin lakes, *Ecology* **7**:445–452.

Kennedy, C. R., 1966, The life history of *Limnodrilus hoffmeisteri* Clap. (Oligochaeta: Tubificidae) and its adaptive significance, *Oikos* **17**:158–168.

Kikuchi, E., and Kurihara, Y., 1977, In vitro studies on the effects of tubificids on the biological, chemical and physical characteristics of submerged ricefield soil and overlying water, Oikos **29**:248–356.

Kleckner, J. F., 1967, The role of the bottom fauna in mixing lake sediments, M. S. thesis, University of Washington, Seattle.

Konstantinov, A. S., 1969, Feeding of chironomid larvae and means of improving the food base of water bodies, in: Transactions of the 6th Conference on Biology of Inland Waters (N. S. Akatora and B. K. Shtegman, eds.), pp. 274–284, Israel Program for Scientific Translation, IPSP No. 5136, Jerusalem.

Kovacik, T. L., and Walters, L. J., Jr., 1973, Mercury distribution in sediment cores from western Lake Erie, in: Proceedings of the 15th Conference on Great Lakes Research, pp. 252–259, International Association for Great Lakes Research, Ann Arbor, Michigan.

Kraft, K. J., 1979, Pontoporeia distribution along the Keweenaw shore of Lake Superior affected by copper tailings, J. Great Lakes Res. **5**:28–35.

Krezoski, J. R., and Robbins, J. A., 1980, Radiotracer studies of solute and particle transport in sediments by freshwater macrofauna, in: Abstracts, International Association for Great Lakes Research Meeting, Kingston, Ontario.

Krezoski, J. R., Mozley, S. C., and Robbins, J. A., 1978, Influence of benthic macroinvertebrates on mixing of profundal sediments in southeastern Lake Huron, Limnol. Oceanogr. **23**:1011–1016.

Ladle, M., 1971, The biology of Oligochaeta from Dorset Chalk streams, Freshwater Biol. **1**:83–79.

Lang, C., and Lang-Dobler, B., 1980, Structure of tubificid and lumbriculid worm communities, and three indices of trophy based on these communities, as descriptors of eutrophication level of Lake Geneva (Switzerland), in: Aquatic Oligochaete Biology (R. O. Brinkhurst and D. G. Cook, eds.), pp. 457–470, Plenum Press, New York.

Lautenschlager, K. P., Kaushik, N. K., and Robinson, J. B., 1978, The peritrophic membrane and fecal pellets of Gammarus lacustris limnoeus Smith, Freshwater Biol. **8**:207–211.

LeFevre, G., and Curtis, W. C., 1912, Studies on the reproduction and artificial propagation of freshwater mussels, Bull. Bur. Fish. **30**:109–201.

Lellak, J., 1969, The regeneration rate of bottom fauna populations of fish ponds after wintering or summering, Verh. Int. Verein. Limnol. **17**:560–569.

Lyman, F. E., 1943, Swimming and burrowing activities of mayfly nymphs of the genus Hexagenia, Ann. Entomol. Soc. Am. **36**:250–256.

Macan, T. T., and Mackereth, J. C., 1951, Notes on Gammarus pulex in the English lake district, Hydrobiologia **9**:1–12.

McCall, P. L., 1977, Community patterns and adaptive strategies of the infaunal benthos of Long Island Sound, J. Mar. Res. **35**:221–266.

McCall, P. L., 1978, Spatial–temporal distributions of Long Island Sound infauna: The role of bottom disturbance in a nearshore marine habitat, in: Estuarine Interactions (M. L. Wiley, ed.), pp. 191–219, Academic Press, New York.

McCall, P. L., 1979, The effects of deposit feeding oligochaetes on particle size and settling velocity of Lake Erie sediments, J. Sediment. Petrol. **49**:813–818.

McCall, P. L., and Fisher, J. B., 1980, Effects of tubificid oligochaetes on physical and chemical properties of Lake Erie sediments, in: Aquatic Oligochaete Biology (R. O. Brinkhurst and D. G. Cook, eds.), pp. 253–317, Plenum Press, New York.

McCall, P. L., Tevesz, M. J. S., and Schwelgien, S. F., 1979, Sediment mixing by Lampsilis radiata siliquoidea (Mollusca) from western Lake Erie, J. Great Lakes Res. **5**:105–111.

Mackey, A. P., 1977, Growth and development of larval chironomidae, Oikos **28**:270–275.

McLachlan, A. J., 1974, Development of some small lake ecosystems in tropical Africa, with special reference to the invertebrates, Biol. Rev. **49**:365–397.

McLachlan, A. J., 1977, Some effects of tube shape on the feeding of Chironomus plumosus L. (Diptera: Chironomidae), J. Anim. Ecol. **46**:139–146.
McLachlan, A. J., and Cantrell, M. A., 1976, Sediment development and its influence on the distribution and tube structure of Chironomus plumosus L. (Chironomidae, Diptera) in a new impoundment, Freshwater Biol. **6**:437–443.
McLachlan, A. J., and McLachlan, J. M., 1976, Development of the mud habitat during the filling of two new lakes, Freshwater Biol. **6**:59–67.
Mackie, G. L., Qadri, S. U., and Clarke, A. H., 1976, Intraspecific variations in growth, birth periods, and longevity of Musculium securis (Bivalvia: Sphaeriidae) near Ottawa, Canada, Malacologia **15**:433–446.
Margalef, R., 1968, Perspectives in Ecological Theory, University of Chicago Press, Chicago.
Martin, J. P., 1946, Microorganisms and soil aggregation. II. Influence of bacterial polysaccharides on soil structure, Soil Sci. **61**:157–166.
Martin, J. P., and Richards, S. J., 1963, Decomposition and binding action of a polysaccharide from Chromobacterium violaceum in soil, J. Bacteriol. **85**:1288–1294.
Marzolf, G. R., 1965, Substrate relations of the burrowing amphipod Pontoporeia affinis in Lake Michigan, Ecology **46**:579–591.
Matteson, M. R., 1948, Life history of Elliptia complanatus (Dillwyn, 1917), Am. Midl. Nat. **40**:690–723.
Meier-Brook, C., 1969, Substrate relations in some Pisidium species (Eulamellibranchiata: Sphaeriidae), Malacologia **9**:121–125.
Milbrink, G., 1973, On the vertical distribution of oligochaetes in lake sediments, Rep. Inst. Freshwater Res. Drottningholm **53**:34–50.
Milbrink, G., 1980, Oligochaete communities in pollution biology: The European situation with special reference to lakes in Scandinavia, in: Aquatic Oligochaete Biology (R. O. Brinkhurst and D. G. Cook, eds.), pp. 433–455, Plenum Press, New York.
Mills, E., 1969, The community concept in marine zoology, with comments on continua and instability in some marine communities: A review, J. Fish. Res. Board Can. **26**:1415–1428.
Mitropolsky, V. I., 1966, Notes on the life cycle and nutrition of Sphaerium corneum L. (Mollusca, Lamellibranchia), Trans. Inst. Biol. Inland Waters Acad. Sci. USSR **12**:125–128.
Monakov, A. V., 1972, Review of studies on feeding of aquatic invertebrates conducted at the Institute of Biology of Inland Waters, Academy of Science, U.S.S.R., J. Fish. Res. Board Can. **29**:363–383.
Moore, D. G., and Scruton, P. C., 1957, Minor internal structures of some recent unconsolidated sediments, A.A.P.G. Bull. **41**:2723–2751.
Moore, J. W., 1979, Ecology of a subarctic population of Pontoporeia affinis Lindstrom (amphipoda), Crustaceana **36(3)**:267–276.
Mozley, S. C., and Alley, W. P., 1973, Distribution of benthic invertebrates in the south end of Lake Michigan, in: Proceedings of the 16th Conference on Great Lakes Research, pp. 87–96, International Association for Great Lakes Research, Ann Arbor, Michigan.
Mozley, S. C., and Garcia, L. C., 1972, Benthic macrofauna in the coastal zone of southeastern Lake Michigan, in: Proceedings of the 15th Conference on Great Lakes Research, pp. 102–116, International Association for Great Lakes Research, Ann Arbor, Michigan.
Mundie, J. H., 1959, The diurnal activity of the larger invertebrates at the surface of Lac la Ronge, Saskatchewan, Can. J. Zool. **37**:945–956.
Myers, A. C., 1977, Sediment processing in a marine subtidal sand bottom community. I. Physical aspects, J. Mar. Res. **35**:609–632.
Nalepa, T. F., and Thomas, N. A., 1976, Distribution of macrobenthic species in Lake Ontario in relation to sources of pollution and sediment parameters, J. Great Lakes Res. **2**:150–163.
Negus, C. L., 1963, A quantitative study of growth and production of unionid mussels in the river Thames at Reading, J. Anim. Ecol. **35**:513–532.

Nittrouer, C. A., and Sternberg, R. W., 1981, The formation of sedimentary strata in an allochthonous shelf environment: The Washington continental shelf, *Mar. Geol.* **42**:201–232.

Okland, J., 1963, Notes on the population density, age distribution, growth, and habitat of *Anodonta piscinalis* rilss. (Moll. Lamellibr.) in a eutrophic Norwegian lake, *Nytt Mag. Zool.* **11**:19–43.

Oliver, D. R., 1971, Life history of the chironomidae, *Annu. Rev. Entomol.* **16**:211–230.

Palmer, M. F., 1968, Aspects of the respiratory physiology of *Tubifex tubifex* in relation to its ecology, *J. Zool.* **154**:463–473.

Parthenaides, E., and Passwell, R., 1968, Erosion of cohesive soil and channel stabilization, Civil Engineering Report No. 19, State University of New York, Buffalo, 119 pp.

Paterson, C. G., and Fernando, C. H., 1969, The macro-invertebrate colonization of a small reservoir in eastern Canada, *Verh. Int. Verein. Limnol.* **17**:126–136.

Pavoni, J. L., Tenney, M. W., and Echelberger, W. F., Jr., 1972, Bacterial exocellular polymers and biological flocculation, *J. Water Poll. Cont. Fed.* **44**:414–431.

Pennak, R. W., 1978, *Freshwater Invertebrates of the United States*, John Wiley and Sons, New York.

Pillai, S. C., 1941, The function of Protozoa in the activated sludge process, *Curr. Sci.* **10**:84–92.

Poddubnaya, T. L., 1961, Data on the nutrition of the prevalent species of tubificids in the Rybinsk basin, *Tr. Inst. Biol. Vodokhran Akad. Nauk SSSR* **4**:219–231.

Poddubnaya, T. L., 1980, Life cycles of mass species of tubificidae (Oligochaeta), in: *Aquatic Oligochaete Biology* (R. O. Brinkhurst and D. G. Cook, eds.), pp. 175–184, Plenum Press, New York.

Postma, H., 1967, Sediment transport and sedimentation in the estuarine environment, in: *Estuaries* (G. H. Lauff, ed.), pp. 158–179, American Association for the Advancement of Science, Washington, D.C.

Potter, D. W. B., and Learner, M. A., 1974, A study of the benthic macro-invertebrates of a shallow euthrophic reservoir in South Wales with emphasis on the Chironomidae (Diptera); their life-histories and production, *Arch. Hydrobiol.* **74**:186–226.

Pringle, C., 1980, Effects of chironomid tube-building activities and resultant floral changes upon the feeding behavior of *Baetis*, in: Titles and Abstracts, North American Benthological Society, 28th Annual Meeting, Savannah, Georgia, p. 36.

Prokopovich, N. P., 1969, Deposition of clastic sediment by clams, *J. Sediment. Petrol.* **39**:891–901.

Ravera, O., 1955, Amount of mud displaced by some freshwater Oligochaeta in relation to depth, *Mem. Ist. Ital. Idrobiol.* **8**(suppl.):247–264.

Rawson, D. S., 1953, The bottom fauna of Great Slave Lake, *J. Fish. Res. Bd. Can.* **10**:486–520.

Reigle, N. J., 1967, An occurrence of Anodonta (Mollusca, Pelecypoda) in deep water, *Am. Midl. Nat.* **78**:530–531.

Rhoads, D. C., 1974, Organism–sediment relations on the muddy sea floor, *Mar. Biol. Annu. Rev.* **12**:263–300.

Rhoads, D. C., 1975, The paleoecologic and environmental significance of trace fossils, in: *The Study of Trace Fossils* (R. W. Frey, ed.), pp. 147–160, Springer-Verlag, New York.

Rhoads, D. C., and Young, D. K., 1970, The influence of deposit-feeding organisms on sediment stability and community trophic structure, *J. Mar. Res.* **28**:150–178.

Rhoads, D. C., Yingst, J. Y., and Ullman, W. J., 1978, Seafloor stability in central Long Island Sound. Part I. Temporal changes in erodibility of fine-grained sediment, in: *Estuarine Interactions* (M. Wiley, ed.), pp. 221–244, Academic Press, New York.

Robbins, J. A., Krezoski, J. R., and Mozley, S. C., 1977, Radioactivity in sediments of the Great Lakes: Post-depositional redistribution by deposit-feeding organisms, *Earth Planet. Sci. Lett.* **36**:325–333.

Robbins, J. A., Edgington, D. N., and Kemp, A. L. W., 1978, Comparative lead-210, cesium-137, and pollen geochronologies of sediments from Lakes Ontario and Erie, *Quat. Res.* **10:**256–278.

Robbins, J. A., McCall, P. L., Fisher, J. B., and Krezoski, J. R., 1979, Effect of deposit feeders on migration of ^{137}Cs in lake sediments, *Earth Planet. Sci. Lett.* **42:**277–287.

Sanders, H. L., 1956, Oceanography of Long Island Sound. X. The biology of marine bottom communities, *Bull. Bingham Oceanogr. Coll.* **15:**245–258.

Sanders, H. L., 1968, Marine benthic diversity in a comparative study, *Am. Nat.* **102:**224–282.

Sanders, H. L., and Hessler, R. R., 1969, Ecology of the deep-sea benthos, *Science* **162:**1419–1424.

Scoffin, T. P., 1970, The trapping and binding of subtidal carbonate sediments by marine vegetation in Bimini Lagoon, Bahamas, *J. Sediment. Petrol.* **40:**249–273.

Segerstrale, S. G., 1967, Observations of summer breeding in populations of the glacial relict *Pontoporeia affinis* Lindstr. (Crustacea, Amphipoda) living at greater depths in the Baltic sea, with notes on the reproduction of *P. femorata* Kröyer, *J. Exp. Mar. Biol. Ecol.* **1:**55–64.

Sephton, T. W., and Paterson, C. G., 1980, Bivalve–non-bivalve relationships in a small reservoir in New Brunswick, Canada, in: Titles and Abstracts, North American Benthological Society, 28th Annual Meeting, Savannah, Georgia, p. 39.

Sly, P. G., 1978, Sedimentary processes in Lakes, in: *Lakes—Chemistry, Geology, Physics* (A. Lerman, ed.), pp. 65–89, Springer-Verlag, New York.

Smith, W. E., 1972, Culture, reproduction, and temperature tolerance of *Pontoporeia affinis* in the laboratory, *Tr. Am. Fish. Soc.* **101:**253–256.

Sorokin, J. I., 1966, Carbon-14 method in the study of the nutrition of aquatic animals, *Int. Rev. Geol. Hydrobiol.* **51:**209–224.

Southard, J. B., 1974, Erodibility of fine abyssal sediments, in: *Deep-Sea Sediments* (A. L. Inderbitzen, ed.), pp. 367–379, Plenum Press, New York.

Stanczykowska, A., 1978, Occurrence and dynamics of *Dreissena polymorpha*, *Verh. Internat. Verein. Limnol.* **20:**2431–2434.

Stansbery, D. H., 1967, Growth and longevity of naiads from Fishery Bay in western Lake Erie, *Am. Malacol. Union. Inc. Bull.* **35:**10–11.

Stansbery, D. H., 1970, A study of the growth rate and longevity of the naiad *Amblema plicata* (Say, 1817) in Lake Erie (Bivalvia: Unionidae), *Am. Malacol. Union. Inc. Bull.* **37:**78–79.

Stober, Q. J., 1972, Distribution and age of *Margaritifera margaritifera* (L.) in a Madison river (Montana, U.S.A.) mussel bed, *Malacologia* **11:**343–350.

Stockner, J. G., and Lund, J. W. G., 1970, Live algae in postglacial lake deposits, *Limnol. Oceanogr.* **15:**41–58.

Sundborg, A., 1956, The River Klavalven: A study of fluvial processes, *Geogr. Ann.* **38:**127–316.

Sweeney, R., Foley, R., Merckel, C., and Wyeth, R., 1975, Impacts of the deposition of dredged spoils on Lake Erie sediment quality and associated biota, *J. Great Lakes Res.* **1:**162–170.

Tenore, K., 1968, Accumulation of detritus and phosphate from the sediments by the brackish water bivalve, *Rangia cuneata*, *U. S. Fish Wildl. Serv. Circ.* **289:**7–8.

Tevesz, M. J. S., and McCall, P. L., 1979, Evolution of substratum preference in bivalves (Mollusca), *J. Paleontol.* **53:**112–120.

Tevesz, M. J. S., Soster, F. M., and McCall, P. L., 1980, The effects of size-selective feeding by oligochaetes on the physical properties of river sediments, *J. Sediment. Petrol.* **50:**561–568.

Thomas, G. J., 1965, Growth in one species of sphaerid clam, *Nautilus* **79:**47–58.

Thut, R. N., 1969, A study of the profundal bottom fauna of Lake Washington, *Ecol. Monogr.* **39**:79–100.

Timm, T., 1975, Zoobenthos of Lake Vortsjarv in 1964–1972, *Hydrobiol. Res. Tartu.* **6**:165–200.

Tudorancea, C., 1969, Comparison of the populations of *Unio tumidus* Philipsson from the complex of Crapina–Dijila marshes, *Ekol. Pol. Ser. A* **27**:185–204.

Tudorancea, C., and Gruia, L., 1969, Observations on the *Unio crassus* Philipsson population from the Nera River, *Muz. Nat. Natur. Grig. Antipa.* **8**:381–394.

Unz, R. F., and Farrah, S. R., 1976, Exopolymer production and flocculation by *Zoogloea* MP6, *Appl. Environ. Microbiol.* **31**:623–626.

Van Cleave, H. J., 1940, Ten years observation on a fresh-water mussel population, *Ecology* **21**:363–370.

Van Valen, L. M., 1980, Patch selection, benefactors, and a revitalization of ecology, *Evol. Theory* **4**:231–233.

Wachs, B., 1967, Die Oligochaeten-Fauna der Fleissgewasser unter besonderer Berücksichtigung der Beziehungen zwischen der Tubificiden-Besiedlung und dem Substrat, *Arch. Hydrobiol.* **63**:310–386.

Walshe, B. M., 1947, Feeding mechanisms of *Chironomus* larvae, *Nature* **16**:474.

Walshe, B. M., 1951, The feeding habits of certain chironomid larvae subfamily Tendipedinae, *Proc. Zool. Soc.* **121**:63–79.

Walters, L. J., Wolery, T. J., and Myser, R. D., 1974, Occurrences of As, Cd, Co, Cr, Cu, Fe, Hg, Ne, Sb, and Zn in Lake Erie sediments, in: *Proceedings of the 17th Conference on Great Lakes Research*, pp. 219–234, International Association for Great Lakes Research, Ann Arbor, Michigan.

Wavre, M., and Brinkhurst, R. O., 1971, Interactions between some tubificid oligochaetes found in the sediments of Toronto Harbour, Ontario, *J. Fish. Res. Board Can.* **28**:335–341.

Wetzel, R. G., 1975, *Limnology*, W. B. Saunders, Philadelphia.

Wiederholm, T., 1978, Long term changes in profundal benthos of Lake Malaren, *Verh. Int. Verein. Limnol.* **20**:818–824.

Wilson, D. S., 1980, *The Natural Selection of Populations and Communities*, Benjamin/Cummings, Menlo Park, California.

Wood, K., 1953, Distribution and ecology of certain bottom-living invertebrates of the western basin of Lake Erie, Ph.D. dissertation, Ohio State University, Columbus, Ohio, 145 pp.

Yingst, J. Y., and Rhoads, D. C., 1978, Seafloor stability in central Long Island Sound. II. Biological interactions and their importance for seafloor erodibility, in: *Estuarine Interactions* (M. Wiley, ed.), pp. 245–260, Academic Press, New York.

Yokley, P., 1972, Life history of *Pleurobema cordatum* (Rafinesque 1820) (Bivalvia: Unionacea), *Malacologia* **11**:351–364.

Young, R. A., and Southard, J. B., 1978, Erosion of fine-grained marine sediments: Sea-floor and laboratory experiments, *Geol. Soc. Am. Bull.* **89**:663–672.

Chapter 4

Effects of Macrobenthos on the Chemical Diagenesis of Freshwater Sediments

J. BERTON FISHER

1. Introduction	177
2. Freshwater Sediment System	178
2.1. Macrobenthos	178
2.2. Chemical Diagenesis of Freshwater Sediments	182
3. Mechanisms by Which Benthos Affect Chemical Diagenesis	186
3.1. Particle Transport	186
3.2. Fluid Transport	189
3.3. Sediment Fabric Alteration	194
3.4. Addition of Reactive Materials	199
4. Observed Effects of Freshwater Macrobenthos on Chemical Diagenesis	200
4.1. Effects of Benthos on the Sedimentary Chemical Milieu	200
4.2. Effects of Benthos on Diagenetic Rates	204
4.3. Effects of Benthos on Materials Flux	205
5. Conclusions	209
References	211

1. Introduction

In this chapter the effects of macrobenthos on chemical changes in sediments occurring during and after burial (i.e., chemical diagenesis) will be considered. The importance of sediments to the biogeochemical cycling of materials is well known (Mortimer, 1941, 1942, 1971; Lee, 1970). Freshwater sediments act as both a source and a sink for biologically important materials such as phosphorus, carbon, nitrogen, sulfur, and silicon. Furthermore, sediments are known to play an active role in regulating cycles of trace metals, radionuclides, and xenobiotics (Jones and Bowser, 1978).

J. BERTON FISHER • Department of Geological Sciences, Case Western Reserve University, Cleveland, Ohio 44106. *Present address*: Research Center, Amoco Production Company, Tulsa, Oklahoma 74102.

Because of this, knowledge of the chemical diagenesis of sediments is essential to an understanding of materials cycling in freshwater environments.

Historically, students of chemical diagenesis have viewed sediments as a homogeneous medium dominated by one-dimensional vertical diffusion. Reactive particles fall onto the sediment–water interface, move into the sediments at the net rate of sedimentation, and react. Solutes derived from these and subsequent reactions either move across the sediment–water interface via molecular diffusion and/or participate in further reactions (e.g., precipitation, adsorption).

Macrobenthos, through their burrowing, feeding, locomotive, respiratory, and excremental activities play an important role in mediating both physical and chemical processes near the sediment–water interface. Sediment and interstitial water are vertically and laterally transported (i.e., advected) by feeding, defecation, and respiratory pumping. Dwelling and feeding burrows may act as conduits for material exchange via fluid advection, solute diffusion, or sediment slumping. Secretion of mucus and excretion of metabolites may create local "hot spots" of bacterial activity (Aller, 1977, 1978; Aller and Yingst, 1978; McCall et al., 1979). Consequently, in the presence of macrobenthos, surfacial sediments are not a homogeneous medium dominated by one-dimensional vertical diffusion. Rather, they are vertically and laterally heterogeneous, strongly mixed, and penetrated by solute exchange conduits.

In the following sections the mechanisms by which macrobenthos affect chemical diagenesis will be examined. Specifically, we will be concerned with how and to what extent the activities of macrobenthos modify (1) sedimentary Eh and pH, (2) rates of chemical diagenesis, and (3) materials flux across the sediment–water interface. Although this chapter is concerned with an overview of how and to what extent macrobenthic organisms affect chemical diagenesis of freshwater sediments, nearly all available information concerns animal–sediment interactions in fine-grained (silt–clay), clastic (detrital), profundal (vegetation-free) lake sediments. As a consequence, examples of animal–sediment interactions in this environment will be used to illustrate the effects of macrobenthos on the diagenesis of freshwater sediments.

2. Freshwater Sediment System

2.1. Macrobenthos

Freshwater sediments are inhabited by a variety of macrobenthos, (operationally defined as those organisms retained by a 500-μm mesh sieve)—principally arthropods (insects and amphipods), annelids (oligochaetes and leeches), and molluscs (bivalves and gastropods) (Brinkhurst, 1974)—all of which interact in some way with their substratum. For any particular organism, the nature of this interaction is determined by the manner in which food is obtained (trophic type), mobility of the organism, and where, in relation to the sediment–water interface, life activities are carried on (life position). The importance of any organism or group of organisms in modifying sedimentary physical and/or chemical properties is also dependent upon individual size (>10 cm for large unionid bivalves), population density (up to >10^5 individuals m^{-2}), and activity rates (which are typically temperature-dependent). Other biotic and abiotic factors such as dissolved oxygen availability and ecological interactions may also be important (McCall and Fisher, 1980).

Benthos obtain food in a number of ways. For marine benthos, these methods of food acquisition have been categorized (Turpaeva, 1957; Savilov, 1957; Newell, 1970; Walker and Bambach, 1974). Although Cummins (1973) has proposed such a classification for aquatic insects, no general classification of trophic types exists for freshwater benthos. However, the general scheme proposed by Walker and Bambach (1974) for marine benthos is applicable to freshwater benthos. This scheme is presented in Table I along with freshwater examples of each trophic type. The feeding activity of an organism may embrace more than one trophic type. For example, some chironomid species engage in browsing, suspension feeding, and deposit feeding (Monakov, 1972; Cummins, 1973). The life habit

Table I. Classification and Terminology of Feeding in Benthic Invertebrates[a]

Trophic type	Description	Freshwater examples
Suspension feeding	Removing food material from suspension in the water mass without the need to subdue or dismember particles.	Unionid bivalves, sphaerid bivalves
Deposit feeding	Removing food from sediment either selectively or nonselectively without the need to subdue or dismember particles	Oligochaetes, amphipods, chironomids
Browsing	Acquiring food by scraping plant materials from environmental surfaces or by chewing or rasping larger plants.	Gastropods, chironomids, crayfish
Carnivory	Actively capturing and subduing prey.	Leeches, chironomids, crayfish
Scavenging	Consuming large particles of dead organisms.	Isopods, chironomids, crayfish
Parasitism	Obtaining nutrition from the fluids or tissues of host organisms.	Leeches

[a] After Walker and Bambach (1974).

Table II. Trophic Type and Life Habit of Dominant Profundal Macrobenthos

Organism	Length scale (cm)	Typical population density (m^{-2})	Primary trophic type	Species referenced	Mobility	Activity zone	Life habit reference
Chironomids	1–2	$10-10^4$	Multiple trophic types	Chironomus plumosus C. riparius C. tentaus	Sedentary (can be mobile)[a]	Infaunal	McLachlan and McLachlan (1976)
Oligochaetes	1–10	$10-10^6$	Subsurface deposit feeding	Limnodrilus hoffmeisteri Stylodrilus heringianus Tubifex tubifex L. socialis	Sedentary (can be mobile)	Infaunal	Alsterberg (1922)
Unionide	3–10	10^0-10^2	Suspension feeding	Branchyura sowerbyi Anodonta radiata silquoidae Elliptio complanata Anodonta sp.	Sedentary (can be mobile)	Infaunal	Coker et al. (1921), Tevesz and McCall (1979)
Amphipods	0.5–1	$10-10^3$	Surface deposit feeding	Pontoporeia hoyi	Mobile	Epifaunal/ shallow infaunal	Marzolf (1965)

[a] Capable of swimming.

of an organism is defined by its mobility and life position. Benthos are mobile, sedentary, or attached, and epifaunal (most activity at or on the sediment–water interface) or infaunal (most activity below the sediment–water interface).

The profundal macrobenthic community of lakes is numerically dominated by oligochaetes, amphipods, insect larvae (notably chironomid larvae), and sphaerid and unionid clams. These organisms are classified using trophic type/habit characteristics in Table II. Shallow infaunal deposit feeders, such as amphipods, and shallow infaunal suspension feeders, such as sphaerid clams, affect only the upper 1–3 cm of deposit. Infaunal forms that construct semipermanent dwelling and feeding burrows, such as chironomid larvae, will not significantly contribute to sediment mixing (at least for one generation of the larvae). Infaunal subsurface deposit feeders, such as tubificid oligochaetes, and deep infaunal, mobile, suspension feeders, such as unionid clams, are capable of thoroughly mixing sediments within their life zone (~10 cm for unionids) (McCall et al., 1979). A generalized diagram showing the life positions of the dominant freshwater benthos of the profundal zone is given in Fig. 1.

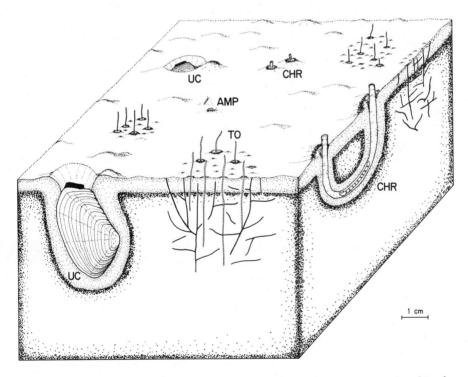

Figure 1. Life positions of dominant profundal freshwater macrobenthos. AMP, Amphipods; CHR, chironomids; TO, tubificid oligochaetes; UC, unionid clams.

Other ways of looking at macrobenthic communities (e.g., biomass, relative sediment reworking activity, relative water pumping) may produce a somewhat different view of "dominance." For example, Fisher and Tevesz (1975) report that, within its life zone, the unionid clam *Elliptio complanata* accounts for only 0.2–2.5% of the total number of benthos but comprises 79–98% of the living biomass (excluding shell mass).

2.2. Chemical Diagenesis of Freshwater Sediments

The chemical diagenesis of sediments comprises those chemical reactions taking place during and after burial of sedimentary materials. These reactions can be divided into two categories, biogenic and abiogenic, based on whether or not the reactions are mediated by bacteria or other microorganisms (Berner, 1976). The early diagenesis of both marine and freshwater sediments is largely carried on by bacterially mediated reactions. Moreover, pore water chemistry is often dominated by these reactions (Manheim and Sayles, 1974), which, in general, control the Eh and pH of sedimentary environments (ZoBell, 1946; Baas Becking et al., 1960; Berner, 1963; Stumm and Morgan, 1970; Ben-Yaakov, 1973; Gardner, 1973). Many geochemically important bacteria are heterotrophs. These organisms gain energy for cell growth and maintenance by oxidizing organic materials using a variety of electron acceptors (Kuznetsov, 1970). A schematic outline of these reactions is given in Table III. For simplicity, organic matter is represented here as CH_2O. The sequence in which these reactions occur closely parallels the order of decreasing free energy change (ΔG^0_{RXN}) of the reactions. Thus, the metabolic activities of microorganisms tend to create a vertical zonation of biogenically mediated diagenetic reactions (see Kuznetsov, 1970; Mortimer, 1971). A schematic diagram of this vertical sequence is given in Fig. 2.

Table III. Free Energy Change (Relative to Aerobic Respiration) and Schematic Chemistry of Important Bacterially Mediated Diagenetic Reactions[a]

Reaction	Schematic chemistry	$\Delta G^0_{RXN}/\Delta G^0_{AR}$ [b]
Aerobic respiration	$\frac{1}{4}CH_2O + \frac{1}{4}O_2 \rightarrow \frac{1}{4}CO_2 + \frac{1}{4}H_2O$	1.00
Denitrification	$\frac{1}{4}CH_2O + \frac{1}{5}NO_3^- + \frac{1}{5}H^+ \rightarrow \frac{1}{4}CO_2 + \frac{1}{10}N_2 + \frac{7}{20}H_2O$	0.95
Nitrate reduction	$\frac{1}{4}CH_2O + \frac{1}{8}NO_3^- + \frac{1}{4}H^+ \rightarrow \frac{1}{4}CO_2 + \frac{1}{8}NH_4^+ + \frac{1}{8}H_2O$	0.66
Fermentation	$CH_2O + \frac{1}{2}H_2O \rightarrow \frac{1}{2}HCOO^- + \frac{1}{2}CH_3OH + H^+$	0.21
Sulfate reduction	$\frac{1}{4}CH_2O + \frac{1}{8}CO_4^{2-} + \frac{1}{8}H^+ \rightarrow \frac{1}{4}CO_2 + \frac{1}{8}HS^- + \frac{1}{4}H_2)O$	0.20
Methane fermentation	$\frac{1}{4}CH_2O \rightarrow \frac{1}{8}CO_2 + \frac{1}{8}CH_4$	0.19

[a] Based on data from Stumm and Morgan (1970).
[b] ΔG^0_{RXN}, free energy change of the reaction; ΔG^0_{AR}, free energy change of aerobic respiration.

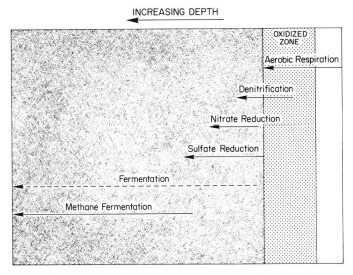

Figure 2. Schematic diagram showing the relative vertical sequence of diagenetic reactions in freshwater sediments.

Solutes produced by organic decay are free to react. These "secondary" reactions further change sediment chemistry. Hydrogen sulfide produced by sulfate reduction and the putrefaction of sulfur-containing organics reacts with iron and other metallics to form various sulfides (Vallentyne, 1961, 1963; Swain, 1965; Doyle, 1967; Nriagu, 1968; Dell, 1971). Hydrogen ions produced by organic decay react with solid carbonates and sulfides to liberate metallics (e.g., Ca, Fe, Mn) and raise alkalinity (Presley and Kaplan, 1968; Weiler, 1973). Phosphorus released from decaying organics can precipitate as authigenic phosphate minerals (D'Anglejan, 1967; Dell, 1973; Shapiro et al., 1971; Otsuki and Wetzel, 1972; Müller and Forstner, 1973; Nriagu and Dell, 1974; Norvell, 1974; Nriagu, 1976; Emerson and Widmer, 1978). Destruction of organic coatings by bacterial decay may promote the abiogenic dissolution of diatoms and other particles of biogenic opaline silica (Berner, 1976).

Diagenetic reactions that are not directly or indirectly controlled by the metabolic activities of microorganisms are less numerous. Examples are the dissolution of opaline silica, $CaCO_3$ dissolution at the sediment–water interface, clay mineral adsorption reactions, and the authigenesis of various clay minerals (Carmouze et al., 1977).

The chemical diagenesis of freshwater sediments has received considerably less attention than that of marine sediments. In lacustrine systems, diagenetic processes can be complicated by seasonal patterns of thermal stratification accompanied by oxygen depletion of hypolimnetic waters. Anoxia in the hypolimnion greatly facilitates the flux of several constituents from sediments to overlying water (e.g., iron, manganese, silica, ammonia, and phosphate) (Mortimer, 1941, 1942, 1971; Hutchinson, 1957; Gorham, 1964; Lee, 1970; Burns and Ross, 1972). For a review of work on the chemistry of interstitial waters from marine and lacustrine sediments, see Glasby (1973).

Is chemical diagenesis in freshwater sediments parallel to that in marine sediments? Are there differences in the process between these environments? Fresh water is, by definition, quite different from oceanic water.

The chemistry of oceanic water represents a "titration" of basic crustal materials (oxides, carbonates, and silicates) by strong acids derived from volcanic gases (Lafon and Mackenzie, 1966), while the composition of fresh water can be viewed as the result of interaction between atmospheric CO_2 and crustal materials. Furthermore, oceanic waters represent a long-time-scale interaction between water and the earth's crust while fresh waters represent a much shorter time of interaction. While the composition of oceanic waters shows little areal variation, the composition of fresh waters exhibits considerable variation among lakes and streams, since it reflects local rock types and other local conditions (Hutchinson, 1957). Even so, it is possible to make some general statements about the differences between oceanic and fresh waters.

In Table IV, the composition of present oceanic water is compared to that of world average river water, median terrestrial water, and water

Table IV. Comparison of the Major Cation and Anion Chemistry and Approximate Ionic Strengths of Present Ocean Water, World Average River Water, Lake Erie Water, Lake Superior Water, and Median Terrestrial Water[a]

Ion	Present oceanic water[b]	World average river water[c]	Lake Erie water[d]	Lake Superior water[d]	Median terrestrial water[e]
Na^+	480.6	0.27	0.52	0.04	1.21
Mg^{2+}	54.5	0.17	0.33	0.01	0.41
Ca^{2+}	10.5	0.37	0.92	0.32	1.10
K^+	10.6	0.06	0.03	0.03	0.03
Cl^-	560.4	0.22	0.71	0.03	0.48
SO_4^{2-}	28.9	0.12	0.27	0.04	0.31
HCO_3^-	2.3	0.96	1.85	0.84	3.77
Ionic strength[f]	0.71	0.002	0.005	0.001	0.006

[a] Values in mM. Ionic strength in M.
[b] Riley (1971). [c] Wetzel (1975). [d] Upchurch (1976). [e] Davies and DeWiest (1966).
[f] Here, ionic strength, I, is calculated using only the major cations and anions listed. $I = \frac{1}{2}\Sigma C_i Z_i^2$, where C_i is the concentration of the ith species and Z_i is the charge of the ith species.

from two Laurentian Great Lakes, Lake Superior and Lake Erie. First, it is immediately apparent that fresh waters are far more dilute than oceanic water. The ionic strength of oceanic water is nearly three orders of magnitude greater than that of fresh waters. Second, the patterns or relative abundances of cations and anions are different. For ocean water the general pattern is $Na^+ > Mg^{2+} > K^+ > Ca^{2+}$, and $Cl^- > SO_4^{2-} > HCO_3^-$. For fresh water the general pattern is $Ca^{2+} > Na^+ > Mg^{2+} > K^+$, and $HCO_3^- > Cl^- > SO_4^{2+}$. With respect to the chemical diagenesis of sediments, the most remarkable difference between marine and fresh waters is the paucity of SO_4^{2-} in fresh waters. In marine systems, the reduction of sulfate largely controls the diagenetic milieu, and it has been the subject of numerous studies (e.g., Berner, 1964a,b, 1970, 1972, 1974; Presley and Kaplan, 1968; Gardner, 1973; Ben-Yaakov, 1973; Rees, 1973; Goldhaber, 1974; Goldhaber and Kaplan, 1975a,b). In freshwater systems (excluding SO_4^{2-}-rich waters), however, the reduction of sulfate is probably unimportant relative to aerobic respiration and methanogenesis (Hayes, 1964). Bacterial reduction of sulfate is independent of sulfate concentration for sulfate concentrations $\leq 10^{-3}$ M (Postgate, 1951; Harrison and Thode, 1958; Kaplan and Rittenberg, 1964; Ramm and Bella, 1974). In most freshwater environments, sulfate concentration (in the overlying water) is approximately two orders of magnitude less than this threshold value. Few detailed measurements of sulfate concentration gradients exist for freshwater sediment. Recent data (T. Tisue, personal communication) from Lake Erie and Lake Michigan indicate that sulfate concentrations fall off quite rapidly with depth in sediments. In Lake Erie, no sulfate was found below a depth of 1 cm, while in Lake Michigan sulfate was found only to a depth of 4–10 cm.

One consequence of this lack of sulfur is high ferrous iron concentrations in the pore waters of freshwater sediments and the formation of reduced authigenic iron minerals [e.g., vivianite, $Fe_3(PO_4)_2 \cdot 8H_2O$; greigite, Fe_3S_4; siderite, $FeCO_3$] rarely found in marine sediments (Berner, 1980). For similar reasons, various authigenic manganese minerals [e.g., rhodochrosite, $MnCO_3$; reddingite, $Mn(PO_4)_2 \cdot 3H_2O$; various forms of MnS] are also found in freshwater sediments (Berner, 1980).

To be sure, sulfate-reducing bacteria are present in freshwater sediments (Kuznetsov, 1970; Dutka et al., 1974; Winfrey and Zeikus, 1979; Kikuchi and Kurihara, 1977), and metallic sulfides, especially FeS, are important constituents of the solid phase of freshwater sediments (Doyle, 1967; Nriagu, 1968, 1975; Jones and Bowser, 1978; Zinder and Brock, 1978; McCall and Fisher, 1980). It appears, however, that the principal source of sulfur to many freshwater sediments is sulfur-containing organic compounds (Nriagu, 1968; Wetzel, 1975; Zinder and Brock, 1978) rather than sulfate. In fact, freshwater sediments with an unusually high content of organic matter (e.g., swamps, marshes, bogs) have an abundant source of sulfide sulfur in their decomposing organic matter. In these sediments, reactive iron is readily converted to pyrite (or marcasite) (Berner, 1980). Except for the lack of a strong sulfate reduction reaction and its mineralogic consequences, diagenetic processes in freshwater sediments parallel those in marine sediments.

3. Mechanisms by Which Benthos Affect Chemical Diagenesis

Benthos can affect the diagenetic environment in four ways: (1) transportation of sediment particles, (2) transportation of interstitial and/or overlying water, (3) alteration of sediment fabric, and (4) addition of reactive materials. Transportation of sediment particles engendered by the burrowing and feeding activity of benthos serves to mix surficial sediments and results in the movement of materials between oxidizing and reducing environments. The biogenic churning of surficial sediments mixes freshly deposited organic matter with older material. Reduced materials brought to the sediment–water interface can react with dissolved oxygen, thereby enhancing sediment oxygen demand. The net effect of biogenic mixing is the alteration of the vertical profile of reactivity in sediments. Advective transport of interstitial and/or overlying water as a consequence of macrobenthic respiration, feeding, and burrowing will aid the movement of dissolved reactants (e.g., O_2, NO_3^-, NH_4^+) across the sediment–water interface. Such enhanced transport may increase the rate at which diagenesis proceeds. Alteration of sediment fabric through the construction of burrows and the pelletization of sediments can also affect solute transport by affecting the porosity and tortuosity of sediments. The addition of reactive materials through the excretion of urine, mucus, and other materials by benthos may locally enrich sediments with microbial nutrients and result in intense local bacterial activity.

3.1. Particle Transport

Movement of sediment particles by benthos, since it can significantly confound stratigraphy, has received a good deal of attention. The particle reworking activities of tubificid oligochaetes have been particularly well studied (Alsterberg, 1922, 1925; Lundbeck, 1926; Ivlev, 1939; Ravera, 1955; Poddubnaya, 1961; Poddubnaya and Sorokin, 1961; Wachs, 1967; Appleby and Brinkhurst, 1970; Davis, 1974a; Aller and Dodge, 1974; Krezoski et al., 1978; Robbins et al., 1979; Fisher et al., 1980; Tevesz et al.,

1980), and the rate and nature of their particle movement activities have been quantified and modeled (Robbins et al., 1979; Fisher et al., 1980). Detailed quantitative data on particle reworking also exist for the lumbriculid oligochaete *Stylodrilus heringianus* (Krezoski, 1981); the amphipod *Pontoporeia hoyi* (Robbins et al., 1979; Krezoski, 1981); the chironomid *Chironomus tentans* (Krezoski, 1981); and the bivalve molluscs *Pisidium*, *Sphaerium*, and *Lampsilis radiata siliquoidea* (McCall et al., 1979; Krezoski, 1981). Knowledge of the particle transport activities of other freshwater benthos is qualitative, but other organisms, especially mayflies, gastropods, crayfish, and various bivalves, are certain to have some impact on sediment mixing (see Chamberlain, 1975).

Several authors have quantified particle mixing by macrobenthos in terms of a particle diffusion coefficient, K, by analogy to eddy diffusion (Goldberg and Koide, 1962; Guinasso and Schink, 1975; Aller and Cochran, 1976; Nozaki et al., 1977; Robbins, 1978; Robbins et al., 1979). Using this approach, the distribution of a particle-associated property, such as utilizable organic matter, microfossils, or particle-bound radionuclides (e.g., ^{137}Cs, ^{210}Pb, ^{234}Th), is determined by particle diffusion, net sedimentation, and chemical reaction (if any) or radioactive decay. For example, a simple model for the distribution of a nonexchangeable particle-bound radionuclide whose source is the sediment–water interface (see Aller, 1978) can be written as

$$\frac{\partial A}{\partial t} = K\left(\frac{\partial^2 A}{\partial z^2}\right) - \omega\left(\frac{\partial A}{\partial z}\right) - \lambda A \qquad (1)$$

Where A is the activity of the radionuclide, t is time, z is depth, λ is the decay constant of the radionuclide, and ω is the sedimentation rate. The particle diffusion coefficient, K, is calculated from observed vertical distributions of radionuclides such as ^{210}Pb. This type of model is an appropriate description of particle reworking by organisms that redistribute particles without any pattern of directionality. Robbins et al. (1979) have shown particle diffusion to be a good representation of particle reworking by a population (16,000 individuals m^{-2}) of the amphipod *P. hoyi*. The effect of *P. hoyi* on the redistribution of ^{137}Cs in an experimental microcosm is shown in Fig. 14 in Chapter 3 (p. 126). In this experiment there was no sedimentation and the decay of ^{137}Cs could be neglected ($t_{1/2}$ for ^{137}Cs $\gg$ experimental time frame) so the distribution of ^{137}Cs by the amphipods was described as

$$\frac{\partial A}{\partial t} = K\left(\frac{\partial^2 A}{\partial z^2}\right) \qquad (2)$$

This approach gave an excellent description of particle reworking by *P. hoyi*. Robbins et al. (1979) give a value of 0.162×10^{-6} cm$^2 \cdot$sec^{-1} for K. This is similar to K values determined for coastal marine sediments (Guinasso and Schink, 1975; Aller and Cochran, 1976) and is comparable to the K inferred from ^{210}Pb profiles reported for a core from Lake Huron, where *P. hoyi* is the principal macrofaunal species (Robbins et al., 1977).

The particle reworking activities of such benthos as amphipods, isopods, unionid and sphaerid clams, gastropods, mayfly larvae, and chironomids—which more or less randomly move particles—should be adequately described by particle diffusion models. Such models, however, do not adequately describe reworking processes that have directionality (Aller, 1978; Robbins et al., 1979; Fisher et al., 1980).

In freshwater environments, tubificid oligochaetes are important agents of bioturbation, by virtue of their life habit, widespread distribution, and high population densities (McCall and Fisher, 1980). The particle reworking activity of tubificid oligochaetes is highly directional. Tubificids ingest sediment at depth in the substratum and expel this material at the sediment–water interface as cylindrical, mucus-bound fecal pellets. Thus tubificids provide a direct link between two distinctly different biogeochemical regimes and can be functionally classified as vertically oriented subsurface deposit feeders, or "conveyor belt" species (Rhoads, 1974). Because of the directionality of tubificid reworking activities, a model incorporating advection is needed to describe adequately their particle transport activities. Such a model has been developed by Fisher et al. (1980) from data on the redistribution of a ^{137}Cs-labeled sediment layer by tubificids in laboratory microcosms (also see Robbins et al., 1979). The effect of tubificid reworking on the redistribution of a ^{137}Cs layer initially at the sediment–water interface is given in Fig. 13 in Chapter 3 (p. 126). For another model of tubificid reworking, see Davis (1974a). For other examples of advective-type bioturbation see Amiard-Triquet (1975) and Aller and Dodge (1974).

As correctly pointed out by Aller (1978), particle transport activities *per se* will not, in general, directly control diffusion rates of interstitial solutes. Effective solute diffusion coefficients, which vary between $\sim 6.75 \times 10^{-6}$ cm$^2 \cdot$sec^{-1} for H$^+$ and $\sim 0.61 \times 10^{-6}$ cm$^2 \cdot$sec^{-1} for S^{2-} (Robinson and Stokes, 1959; Chapman, 1967; Manheim, 1970; Li and Gregory, 1974; Reid et al., 1977), are approximately one to two orders of magnitude greater than particle diffusion coefficients. It should be mentioned, however, that the redistribution of reactivity owing to particle transport phenomena can strongly affect the reactivity of the sediment–water interface. Robbins (1978) showed that, under conditions of constant sedimentation rate, constant ^{210}Pb supply to the sediment–water interface, and rapid biological mixing, the activity of ^{210}Pb at the sediment surface, $A_I(0)$, is given by

$$A_I(0) = A_S \left(\frac{\omega}{\omega + \lambda S}\right) \qquad (3)$$

where A_S is the activity (i.e., concentration) of ^{210}Pb in material falling to the sediment–water interface, ω is the sedimentation rate, λ is the ^{210}Pb decay constant, and S is the depth of the mixed layer. This concept can be extended to other materials subject to decomposition (e.g., organic N, C, P) that fall to the sediment–water interface. The amount of reactive matter at the sediment–water interface is decreased by increased depth of biological mixing or an increased decay constant (decreased half-life). High sedimentation rates or changes in the supply of reactive materials to the sediment–water interface may mask this effect.

Movement of materials from depth to the sediment–water interface can also enhance the reactivity of the sediment surface. In a study of the effects of tubificids on sediment oxygen consumption, McCall and Fisher (1980) found that the oxygen consumption of tubificid-inhabited sediments (T. tubifex, simulated population density 100,000 individuals m^{-2}) exceeded the simple sum of tubificid respiration and the oxygen demand of sediments lacking a tubificid population by a factor of ~2. They found that 50–70% of the enhanced demand could be accounted for by the oxidation of FeS brought to the sediment–water interface by tubificid feeding.

3.2. Fluid Transport

Advective transport of pore and/or overlying water across the sediment–water interface is potentially the most significant way in which macrobenthos might affect chemical diagenesis and materials balance in freshwater systems. Although nearly all benthic organisms are capable of causing some fluid advection, infaunal organisms that engage in active burrow irrigation, such as chironomids and burrowing mayflies (Lyman, 1943; Walshe, 1947), or that might inject water to aid burrowing, such as unionid clams (McCall et al., 1979), are probably the largest contributors to biogenic fluid advection.

Despite the potential importance of biogenic fluid transport, few experimental data are available. Moreover, the published data are entirely limited to chironomids. How can the effects of chironomid activities on the transport of pore and overlying water and their associated solutes be viewed? The simplest approach would be to model the effects of chironomids on solute transport as "enhanced diffusion." By way of example, this interpretation was applied to Tessenow's (1964) experimental data on interstitial silica concentrations in the presence of various chironomid population densities. A very simple one-dimensional one-layer diffusional transport model was used:

$$\frac{\partial C}{\partial t} = D\left(\frac{\partial^2 C}{\partial z^2}\right) \qquad (4)$$

where D is the effective solute diffusion coefficient. An analytical solution was derived for the following boundary conditions:

$$C(z = 0, t) = C_o$$
$$C(z \to \infty, t) = C_i$$

and the initial condition

$$C(z, t = 0) = C_i$$

where C_o is the silica concentration in the overlying water (assumed constant), and C_i is the initial interstitial water silica concentration. Both C_o and C_i were estimated from Tessenow's (1964) data. The results of computations assuming various values for D are shown in Fig. 3. As can be seen, there is reasonable agreement between the experimental data and the theoretical computations. The calculated flux of silica using $D = 10 \times 10^{-6}$ cm$^2 \cdot$sec^{-1} is approximately three times that calculated with $D = 1 \times 10^{-6}$ cm$^2 \cdot$sec^{-1}. Of course, the model used here is not really appropriate since silica is being produced in the sediments. Nevertheless, it does appear that a diffusion analogue adequately describes the data, and that the presence of a chironomid population density of 770 individuals m^{-2} produced roughly a tenfold increase in diffusivity over that observed in sediments unaffected by chironomids.

Edwards and Rolley (1965) studied the migration of Li$^+$ from overlying water into chironomid-inhabited sediments (*Chironomous riparius*, simulated population densities 0, 25, 50, and 100 $\times$ 10^3 individuals m^{-2}). These workers found that in sediments containing chironomid populations Li$^+$ concentrations in the top ~2 cm of sediment were equivalent to those of the overlying water after only 6 hr. Diffusion alone cannot account for this finding (see Manheim, 1970). The data presented by Edwards and Rolley (1965) are not extremely detailed, but at the two highest chironomid population densities examined, it appears that Li$^+$-

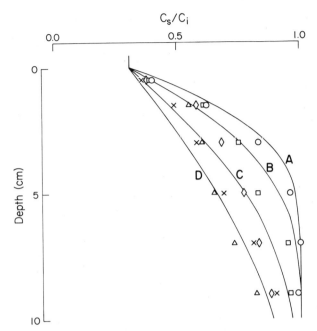

Figure 3. Comparison of silica concentration profiles calculated using a one-dimensional, one-layer diffusion model for various values of D (A: $D = 1 \times 10^{-6}$ cm^2 sec^{-1}; B: $D = 2 \times 10^{-6}$ cm^2 sec^{-1}; C: $D = 5 \times 10^{-6}$ cm^2 sec^{-1}; D: $D = 10 \times 10^{-6}$ cm^2 sec^{-1}) to the silica concentration profile data of Tessenow (1964) in the presence of various chironomid population densities (○, 0 individuals m^{-2}; □, 193 individuals m^{-2}; ◇, 385 individuals m^{-2}; ×, 580 individuals m^{-2}; △, 770 individuals m^{-2}). Plotted concentrations are nondimensionalized (C_s/C_i). C_s is the silica concentration at depth z. C_i is the silica concentration at $z = \infty$.

laden overlying water is moving into the sediment at a velocity of ~5 × 10^{-5} cm·sec^{-1}. This translates to a "water exchange rate" of ~4.3 × 10^4 cm^3·m^{-2}·day^{-1}. Although this value is high compared to the biopumping rate of 7 × 10^3 cm^3·m^{-2}·day^{-1} given by McCaffrey et al. (1980) for Narragansett Bay sediments, it is not unreasonable, since the activities of each chironomid need result in only ~1–0.5 cm^3·day^{-1} of water exchange.

In experiments conducted in our laboratory, the effect of chironomids (Chironomus plumosus, simulated population densities 0, 1096, 3289, and 6798 individuals m^{-2}) on pore water chemical profiles and materials flux across the sediment–water interface was investigated. Pore water concentration profiles for Fe^{2+} and H$_4$SiO$_4$ observed in this experiment are given in Figs. 4 and 5. The presence of chironomids was found to suppress the concentration of both Fe^{2+} and H$_4$SiO$_4$ in the top 12 cm of the sediment column. Soluble reactive phosphorus (SRP), NH$_4^+$, and

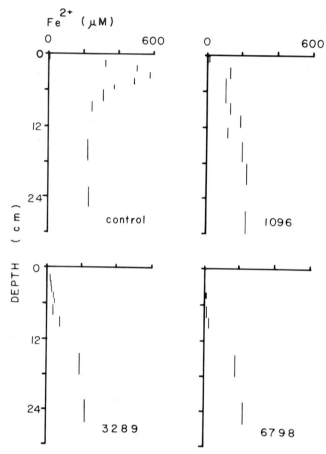

Figure 4. Vertical profiles of Fe^{2+} in control and experimental (with *Chironomus plumosus*) sediments. Simulated population densities (individuals m^{-2}) indicated on figures.

HCO_3^- concentrations were also strongly suppressed in this zone. The suppression of Fe^{2+} and SRP concentrations is probably the result of deeper O_2 penetration into chironomid-inhabited sediments (evidenced by their oxidized burrow walls). The suppression of HCO_3^-, H_4SiO_4, and NH_4^+ concentrations is probably the result of chironomid water pumping activities. By way of illustration, direct flux measurements (made by measuring overlying water concentration change) and indirect flux estimates (made by applying Fick's law to pore water concentration gradients) for H_4SiO_4 are compared in Table V. Direct and indirect flux estimates for H_4SiO_4 give comparable results when no chironomids are present, but these two estimates deviate progressively as chironomid population density increases. At the highest density used (6798 individuals m^{-2}) the

Effects of Macrobenthos on Freshwater Sediment Diagenesis

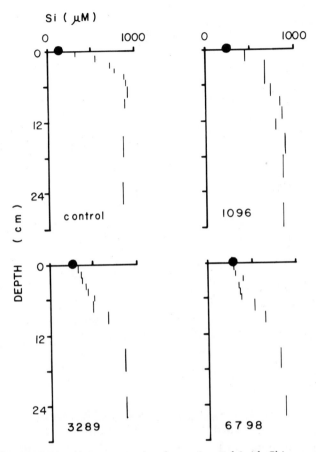

Figure 5. Vertical profiles of Si in control and experimental (with *Chironomus plumosus*) sediments. Simulated population densities (individuals m^{-2}) indicated on figures.

Table V. Direct and Indirect Estimates of Silica Flux from Control and Experimental (Chironomid-Inhabited) Sediments

Chironomus plumosus (individuals m^{-2})	Direct flux estimate ($\times 10^{-6}$ mol·m^{-2}·day^{-1})	Indirect flux estimate ($\times 10^{-6}$ mol·m^{-2}·day^{-1})
0	1393	1651
1096	2644	855
3289	4191	150
6798	4335	76

direct flux estimate is nearly 60 times larger than the indirect flux estimate because the activities of the chironomids have greatly decreased the H_4SiO_4 concentration gradient at the sediment–water interface. Furthermore, the presence of chironomids has increased silica production, since the amount of silica transferred to the overlying water is approximately four times greater than can be accounted for by simply flushing interstitial water from the sediments. The activity of the chironomids, rather than just the presence of their burrows, is necessary for enhanced materials flux across the sediment–water interface. In recent experiments conducted at the University of Michigan (M. A. Quigley, personal communication), it was found that silica flux from sediments was reduced by a factor of 6.7 when chironomids were inactivated by physostygmine, a neurotoxin.

More work is needed to describe adequately the effects of fluid-advecting benthos on materials movement across the sediment–water interface. Carefully controlled experiments using tracers with well-known properties that can be followed nondestructively and with high resolution (e.g., ^{22}Na) need to be conducted. In particular, the assumption of one-dimensionality needs to be critically examined, since the presence of burrows creates an intrinsically two-dimensional situation (see Aller, 1977, 1978; Aller and Yingst, 1978; Berner, 1980).

3.3. Sediment Fabric Alteration

The feeding and burrowing activities of macrobenthos alter sediment fabric. Their construction of feeding and dwelling burrows, movement through the sediment, and production of fecal pellets radically alter the textural properties of surface sediment (Rhoads, 1970; Rhoads and Young, 1970; McCall, 1979; McCall and Fisher, 1980). Collectively, these activities increase the porosity and decrease the tortuosity of the substratum. Such modifications of porosity and tortuosity affect diffusive flux across the sediment–water interface, since

$$J_z = D_w \left(\frac{\phi}{\theta^2}\right)\left(\frac{\partial C}{\partial z}\right), \quad z = 0 \qquad (5)$$

where ϕ is porosity, θ is tortuosity, D_w is the Fick's law diffusion coefficient in free solution, and J_z is the diffusional flux in $mass \cdot area^{-1} \cdot time^{-1}$. Tortuosity is defined as

$$\theta = \frac{dl}{dz} \qquad (6)$$

where dl is the length of the actual sinuous path over a depth interval dz. Porosity is the ratio of the volume of contained water to the total volume:

$$\phi = \frac{V_{water}}{V_{water} + V_{solids}} \tag{7}$$

In general, porosity and tortuosity are not treated explicitly. Rather, their effect on diffusivity is taken into account using an effective or apparent diffusion coefficient, D_e. The maximum value of D_e is, of course, D_w. Typically, values of D_e for chemical species that do not participate in exchange reactions are on the order of $0.55D_w$ (Li and Gregory, 1974), but values of $0.1D_w$ are commonly observed (Manheim, 1970). Thus the maximum effect of sediment fabric alteration on diffusivity is about a factor of ten. McCall et al. (1979) showed that the activities of unionid bivalves significantly increased sediment water content in the upper 4 cm of their experimental sediments. The change in porosity (and, consequently, diffusional flux) owing to unionid activity calculated from their water content data is fairly modest, however, varying from 10% to 20%. Krezoski (1981) found that the presence of *Stylodrilus heringianus* (a lumbriculid oligochaete, simulated population density 67,000 individuals m^{-2}) produced a zone of enhanced constant porosity 5.5 cm in thickness, while the porosity of control cells exhibited an exponential decrease to a constant value at 4 cm. His data are given in Fig. 6. In experiments with *Pontoporeia hoyi* these workers found no difference in porosity between

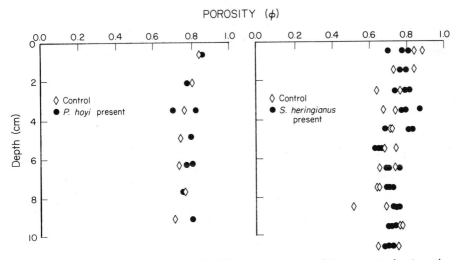

Figure 6. Effect of the presence of *Stylodrilus heringianus* and *Pontoporeia hoyi* on the porosity of lake sediments. Data from Krezoski (1981).

sediments with a *P. hoyi* population (simulated population density 47000 individuals m^{-2}) and control sediments (see Fig. 6).

McCall and Fisher (1980), in a study of the effects of *Tubifex tubifex* on the movement of dissolved material across the sediment–water interface, found that the movement of chloride from sediments to overlying water was enhanced by the presence of tubificids (simulated population density 100,000 individuals m^{-2}), but found no evidence to support the contention (Wood, 1975) that tubificids pump water. The effects of the tubificid population on chloride movement were found to be well-approximated by assuming that the diffusivity of chloride in the tubificid-affected sediments decreased linearly from 11×10^{-6} cm$^2 \cdot$sec^{-1} at the sediment–water interface to 5.5×10^{-6} cm$^2 \cdot$sec^{-1} (the diffusivity of chloride in control sediments) at a depth of 0.75 cm and remained at this value to the base of the sediment column. McCall and Fisher (1980) attributed this effect to the tubificids' generation of a porous fecal pellet zone in the upper 1 cm of sediment. Chatarpaul *et al.* (1980) observed a

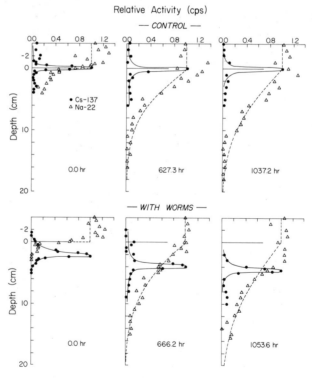

Figure 7. Effect of the presence of *Stylodrilus heringianus* on the migration of ^{22}Na and ^{137}Cs through lake sediments. Data from Krezoski (1981).

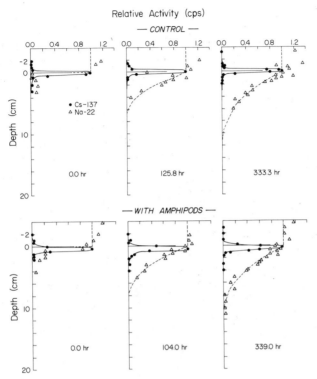

Figure 8. Effect of the presence of *Pontoporeia hoyi* on the migration of ^{22}Na and ^{137}Cs through lake sediments. Data from Krezoski (1981).

pattern of chloride loss from overlying water to sediment in the presence of tubificids, which could also be explained as the result of enhanced diffusion in the upper 1 cm of sediment.

Krezoski (1981) studied the effects of S. *heringianus* and P. *hoyi* on the migration of ^{22}Na through sediments. Data were collected by non-destructively scanning the experimental cells with a gamma scanner (Robbins et al., 1979). Krezoski found that the presence of both S. *heringianus* (simulated population density 67,000 individuals m^{-2}) and P. *hoyi* (simulated population density 47,000 individuals m^{-2}) enhanced the apparent effective diffusivity, D_e, of ^{22}Na. In control sediments the D_e of ^{22}Na was found to be 0.92×10^{-5} cm$^2 \cdot$sec^{-1}. In three replicate experiments with S. *heringianus* populations the average D_e of ^{22}Na was found to be 1.33×10^{-5} cm$^2 \cdot$sec^{-1}. In two replicate experiments with P. *hoyi* the average D_e of ^{22}Na was 1.45×10^{-5} cm$^2 \cdot$sec^{-1}. Data from this study are given in Figs. 7 and 8. Krezoski (1981) treats the sediment as a one-layer system (i.e., D_e is the same at all depths in the sediment).

It appears (at least for S. heringianus) that a two-layer model would be more appropriate. In the presence of S. heringianus the activity of ^{22}Na in the sediment is equivalent to that in the overlying water to a depth of ~3 cm after 1053.6 hr. P. hoyi does not produce a similar pattern of ^{22}Na activity. In the presence of the amphipod, the observed ^{22}Na profiles appear to be well-approximated by a one-layer diffusion model.

Despite the relatively modest increase in D_e that can be attributed to macrobenthic sediment fabric alteration, it should be remembered that these alterations are made in the most biologically active zone of the sediment column. As a result, the biogeochemical consequences of such enhanced diffusion may be large. For example, such enhanced diffusion could greatly modify sediment oxygen demand, speed the mineralization of organic matter entering sediments, and facilitate silica release.

In the presence of lateral pressure gradients resulting from waves and currents, changes in porosity and tortuosity may enhance passive (i.e., not the result of biological pumping) ventilation of sediments (Vogel and Bretz, 1972), since porosity and tortuosity affect the hydraulic conductivity (i.e., permeability) of sediments. Using falling head permeameters, McCall and Fisher (1980) found that the presence of tubificids significantly increased the permeability of sediments and that the permeability

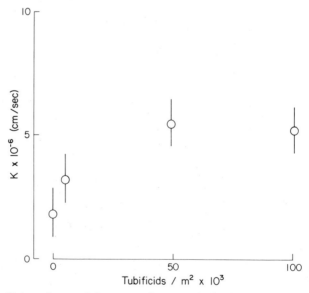

Figure 9. Coefficient of permeability, K, as a function of tubificid (*Tubifex tubifex*) population density. Confidence intervals are 95% Scheffé-type intervals. Data from McCall and Fisher (1980).

increased with increasing tubificid population density. It appeared that the increased permeability was principally due to decreased tortuosity (owing to tubificid burrows), since no significant difference in porosity was observed between control and experimental sediments. Their data are presented in Fig. 9. The significance of biogenically enhanced permeability in affecting chemical diagenesis has not been evaluated. It is likely, however, to have an impact similar in nature to that of fluid advection.

Aside from modifying tortuosity and porosity, the production of fecal pellets may significantly affect the diagensis of their contained organic matter. Iovino and Bradley (1969) showed that Chironomus spp. in a small Florida lake produced strongly compacted fecal pellets rich in undigested blue-green algae. They suggest that the copropel comprised of these pellets appears to be a modern analogue to the precursor of oil shale in the Green River Formation.

3.4. Addition of Reactive Materials

Particle transport, fluid transport, and sediment fabric alteration can all change the distribution of reactivity in sediments and the flux of materials across the sediment–water interface, but none of these processes alters the amount of reactive material in sediments. Aquatic macrobenthos can also add reactive materials to sediment through the excretion of urine (Potts, 1954a,b), mucus, and various sediment-agglutinating materials (Chamberlain, 1975). Nearly all aquatic invertebrates excrete copious amounts of hypoosmotic urine (Potts, 1954a,b; Parry, 1960; Staddon, 1969; Stobbart and Shaw, 1974). In addition to large quantities of ammonia and bicarbonate, the urine of macrobenthos contains bacterial nutrients such as sugars and amino acids. When excreted reactive materials are associated with burrows, dramatic local chemical changes can occur. Aller (1977) and Aller and Yingst (1978), working in the marine environment, showed that organic enrichment of Amphitrite burrow walls increased microbiological activity in the burrow walls (particularly sulfate reduction); produced an increase in the concentration of iron sulfides immediately around the burrows; and enriched the burrow walls in Fe, Mn, and other metals. McCall et al. (1979) found that the burrows of the freshwater mussel Lampsilis radiata siliquoidea were surrounded by a zone enriched in iron sulfide. This enrichment was surmised to be a result of intense microbiological activity owing to local nutrient enrichment and enhanced availability of electron acceptors near the clam. Injection of oxygen-rich water by the clams may result in a local SO_4^{2-} source through the oxidation of FeS and other easily oxidized metallic sulfides.

4. Observed Effects of Freshwater Macrobenthos on Chemical Diagenesis

Data on the effects of freshwater macrobenthos on chemical diagenesis are limited to their effects on sedimentary Eh and pH, sediment oxygen consumption, and the flux of various materials across the sediment–water interface. Since biogenically mediated diagenetic reactions in general produce H^+ and reduce Eh, it is appropriate that this discussion begin with a consideration of the effects of benthos on sedimentary Eh and pH.

4.1. Effects of Benthos on the Sedimentary Chemical Milieu

Knowledge of the Eh and pH of an environment furnishes a synoptic view of chemical processes taking place, but does not provide detailed information about the specific chemistry of the environment (Baas Becking et al., 1960). Eh data provide information about the relative abundance of electron acceptors (e.g., O_2, NO_3^-, SO_4^{2-}) and electron donors (e.g., HS^-, NO_2^-, NH_4^+), and pH data reveal the abundance of H^+. Neither parameter provides specific knowledge concerning the chemical reactions that produced these conditions. Consequently, information on sedimentary Eh and pH is useful in the examination of the effects of benthos on chemical diagenesis but, taken alone, provides no knowledge concerning the exact mechanisms whereby diagenesis is modified. Even so, it is useful to begin our examination of the effects of benthos on sediment diagenesis with a discussion of the patterns of Eh and pH modification noted in their presence.

Numerous workers have observed that the presence of macrobenthos is correlated with a thickened oxidized (i.e., light-colored) layer [e.g., Milbrink, 1969 (tubificids); Hargrave, 1972, 1975; Kirchner, 1975 (mixed natural populations)]. Davis (1974b), in an experimental study, found that the presence of a tubificid population of 800 individuals m^{-2} increased the redox potential of profundal lake sediment between depths of 1 and 4 cm and thickened the oxidized zone from 0.3 to 1.6 cm. In another experimental study, Edwards (1958) found that the introduction of a *Chironomus riparius* population (simulated population densities of 12,500 and 50,000 individuals m^{-2}) into settled activated sludge displaced the redox potential originally defining the redox color boundary downward to 8 cm from its initial depth of 3 cm. Furthermore, he noted that the higher density (50,000 individuals m^{-2}) of chironomids produced more extensive shifting of Eh and that redox conditions returned to their original distribution after the chironomids pupated and emerged. Schumacher

(1963) observed that the oxidized zone extended to at least 15 cm with a tubificid population of 50,000 individuals m^{-2} and that the depth of the oxidized zone was correlated with the population density of worms. McCall and Fisher (1980) found that homogenized Lake Erie sediments lacking macrobenthos developed an oxidized zone 0.1–0.3 cm in thickness; the same sediments inhabited by tubificids (100,000 individuals m^{-2}) developed an oxidized zone 1–3 cm thick. McCall et al. (1979) observed that sediment immediately adjacent to the unionid *Lampsilis radiata silquoidea* were oxidized.

Simply stated, an oxidizing environment contains an excess of electron acceptors (e.g., O_2, NO_3^-, SO_4^{2-}), while a reducing environment contains an excess of electron donors (e.g., Fe^{2+}, HS^-, NH_4^+). From the above data it is clear that the actions of chironomids, tubificids, and unionids aid the transport of electron acceptors into sediments and thus increase the depth over which more energetic diagenetic reactions occur (see Table II). Chironomids more or less continuously pump overlying water through their burrows, tubificids enhance porosity in the upper 1 cm of substratum, and unionids episodically pump water into sediments to aid burrowing. It is likely that such activity increases the rate of organic decomposition.

The actions of some macrobenthos can also affect sedimentary pH. Davis (1974b) showed that the presence of a tubificid population decreased pH in the upper 1–2 cm and increased pH below 2 cm. Edwards (1958) found that the presence of chironomids increased sedimentary pH in settled activated sludge over a depth range of 8 cm. Fisher and Matisoff (1981) found that the presence of tubificids (100,000 individuals m^{-2}) produced a local pH minimum at a depth of 4–5 cm (see Fig. 10). They also noted that no pH gradient occurred below 4–6 cm in control cells, while in cells containing worms a continuous decrease in pH occurred below the local pH minimum. Profiles of pH in both with- and without-worm cases did not vary substantially over the course of the experiment (89 days) and were comparable to the pH profiles reported by Davis (1974b). In further experiments, (see Fig. 10) a thin layer (0.2 cm) of Milorganite (Milwaukee sewage sludge) was placed at a depth of 2 cm. In Milorganite-treated sediments without worms, a highly systematic time series of pH profiles evolved, while in similarly treated sediments with worms (100,000 individuals m^{-2}) no change in pH profile occurred during the course of the experiment (89 days). In fact, pH profiles in Milorganite-treated sediments with worms were equivalent to those observed in untreated sediments with worms. The activities of macrobenthos obviously affect sedimentary pH. These modifications of pH have been attributed to enhanced exchange (presumably of acid metabolites) across the sediment–water interface (Edwards, 1958; Davis, 1974b). This may well be

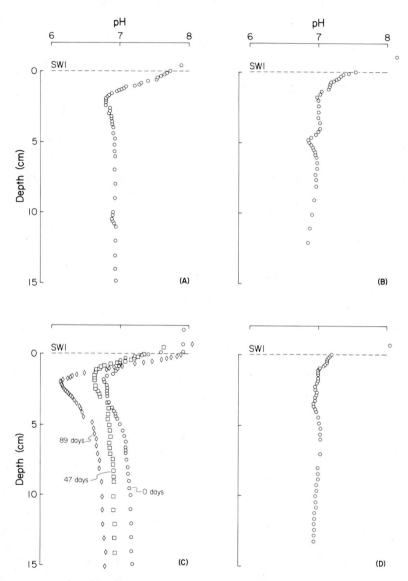

Figure 10. Vertical profiles of pH observed in homogenized Lake Erie sediments. (A) Control; (B) with tubificids; (C) without tubificids, but with a Milorganite layer (2 cm depth); (D) with tubificids and a Milorganite layer (2 cm depth). SWI, sediment–water interface. Data from Fisher and Matisoff (1981).

Effects of Macrobenthos on Freshwater Sediment Diagenesis

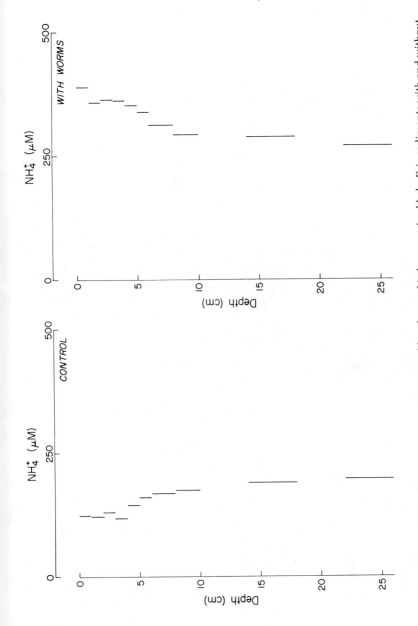

Figure 11. Comparison of ammonia concentration profiles observed in homogenized Lake Erie sediments with and without tubificids. Population density: 10^5 individuals m^{-2}.

true in the case of chironomids, but the strong alteration of pH profiles by tubificids in the Milorganite-treated sediments and their effects on pH deep in sediments are difficult to explain in terms of solute exchange alone, because aquatic oligochaetes do not pump water (McCall and Fisher, 1980; Krezoski, 1981).

In experiments in our laboratory, it was found that the presence of tubificids (100,000 individuals m^{-2}) resulted in a local ammonium maximum at a depth of ~4 cm (see Fig. 11). This result indicates that an additional source of ammonia is present in the upper 3–5 cm of worm-inhabited sediments and suggests a possible explanation for the observed effects of tubificids on interstitial water pH: The worms may be excreting enough ammonia to "buffer" pore waters against changes in pH.

4.2. Effects of Benthos on Diagenetic Rates

The actions of freshwater macrobenthos affect the transport of reactants and reaction products (e.g., NH_4^+, HCO_3^-) across the sediment–water interface. It should follow, therefore, that their activities should also affect the rate at which diagenetic reactions take place. One index of diagenetic rate is the rate at which electron acceptors are consumed by sediments.

Edwards and Rolley (1965), Hargrave (1975), and Granéli (1982) all noted a significant increase in oxygen consumption by sediments inhabited by various chironomid species (*Chironomus riparius, C. anthricinus,* and *C. pulmosus,* respectively) that could not be explained as the simple sum of chironomid respiration and sediment oxygen demand in the absence of chironomids. Andersen (1977) found that the rate at which oxygen was consumed by sediments was correlated with the wet weight of *C. plumosus* present, and that the oxygen demand of sediment containing chironomids was approximately twice the simple sum of chironomid respiration and sediment oxygen demand in the absence of chironomids. The relationship between the wet weight of chironomids and reported sediment oxygen demand (SOD) (Andersen, 1977) was as follows:

$$SOD[mmol \cdot m^{-2} \cdot day^{-1}] = (14.97 \text{ mmol} \cdot m^{-2} \cdot day)$$
$$+ (0.18 \text{ mmol} \cdot m^{-2} \cdot day^{-1} \cdot mg \text{ wet biomass}^{-1})*B$$

where *B* is wet chironomid biomass in milligrams.

The actions of tubificids also affect the rate of electron acceptor uptake by sediments. McCall and Fisher (1980) showed that the presence of *Tubifex tubifex* (100,000 individuals m^{-2}) enhanced sediment oxygen consumption by a factor of 2. Approximately 20% of the enhanced demand was the result of tubificid respiration, while 50–70% of the enhanced demand was due to the oxidation of FeS brought to the sediment–water interface by tubificid feeding activities. The remaining 10–30% of the enhanced demand was apparently due to increased microbial activity. Thus, tubificids enhance the rate of organic decay not only through aiding the transport of dissolved oxygen, but also by recycling sulfur to produce SO_4^{2-} at the sediment–water interface.

Macrobenthos also affect denitrification. Andersen (1977) reported that the presence of *C. plumosus* increased the rate of nitrate consumption by sediments. Nitrate consumption by sediments was strongly correlated with biomass of *C. plumosus* present. The consumption of nitrate was dependent on nitrate concentration in the overlying water and *C. plumosus* enhanced nitrate consumption by 0.11×10^{-3} mol·m^{-2}·day^{-1}·mg wet biomass^{-1} at nitrate concentrations of between 0.14 and 0.21 mM and by 0.39×10^3 mol·m^{-2}·day^{-1}·mg wet biomass^{-1} at nitrate concentrations of between 0.42 and 0.71 mM.

Tubificids have also been shown to enhance the consumption of nitrate by sediments. Chatarpaul et al. (1979, 1980) showed that the presence of tubificids (mixed population of *T. tubifex* and *Limnodrilus hoffmeisteri*, 10,000 to 13,000 individuals m^{-2}) increased the rate of both nitrification and denitrification in sediments. In the absence of worms, the measured rates of nitrification and denitrification were respectively 1.9 and 3.3×10^{-3} mol·m^{-2}·day^{-1}. Chatarpaul et al. (1980) further showed that, even when clean glass beads were used as a substratum for the tubificids, the concentration of nitrate in overlying water decreased at a rate of 0.5 mmol·day^{-1} (no data given to convert to mol·m^{-2}·day^{-1}). Furthermore, the decrease in nitrate concentration could be stopped by removing the worms. This finding suggests that denitrifying bacteria were associated with the oligochaetes. Indeed, Chatarpaul et al. (1980) isolated nitrifying and denitrifying bacteria from the body walls and guts of tubificids.

4.3. Effects of Benthos on Materials Flux

The effects of chironomids on the flux of materials across the sediment–water interface have received considerable attention. Ganapati (1949) found that *C. plumosus* larvae increased the flux of both ammonia and phosphorus from sand filters of the Madras waterworks. Edwards (1958) observed that *C. riparius* enhanced ammonia flux from sediments and that the observed flux was greater by a factor 5 than that expected from chironomid ammonia excretion. Tessenow (1964) showed that the presence of *C. plumosus* (100 individuals m^{-2}) increased silica flux by 2.28–2.88 mmol·m^{-2}·day^{-1}, while Weissenbach (1974) found that the

Table VI. Chironomid-Induced Flux Increase Factors[a] for Various Materials in Three Lake Sediments[b]

Lake	Total P	SiO$_z$	NH$_4$-N	NO$_3$-N	Fe	Mn
Trummen						
Aerobic conditions	2.27	2.86	0.61	1.97	3.43	2.20
Nitrogen atmosphere	0.92	1.29	1.18	1.00	0.81	0.71
Årungen						
Aerobic conditions	2.08	2.07	154.90	1.10	13.71	4.01
Nitrogen atmosphere	1.08	1.14	0.94	1.10	0.67	0.91
Vombsjön						
Aerobic conditions	4.43	2.06	33.40	0.86	1.03	1.90
Nitrogen atmosphere	1.26	0.98	1.25	1.00	0.97	0.89

[a] Flux after chironomid addition/flux before chironomid addition.
[b] Based on data from Granéli (1979). Chironomid population density: 1000 individuals m^{-2}.

presence of chironomids (species not given, ~1000 individuals m^{-2}) increased silica flux by 2.52–2.88 mmol·m^{-2}·day^{-1}. In a study of sediments from three Swedish entrophic lakes (Trummen, Årungen, and Vombsjön), Granéli (1979) observed silica flux increases of 0.25, 0.21, and 0.64 mmol·m^{-2}·day^{-1} (respectively) per 100 chironomids m^{-2}, which corresponds to a flux increase by a factor of 2–3 over sediments not having a chironomid population (1000 individuals m^{-2}). Granéli (1979) found that the presence of chironomids affected the flux of total phosphorous ($\times 2$–$\times 4$), ammonia ($\times 0.61$–$\times 155$), iron ($\times 1$–$\times 14$), and manganese ($\times 2$–$\times 4$) when the overlying water was aerobic, but had virtually no effect on flux under anoxic conditions (see Table VI). Gallepp (1978, 1979) found that both C. riparius and C. tentans caused increased phosphorus flux from sediments. The larger species, C. tentans, produced the greater effect. Release of total phosphorus at 20°C increased in a linear manner from 9.69 to 30.35 mmol·m^2·day^{-1} over a range of six chironomid densities from 0 to 6585 larvae m^{-2}. Increasing temperature from 10° to 20°C had little effect on phosphorus release rates from sediments without chironomids, but release from sediments with C. tentans populations was increased tenfold at the higher temperature. Gallepp, however, could not differentiate between phosphorus translocated from the sediments and phosphorus excreted by the chironomids, and felt that the increased phosphorus flux might be almost entirely caused by chironomid excretion. Chironmids may also transport materials, such as phosphorus and iron, from lakes when the larvae emerge as adults (Neame, 1975, 1977).

Macrobenthos aside from chironomids are also known to enhance materials flux across the sediment–water interface. Jernelöv (1970) noted that the transfer of Hg from sediments was affected by both tubificids and the unionid clam *Anodonta* sp. Tubificids (size and species not given,

820,000 individuals m^{-2}) were found to cause Hg release from 0–2.5 cm, while *Anodonta* sp. (126 individuals m^{-2}) caused release of Hg from a depth of 0.9 cm. In the absence of macrobenthos, Hg was released only from the upper 1 cm. Davis et al. (1975) showed that the presence of tubificids (mixed population, *Tubifex* spp. ~90%, 2500 individuals m^{-2}) accelerated the removal of phosphorus from overlying water, increased the depth of phosphorus penetration, and reduced the proportion of SRP in the interstitial water of the uppermost sediments. These workers also noted that the presence of tubificids did not have an effect on phosphorus release from sediments under low-oxygen conditions in the overlying water. Kikuchi and Kurihara (1977), in a study of the effects of tubificids on the chemical and physical characteristics of submerged ricefield soil, found that the presence of tubificids (*L. socialis* and *Branchyura sowerbyi*, 8842 individuals m^{-2}) kept the activity of Fe^{2+} high in the upper 1 cm of the sediment, increased the movement of Fe^{2+} into the overlying water, enhanced ammonia concentration in the sediment and ammonia flux across the sediment–water interface, decreased the number of aerobic bacteria in the upper 1 cm of sediment, increased the amount of hexose in the overlying water, and increased the number of sulfate-reducing bacteria throughout the sediment column.

In experiments in our laboratory, control sediments and sediments preconditioned by tubificid activity (mixed population of *L. hoffmeisteri* and *Limnodrilus cervix*, 100,000 individuals m^{-2}; 30 days of tubificid activity; homogenized Lake Erie Sediment) were sealed and their overlying water permitted to go anoxic. Measurements of the concentration changes in bicarbonate, ammonia, ferrous iron, SRP, and dissolved silica were made in the water overlying both control and worm-affected sediments. In addition, control and experimental microcosms were sacrificed at intervals to follow the concentration profiles of these materials in the interstitial water. It was found that the presence of tubificids had no effect on bicarbonate flux; enhanced ammonia flux ($\times$ ~1.4); and suppressed the flux of silica ($\times$ ~1.4), phosphorus ($\times$ ~6), and ferrous iron ($\times$ ~2). A tentative explanation for the difference in phosphorus flux between bioturbated and nonbioturbated sediments is that in nonbioturbated sediments a ferric-hydroxide-rich surface layer forms owing to upward diffusion of Fe^{2+} and its oxidation at the sediment–water interface. This iron-hydroxide-rich layer is the site for the absorption of upward-diffusing phosphorus. Phosphorus entering this surface layer forms a ferric hydroxide–orthophosphate complex $[Fe_x(OH)_{3(x-y)}(PO_4)_y \cdot 2H_2O]$ (Stamm and Kohlschutter, 1965; Williams et al., 1976a,b). Phosphorus thus bound will be readily available for release to the overlying water when the ferric hydroxide dissolves following anoxia. In sediments bioturbated by tubificids such an iron-rich surface layer is prevented from forming because

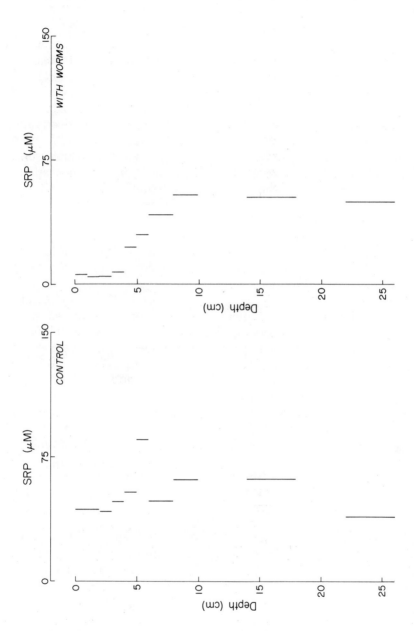

Figure 12. Phosphorus concentration profiles observed in homogenized Lake Erie sediments with and without tubificids. Population density: 10^5 individuals m^{-2}. SRP, soluble reactive phosphorus.

the worms continuously subduct material from the sediment–water interface. This action of tubificids during oxic conditions reduces the initial flux of phosphorus (and iron) when the overlying water is driven anoxic. Furthermore, examination of interstitial water SRP concentration profiles from a strongly bioturbated experimental core and its companion control core sacrificed near the end of the experiment (Fig. 12) shows that the worms prevented a strong phosphorus concentration gradient from forming at the sediment–water interface. Indeed, quite a long period of worm inactivity would be required before enough phosphorus could diffuse to establish a phosphorus gradient in the bioturbated core equivalent to that in the nonbioturbated sediment. The enhancement of ammonia flux in the presence of worms appears to be the result of tubificid-induced ammonia production in the upper few centimeters of microcosm sediments (see Fig. 11). The suppression of silica flux in the presence of worms is not currently explicable.

5. Conclusions

Freshwater macrobenthos affect the chemical diagenesis of sediments in a number of ways: advection of sediment particles, "diffusion" of sediment particles, pumping of interstitial and overlying water, alteration of sediment fabric, and injection of reactive materials. What is the relative importance of these processes to the chemical diagenesis of freshwater sediments? The pumping of water through sediments is almost certainly the most significant of these activities. Data on the effects of chironomids and tubificids on materials flux across the sediment–water interface indicate that water pumping by chironomids is approximately twice as important as sediment fabric alteration by tubificids. The advection of sediment particles from depth to the sediment–water interface by tubificids links two distinctly different biogeochemical regimes and certainly enhances sediment oxygen consumption. The advection of sediment particles does not appear to increase materials flux out of sediments appreciably, but possibly enhances the movement of some redox-sensitive materials from the overlying water into sediments, preconditions sediments so as to inhibit the release of materials such as iron and phosphorus, and modifies the reactivity of the sediment–water interface. As mentioned earlier, the alteration of sediment fabric through the construction of burrows and the generation of surficial fecal pellet layers are inherently less important than water pumping in enhancing the movement of material across the sediment–water interface, but can, at least theoretically, enhance the diffusivity of sediments by as much as a factor of 10. Empirical observations have shown an increase in diffusivity by approximately factors of 2–10. The importance of particle "diffusion" to chemical diagenesis probably depends on the length scale (i.e., size of the particle-diffusing organism) of the process, local sedimentation rate, and mixing rate. For example, the actions of P. hoyi are confined to uppermost 1 cm of sediment, while those of large unionid clams may extend to 30 cm (Robbins et al., 1979; McCall et al., 1979). Consequently, the effects of these two organisms on surface reactivity are vastly different. P. hoyi could reduce surface reactivity by a factor of $\omega/(\omega + \lambda)$, while large unionids could reduce surface reactivity by $\omega/(\omega + 30\lambda)$. The significance of reactant injection by macrobenthos has not yet been empirically evaluated in freshwater environments. Indeed, basic data concerning the chemical nature of materials excreted by freshwater macrobenthos are almost wholly lacking. There is some evidence that suggests that the excretion of ammonia by tubificids may aid in controlling interstitial water pH (Fisher and Matisoff, 1981). Chatarpaul et al. (1980) suggest that the bodies of at least some benthos (tubificid oligochaetes) may actually be sites of certain diagenetic reactions (nitrification and denitrification). The following ranking, in order of relative importance, of the above mechanisms is suggested:

1. Water pumping
2. Particle advection and diffusion
3. Fabric alteration
?. Reactant injection
?. Reaction site

The above mechanisms are, of course, linked to tropic group/life habit type. Water pumping through sediments is carried on by sedentary infaunal suspension feeders and may be continuous or episodic. Sediment particle advection is carried on by sedentary infaunal subsurface deposit feeders. Mobile epifaunal and infaunal deposit feeders will cause particle diffusion. Nearly all tropic group/life habit types can alter sediment fabric and inject reactants. Furthermore, the intensity of all of the mechanisms is likely to be related to organism size, population density, temperature, availability of dissolved oxygen, and other biotic and abiotic variables.

From the foregoing discussion, it is clear that the actions of macrobenthos affect the chemical diagenesis of freshwater sediments. Current knowledge of the mechanisms whereby these effects are generated is sketchy. In particular, the significance of various transport-modifying mechanisms should be investigated. For example, detailed time series experiments should be conducted to examine the effects of water pumping by chironomids on the migration of solutes through sediments. Ideally, radiotracer experiments would be used to determine the behavior of various materials in sediments populated by chironomids. Particle and solute transport measurements for organisms other than tubificids and chironomids should be made. In particular, the effects of unionid bivalves and burrowing mayflies on particle and solute transport should be investigated. The significance of reactant injection to sediment diagenesis needs investigation. What materials are excreted by various macrobenthos? How much material is injected? How rapidly does this material decay? What are the decomposition products? How does this process affect overall chemical mass transfer in surficial sediments? Do macrobethos provide reaction sites (as suggested by Chatarpaul et al., 1980)? What is the significance of this phenomenon? Finally, the problem of dimensionality needs to be considered in detail. For relatively large organisms, which occur at lower population densities (e.g., chironomids, unionid clams), two-dimensional models are probably necessary. Two-dimensional models require two-dimensional data for development and verification. New and far more advanced data-gathering techniques (e.g., multiple microelectrode probes, two-dimensional gamma scanning techniques) need to be developed.

In addition to the above work, the relationships between particle movement, water pumping, sediment fabric alteration, and reactant injection by macrobenthos and various biotic and abiotic variables need exploration. What are the effects of such factors as seasonal temperature variations, episodic anoxic interludes, organism population density, inter- and intraspecific interactions, and predation on interactions between benthos and their substratum?

The work suggested here, because of the detail required, must largely be carried out in the laboratory. How can the results of these suggested investigations be extended to the field? What field experiments might be conducted to answer the questions posed? Investigation of the effects of animal–sediment interactions on the chemical diagenesis of freshwater sediments should prove a fertile and productive area for future research.

ACKNOWLEDGMENTS. Funding from NOAA and EPA supported some of the work described in this chapter.

References

Aller, R. C., 1977, The influence of deposit feeding benthos on chemical diagenesis of marine sediments, Ph.D thesis, Yale University, New Haven, Connecticut.
Aller, R. C., 1978, The effects of animal–sediment interactions on geochemical processes near the sediment–water interface, in: *Estuarine Interactions* (M. L. Wiley, ed.), pp. 157–172, Academic Press, New York.
Aller, R. C., and Cochran, J. K., 1976, 234-Th/238-U disequilibrium in near shore sediments: Particle reworking and diagenetic time scales, *Earth Planet. Sci. Lett.* **20**:37–50.

Aller, R. C., and Dodge, R. E., 1974, Animal–sediment relations in a tropical lagoon— Discovery Bay, Jamaica, *J. Mar. Res.* **32**:209–232.

Aller, R. C., and Yingst, J. Y., 1978, Biogeochemistry of tube-dwellings: A study of the sedentary polychaete *Amphitrite ornata* (Leidy), J. Mar. Res. **36**:201–254.

Alsterberg, G., 1922, Die respiratorischen Mechanismen der Tubificiden, *Lunds Univ. Arsskr.* **18**:1–176.

Alsterberg, G., 1925, Die Nahrungszirkulation einiger Binnensetypen, *Archiv. Hydrobiol.* **15**:291–338.

Amiard-Triquet, C., 1975, Etude experimentale de la contamination par le cerium 144 et le fer 59 d'un sédiment à *Arenicola marina* L. (Annelide Polychete), *Cah. Biol. Mar.* **15**:483–494.

Andersen, J. M., 1977, Importance of the denitrification process for the rate of degradation of organic matter in lake sediments, in: *Interactions between Sediments and Fresh Water* (H. L. Golterman, ed.), pp. 357–362, Junk, The Hague.

Appleby, A. G., and Brinkhurst, R. O., 1970, Defecation rate of three tubificid oligochaetes found in the sediment of Toronto Harbour, Ontario, *J. Fish. Res. Board Can.* **27**:1971–1982.

Baas Becking, L. G. M., Kaplan, I. R., and Moore, D., 1960, Limits of the natural environment in terms of pH and oxidation–reduction potentials, *J. Geol.* **68**:243–284.

Ben-Yaakov, S., 1973, pH buffering of pore water of recent anoxic marine sediments, *Limnol. Oceanogr.* **18**:86–94.

Berner, R. A., 1963, Electrode studies of hydrogen sulfide in marine sediments, *Geochim. Gosmochim. Acta* **27**:563–575.

Berner, R. A., 1964a, An idealized model of dissolved sulfate distribution in recent sediments, *Geochim. Cosmochim. Acta* **28**:1497–1503.

Berner, R. A., 1964b, Distribution and diagenesis of sulfur in some sediments from the Gulf of California, *Mar. Geol.* **1**:117–140.

Berner, R. A., 1970, Sedimentary pyrite formation, *Am. J. Sci.* **268**:1–23.

Berner, R. A., 1972, Sulfate reduction, pyrite formation, and the oceanic sulfur budget, in: *Nobel Symposium 20: The Changing Chemistry of the Oceans* (D. Dyrssen and D. Jagner, eds.), pp. 347–361, Almquist and Wiksell, Stockholm.

Berner, R. A., 1974, Kinetic models for the early diagenesis of nitrogen, sulfur, phosphorus, and silicon in anoxic marine sediments, in: *The Sea*, Volume 5: *Marine Chemistry* (E. D. Goldberg, ed.), pp. 427–450, John Wiley and Sons, New York.

Berner, R. A., 1976, The benthic boundary layer from the viewpoint of a geochemist, in: *The Benthic Boundary Layer* (I. N. McCave, ed.), pp. 33–55, Plenum Press, New York.

Berner, R. A., 1980, *Early Diagenesis: A Theoretical Approach*, Princeton University Press, Princeton, N.J.

Brinkhurst, R. O., 1974, *The Benthos of Lakes*, St. Martin's Press, New York.

Burns, N. M. and Ross, C., 1972, Oxygen–nutrient relationships in the Central Basin of Lake Erie, in: *Project Hypo* (N. M. Ross and C. Ross, eds.), pp. 85–119, Canadian Center for Inland Waters Paper 6, Canadian Center for Inland Waters, Hamilton, Ontario.

Carmouze, J. P., Golterman, H. L., and Pedro, C., 1977, The neoformation of sediments in Lake Chad; their influence on salinity control, in: *Interactions between Sediments and Fresh Water* (H. L. Golterman, ed.), pp. 33–39, Junk, The Hague.

Chamberlain, C. K., 1975, Recent lebensspuren in nonmarine aquatic environments, in: *The Study of Trace Fossils* (R. W. Frey, ed.), pp. 431–458, Springer-Verlag, New York.

Chapman, T. W., 1967, The transport properties of concentrated electrolytic solutions, Ph.D. thesis, University of California, Berkeley.

Chatarpaul, L., Robinson, J. B., and Kaushik, N. K., 1979, Role of tubificid worms in nitrogen transformations in stream sediment, *J. Fish. Res. Board Can.* **36**:673–678.

Chatarpaul, L., Robinson, J. B., and Kaushik, N. K., 1980, Effects of tubificid worms on denitrification and nitrification in stream sediment, *Can. J. Fish. Aquat. Sci.* **37**:656–663.

Coker, R. E., Shira, A. R., Clark, H. W., and Howard, A. D., 1921, Natural history and propagation of fresh-water mussels, *Bull. U.S. Bur. Fish.* **37**:77–181.

Cummins, K. W., 1973, Trophic relationships of aquatic insects, *Annu. Rev. Entomol.* **18**:183–206.

D'Anglejan, B. F., 1967, Origin of marine phosphorites off Baja California, Mexico, *Mar. Geol.* **5**:15–44.

Davies, S. N., and DeWiest, R. C. M., 1966, *Hydrogeology*, John Wiley and Sons, New York.

Davis, R. B., 1974a, Stratigraphic effects of tubificids in profundal lake sediments, *Limnol. Oceanogr.* **19**:466–488.

Davis, R. B., 1974b, Tubificids alter profiles of redox potential and pH in profundal lake sediments, *Limnol. Oceanogr.* **19**:342–346.

Davis, R. B., Thurlow, D. L., and Brewster, R. E., 1975, Effects of burrowing tubificids on the exchange of phosphorus between lake sediment and overlying water, *Verh. Int. Verein. Limnol.* **19**:382–394.

Dell, C. I., 1971, Late quaternary sedimentation in Lake Superior, Ph.D. thesis, University of Michigan, Ann Arbor.

Dell, C. I., 1973, Vinianite: An authigenic phosphate mineral in Great Lake sediments, in: *Proceedings of the 16th Conference on Great Lakes Research*, pp. 1027–1028, International Association for Great Lakes Research, Ann Arbor, Michigan.

Doyle, R. W. S., 1967, Eh and thermodynamic equilibrium in environments containing dissolved ferrous iron, Ph.D. thesis, Yale University, New Haven, Connecticut.

Dutka, B. J., Bell, J. B., and Liu, D. L. S., 1974, Microbiological examination of offshore Lake Erie sediments, *J. Fish. Res. Board Can.* **31**:299–308.

Edwards, R. W., 1958, The effect of larvae of *Chironomus riparius* Meigen on the redox potentials of settled activated sludge, *Ann. Appl. Biol.* **46**:457–464.

Edwards, R. W., and Rolley, H. L. J., 1965, Oxygen consumption of river muds, *J. Ecol.* **53**:1–19.

Emerson, S., and Widmer, G., 1978, Early diagenesis in anaerobic lake sediments. II. Thermodynamic and kinetic factors controlling the formation of iron phosphate, *Geochim. Cosmochim. Acta* **42**:1307–1316.

Fisher, J. B., and Matisoff, G., 1981, High resolution vertical profiles of pH in recent sediments, *Hybrobiologia* **79**:277–284.

Fisher, J. B., and Tevesz, M. J. S., 1975, Distribution and population density of *Elliptio complanata* (Mollusca) in Lake Pocotopaug, Connecticut, *Veliger* **18**:332–338.

Fisher, J. B., Lick, W., McCall, P. L., and Robbins, J. A., 1980, Vertical mixing of lake sediments by tubificid oligochaetes, *J. Geophys. Res.* **85**:3997–4006.

Gallepp, G. W., 1979, Chironomid influence on phosphorous release in sediment–water microcosms, *Ecology* **60**:547–556.

Gallepp, G. W., Kitchell, J. F., and Bartell, S. M., 1978, Phosphorus release from lake sediments as affected by chironomids, *Verh. Int. Verein. Limnol.* **20**:458–465.

Ganapati, S. V., 1949, The role of the bloodworm, *Chironomus plumosus*, in accounting for the presence of phosphorus and excessive free ammonia in the filtrates from the slow sand filters of the Madras water works, *Zool. Soc. India J.* **6**:41–43.

Gardner, R. L., 1973, Chemical models for sulfate reduction in closed anaerobic marine environments, *Geochim. Cosmochim. Acta* **37**:53–68.

Glasby, G. P., 1973, Interstitial waters in marine and lacustrine sediments: A review, *J. R. Soc. N. Z.* **3**:43–59.

Goldberg, E. D., and Koide, M., 1962, Geochronological studies of deep sea sediments by the ionium/thorium method, *Geochim. Cosmochim. Acta* **26**:417–450.

Goldhaber, M. B., 1974, Equilibrium and dynamic aspects of the marine geochemistry of sulfur, Ph.D. thesis, University of California, Los Angeles.
Goldhaber, M. B., and Kaplan, I. R., 1975a, Apparent dissociation constants of hydrogen sulfide in chloride solutions, *Mar. Chem.* **3**:83–104.
Goldhaber, M. B., and Kaplan, I. R., 1975b, Controls and consequences of sulfate reduction rates in recent marine sediments, *Soil Sci.* **119**:42–55.
Gorham, E., 1964, Molybdenum, manganese, and iron in lake muds, *Verh. Int. Verein. Limnol.* **15**:330–332.
Granéli, W., 1979, The influence of *Chironomus plumosus* larvae on the exchange of dissolved substances between sediment and water, *Hydrobiologia* **66**:149–159.
Granéli, W., 1982, The influence of *Chironomus plumosus* on oxygen uptake of sediment, *Arch. Hydrobiol.* (in press).
Guinasso, N. L., and Schink, D. R., 1975, Quantitative estimates of biological mixing rates in abyssal sediments, *J. Geophys. Res.* **80**:3032–3043.
Hargrave, B. T., 1972, Oxidation–reduction potentials, oxygen concentration and oxygen uptake of profundal sediments in a eutrophic lake, *Oikos* **23**:167–177.
Hargrave, B. T., 1975, Stability in structure and function of the mud–water interface, *Verh. Int. Verein. Limnol.* **19**:1073–1079.
Harrison, A. G., and Thode, 1958, The kinetic isotope effect in the chemical reduction of sulfate, *Tr. Faraday Soc.* **53**:1648–1651.
Hayes, F. R., 1964, The mud–water interface, *Oceanogr. Mar. Biol. Annu. Rev.* **2**:121–145.
Hutchinson, G. E., 1957, *A Treatise on Limnology*, John Wiley and Sons, New York.
Iovino, A., and Bradley, W., 1969, The role of larval chironomidae in the production of lacustrine copropel in Mud Lake, Marion County, Florida, *Limnol. Oceanogr.* **14**:898–900.
Ivlev, V. S., 1939, Transformation of energy by aquatic animals. Coefficient of energy consumption by *Tubifex tubifex* (Oligochaeta), *Int. Rev. Ges. Hydrobiol. Hydrogr.* **38**:449–458.
Jones, B. F., and Bowser, C. J., 1978, The mineralogy and related chemistry of lake sediments, in: *Lakes: Chemistry, Geology, Physics* (A. Lerman, ed.), pp. 179–235, Springer-Verlag, New York.
Jernelöv, A., 1970, Release of methyl mercury from sediments with layers containing inorganic mercury at different levels, *Limnol. Oceanogr.* **15**:958–960.
Kaplan, I. R., and Rittenberg, S. C., 1964, Microbiological fractionation of sulfur isotopes, *J. Gen. Microbiol.* **34**:195–212.
Kikuchi, E., and Kurihara, Y., 1977, In vitro studies on the effects of tubificids on the biological, chemical, and physical characteristics of submerged ricefield soil and overlying water, *Oikos* **29**:348–356.
Kirchner, W. B., 1975, The effect of oxidized material on the vertical distribution of freshwater benthic fauna, *Freshwater Biol.* **5**:423–429.
Krezoski, J. R., 1981, The influence of zoobenthos on fine-grained particle reworking and benthic solute transport in Great Lakes sediments. Ph.D. thesis, University of Michigan, Ann Arbor.
Krezoski, J. R., Mozley, S. C., and Robbins, J. A., 1978, Influence of benthic macroinvertebrates on mixing of profundal sediments in southeastern Lake Huron, *Limnol. Oceanogr.* **23**:1011–1016.
Kuznetsov, S. I., 1970, *The Microflora of Lakes and its Geochemical Activity*, University of Texas Press, Austin.
Lafon, G. M., and Mackenzie, F. T., 1966, Early evolution of the oceans—A weathering model, in: *Studies in Paleo-Oceanography* (W. W. Hay, ed.), *Soc. Econ. Paleontol. Mineral. Spec. Publ.* **20**:205–218.
Lee, G. F., 1970, Factors affecting the transfer of materials between water and sediments,

University of Wisconsin Water Resources Center, Eutrophication Information Program, Literature Review 1, University of Wisconsin, Madison.

Li, Y.-H., and Gregory, S., 1974, Diffusion of ions in sea water and in deep sea sediments, Geochim. Cosmochim Acta 38:703–714.

Lundbeck, J., 1926, Die Bodentierwelt Norddeutschen Seen, Arch. Hydrobiol. Suppl 7.

Lyman, F. E., 1943, Swimming and burrowing activities of mayfly nymphs of the genus Hexigenia, Ann. Entomol. Soc. Am. 26:250–256.

McCaffrey, R. J., Meyers, A. C., Davey, E., Morrison, G., Bender, M., Luedtke, N., Cullen, D., Froelich, P., and Klinkhammer, G., 1980, The relation between pore water chemistry and benthic fluxes of nutrients and manganese in Narragansett Bay, Rhode Island, Limnol. Oceanogr. 25:31–44.

McCall, P. L., 1979, Effects of deposit feeding oligochaetes on particle size and settling velocity of Lake Erie sediments, J. Sediment. Petrol. 49:813–818.

McCall, P. L., and Fisher, J. B., 1980, Effects of tubificid oligochaetes on physical and chemical properties of Lake Erie sediments, in: Aquatic Oligochaete Biology (R. O. Brinkhurst and D. G. Cook, eds.), pp. 253–318, Plenum Press, New York.

McCall, P. L., Tevesz, M. J. S., and Schwelgien, S. F., 1979, Sediment mixing by Lampsilis radiata siliquoidea (Mollusca) from western Lake Erie, J. Great Lakes Res. 5:105–111.

McLachlan, A. J., and McLachlan, S. M., 1976, Development of the mud habitat during the filling of two new lakes, Freshwater Biol. 6:59–67.

Manheim, F. T., 1970, The diffusion of ions in unconsolidated sediments, Earth Planet. Sci. Lett. 9:307–309.

Manheim, F. T., and Sayles, F. L., 1974, Composition and origin of interstitial waters of marine sediments, in: The Sea, Volume 5: Marine Chemistry (E. D. Goldberg, ed.), pp. 527–568, John Wiley and Sons, New York.

Marzolf, G. R., 1965, Substrate relations of the burrowing amphipod Pontoporeia affinis in Lake Michigan, Ecology 46:579–592.

Milbrink, G., 1969, Microgradient at the mud–water interface, Rep. Inst. Freshwater Res. Drottningholm 49:129–148.

Monakov, A. K., 1972, Review of studies on feeding of aquatic invertebrates conducted at the Institute of Biology of Inland Waters, Academy of Sciences, USSR, J. Fish. Res. Board Can. 29:363–383.

Mortimer, C. H., 1941, The exchange of dissolved substances between mud and water in lakes, J. Ecol. 29:280–329.

Mortimer, C. H., 1942, The exchange of dissolved substances between mud and water in lakes, J. Ecol. 30:147–201.

Mortimer, C. H., 1971, Chemical exchanges between sediments and water in the Great Lakes—Speculations on probable regulatory mechanisms, Limnol. Oceanogr. 16:387–404.

Müller, G., and Forstner, U., 1973, Recent iron ore formation in Lake Malawi, Africa, Miner. Deposita 8:278–290.

Neame, P. A., 1975, Benthic oxygen and phosphorus dynamics in Castle Lake, California, Ph.D. thesis, University of California, Davis.

Neame, P. A., 1977, Phosphorus flux across the sediment–water interface, in: Interactions between Sediments and Fresh Water (H. L. Golterman, ed.), pp. 307–312, Junk, The Hague.

Newell, R. C:, 1970, Biology of Intertidal Animals, Elsevier, New York.

Norvell, W. A., 1974, Insolubilization of inorganic phosphate by anoxic lake sediment, Soil Sci. Soc. Am. Proc. 38:441–445.

Nozaki, Y., Cochran, J. K., Turekian, K. K., and Keller, G., 1977, Radiocarbon and (210) Pb distribution in submersible taken deep-sea cores from project FAMOUS, Earth Planet. Sci. Lett. 34:167–173.

Nriagu, J. O., 1968, Sulfur metabolism and sedimentary environment: Lake Mendota, Wisconsin, *Limnol. Oceanogr.* **13**:430–439.
Nriagu, J. O., 1975, Sulphur isotopic variations in relation to sulphur pollution of Lake Erie, in: *Isotope Ratios as Pollutant Source and Behavior Indicators*, pp. 77–93, IAEA-SM-191/28, International Atomic Energy Agency, Vienna.
Nriagu, J. O., 1976, Phosphate–clay mineral relations in soils and sediments, *Can. J. Earth Sci.* **13**:717–736.
Nriagu, J. O., and Dell, C. I., 1974, Diagenetic formation of iron phosphates in recent lake sediments, *Am. Mineral.* **59**:934–946.
Otsuki, A., and Wetzel, R. G., 1972, Coprecipitation of phosphate with carbonates in a marl lake, *Limnol. Oceanogr.* **17**:763–767.
Parry, G., 1960, Excretion, in: *The Physiology of Crustacea* (T. H. Waterman, ed.), pp. 341–363, Academic Press, New York.
Poddubnaya, T. L., 1961, Concerning the feeding of the high density species of tubificids in Rybinsk Reservoir, *Tr. Inst. Biol. Vodokhran. Akad. Nauk SSSR* **4**:219–231.
Poddubnaya, T. L., and Sorokin, Yu. I., 1961, The depth of the layer of optimal feeding of tubificids in connection with their movements in the sediment, *Izv. Inst. Biol. Vodokhran. AN SSSR* **10**:14–17.
Postgate, J. R., 1951, The reduction of sulphur compounds by *Desulphovibrio desulphuricans*, *J. Gen. Microbiol.* **5**:725–738.
Potts, W. T. W., 1954a, The energetics of osmotic regulation in brackish and freshwater animals, *J. Exp. Biol.* **31**:618–630.
Potts, W. T. W., 1954b, The rate of urine production of *Anodonta cygnea*, *J. Exp. Biol.* **31**:614–617.
Presley, B. J., and Kaplan, I. R., 1968, Changes in dissolved sulfate, calcium, and carbonate from interstitial water of near-shore sediments, *Geochim. Cosmochim. Acta* **32**:1037–1048.
Ramm, A. E., and Bella, D. A., 1974, Sulfide production in anaerobic microcosms, *Limnol. Oceanogr.* **19**:110–118.
Ravera, O., 1955, Amount of mud displaced by some freshwater oligochaeta in relation to depth, *Mem. Ist. Ital. Idrobiol.* **8**(suppl.):247–264.
Rees, C. E., 1973, A steady state model for sulfur isotope fractionation in bacterial reduction process, *Geochim. Cosmochim. Acta* **37**:1141.
Reid, R. C., Prausnitz, J. M., and Sherwood, T. K., 1977, *The Properties of Gases and Liquids*, McGraw-Hill, New York.
Rhoads, D. C., 1970, Mass properties, stability, and ecology of marine muds related to burrowing activity, in: *Trace Fossils* (T. P. Crimes and J. C. Harper, eds.), pp. 391–406, *Geol. J. Spec. Issue* **3**.
Rhoads, D. C., 1974, Organism–sediment relations on the muddy sea floor, *Oceanogr. Mar. Biol. Annu. Rev.* **12**:263–300.
Rhoads, D. C., and Young, D. K., 1970, The influence of deposit feeding organisms on sediment stability and community trophic structure, *J. Mar. Res.* **28**:150–178.
Riley, J. P., 1971, The major and minor elements in seawater, in: *Introduction to Marine Chemistry* (J. P. Riley and R. Chester, eds.), pp. 60–101, Academic Press, New York.
Robbins, J. A., 1978, Geochemical and geographical applications of radioactive lead, in: *Biogeochemistry of Lead* (J. O. Nriagu, ed.), pp. 285–337, Elsevier, Amsterdam.
Robbins, J. A., Krezoski, J. R., and Mozley, S. C., 1977, Radioactivity in sediments of the Great Lakes: Post-depositional redistribution by deposit feeding organisms, *Earth Planet. Sci. Lett.* **36**:325–333.
Robbins, J. A., McCall, P. L., Fisher, J. B., and Krezoski, J. R., 1979, Effects of deposit feeders on migration of cesium-137 in lake sediments, *Earth Planet. Sci. Lett.* **42**:277–287.
Robinson, R. A., and Stokes, R. H., 1959, *Electrolyte Solutions*, Butterworths, London.

Savilov, A. I., 1957, Biolgical aspects of the bottom fauna grouping of the North Okhotsk Sea, *Tr. Inst. Okeanol. Mar. Biol.* **20**:67–136.
Schumacher, A., 1963, Quantitative Aspekte der Beziehung zwischen Stärke der Tubificidenbesiedlung und Schichtdicke der Oxydationzone in den Süsswasserwatten der Unterelbe, *Arch. Fischwiss.* **14**:48–51.
Shapiro, J., Edmondson, W. T., and Allison, D. E., 1971, Changes in the chemical composition of sediments of Lake Washington, *Limnol. Oceanogr.* **16**:437–452.
Staddon, B. W., 1969, Water balance in *Ilyocoris*, Naucoridae, *J. Exp. Biol.* **51**:643–665.
Stamm, H. H., and Kohlschutter, H. W., 1965, Die Sorption von Phosphationene an Eisen (III)-Hydroxide, *J. Inorg. Nucl. Chem.* **27**:2103–2108.
Stobbart, R. H., and Shaw, J., 1974, Salt and water balance: Excretion, in: *The Physiology of Insects* (M. Rockstein, ed.), pp. 361–446, Academic Press, New York.
Stumm, W., and Morgan, J. J., 1970, *Aquatic Chemistry*, John Wiley and Sons, New York.
Swain, F. M., 1965, Geochemistry of some quaternary lake sediments of North America, in: *The Quaternary of the United States* (H. E. Wright and D. G. Frey, eds.), pp. 765–781, Princeton University Press, Princeton, New Jersey.
Tessenow, U., 1964, Experimentaluntersuchungen zur Kieselsäurerückführung aus dem Schlamm der See durch Chironomidenlarven (Plumosus-Gruppe), *Arch. Hydrobiol.* **60**:497–504.
Tevesz, M. J. S., and McCall, P. L., 1979, Evolution of substratum preference in bivalves (Mollusca), *J. Paleontol.* **53**:112–120.
Tevesz, M. J. S., Soster, F., and McCall, P. L., 1980, The effects of size selective feeding by oligochaetes on the physical properties of river sediments, *J. Sediment. Petrol.* **50**:561–568.
Turpaeva, E. P., 1957, Food interrelationships of dominant species in marine benthic biocoenoses, *Tr. Inst. Okeanol. Mar. Biol.* **20**:137–148.
Upchurch, S. B., 1976, Chemical characteristics of the Great Lakes, in: *Great Lakes Basin Framework Study*, Appendix 4: *Limnology of Lakes and Embayments*, pp. 151–238, Public Information Offices, Great Lakes Basin Commission, Ann Arbor, Michigan.
Vallentyne, J. R., 1961, On the rate of formation of black spheres in recent sediments, *Verh. Int. Verein. Limnol.* **14**:291–295.
Vallentyne, J. R., 1963, Isolation of pyrite spherules from recent sediments, *Limnol. Oceanogr.* **8**:16–30.
Vogel, S., and Bretz, W. L., 1972, Interfacial organisms: Passive ventilation in the velocity gradients near surfaces, *Science* **175**:210.
Wachs, B., 1967, Die Oligochaeten-Fauna der Fleissgewasser unter besonderer Berücksichtigung der Beziehungen zwischen der Tubificiden-Besiedlung und dem Substrat, *Arch. Hydrobiol.* **63**:310–386.
Walker, K. R., and Bambach, R. K., 1974, Feeding by benthic invertebrates: Classification and terminology for paleoecological analysis, *Lethaia* **7**:67–78.
Walshe, B. M., 1947, Feeding mechanisms of Chironomous larvae, *Nature* **160**:474.
Weiler, R. R., 1973, The interstitial water composition in the sediments of the Great Lakes. I. Western Lake Ontario, *Limnol. Oceanogr.* **18**:918–931.
Weissenbach, H., 1974, Untersuchungen zum Phophorhaushalt eines Hochgebirgsees (Vorder Finstertaler See, Kühtai, Tirol) unter besonderer Berücksichtigung der Sedimente, Dissertation, Leopold Franzens-Universität, Innsbruck.
Wetzel, R. G., 1975, *Limnology*, W. B. Saunders, Philadelphia.
Williams, J. D. H., Jaquet, J. M., and Thomas, R. L., 1976a, Forms of phosphorus in the surficial sediments of Lake Erie, *J. Fish. Res. Board Can.* **33**:413–429.
Williams, J. D. H., Murphy, T. P., and Mayer, T., 1976b, Rates of accumulation of phosphorus forms in Lake Erie sediments, *J. Fish. Res. Board Can.* **33**:430–439.

Winfrey, M. R., and Zeikus, J. G., 1979, Anaerobic metabolism of immediate methane precursors in Lake Mendota, *Appl. Environ. Microbiol.* **37:**244–253.

Wood, L. W., 1975, Role of oligochaetes in the circulation of water and solutes across the mud–water interface, *Verh. Int. Verein. Limnol.* **19:**1530–1538.

Zinder, S. H., and Brock, T. D., 1978, Methane, carbon dioxide, and hydrogen sulfide production from the terminal methiol group of methionine by anaerobic lake sediments, *Appl. Environ. Microbiol.* **35:**344–352.

ZoBell, C. E., 1946, Studies on redox potential of marine sediments, *Bull. Am. Assoc. Petrol. Geol.* **30:**477–513.

III

Ancient Environments

Chapter 5

Geological Significance of Marine Biogenic Sedimentary Structures

CHARLES W. BYERS

1. Introduction 221
2. Traces and Sedimentology 222
   2.1. Bioturbation in Recent Muds 223
   2.2. Rate and Continuity of Sedimentation 225
   2.3. Substrate Consistency 227
   2.4. Diagenesis 230
3. Paleobathymetry 233
   3.1. Introduction 233
   3.2. Problems with Seilacher's Model 234
   3.3. Evidence from the Oceans 238
   3.4. The Graphoglyptid Controversy 241
4. Precambrian Traces 243
   4.1. The Precambrian–Cambrian Boundary 243
   4.2. The Oldest Trace Fossil 248
   References 252

1. Introduction

The "geological significance" of traces encompasses a huge range of topics; to survey the field even briefly would require an entire book and probably multiple authors. Fortunately a major review volume was published within the last decade (Frey, 1975), and a diversity of topics were addressed. In addition, the revised Volume W of the Treatise on Invertebrate Paleontology (Häntzschel, 1975), which covers the systematics of trace fossils, also appeared recently. These two books have served to provide a solid base for research in trace fossils in terms of current systematics and concepts of classification, how traces are made and pre-

CHARLES W. BYERS • Department of Geology and Geophysics, University of Wisconsin, Madison, Wisconsin 53706.

served, and the relation of traces to sedimentology, biostratigraphy, paleontology, and paleoecology.

My intent in this chapter is twofold: I will review some of the recent literature on trace fossils, and I will focus on what seem to me to be the most exciting and significant areas of trace fossil research. Gould (1980) recently reviewed the status of paleontology in general and concluded that the science has been overly concerned with performing as a "service industry" for stratigraphic ordering and environmental reconstruction. This has been especially true in the realm of trace fossils. The paramount interest in traces has been for their utility as bathymetric indicators, and the inductivist methodology Gould deplored has been pervasive. Nearly all trace fossil studies have addressed one of these questions:

1. What traces are found in a particular depositional environment (or in a particular *inferred* environment, as interpreted from the rock record)?
2. What traces are found in rocks of a particular age?
3. What was the animal that made a particular trace, and what was it doing?

Each of these questions focuses on the particular. I find Gould's thesis quite applicable to ichnology—students of traces have hardly begun the task of formulating and testing hypotheses. Although data have been gathered and a few generalities have been formulated, the predictive and explanatory aspects of science have been lacking. Perhaps this is simply the result of the youthfulness of the discipline; not many years ago Seilacher needed to assert "It is generally agreed that biogenic sedimentary structures are true fossils and that their study . . . is part of paleontology" (Seilacher, 1964). I will discuss below some encouraging signs of new approaches to ichnology.

2. Traces and Sedimentology

Trace fossils are a special kind of sedimentary structure. They are formed, as a rule, within the depositional environment and soon after the physical deposition of the sediment. Therefore they typically reflect conditions in that environment that are of interest to sedimentologists: water energy, rate of deposition, uniformity of deposition, consistency of substrata. Traces may also be important in the interpretation of the postdepositional history of a sedimentary rock. Because burrows usually have fill that is chemically or physically different from the surrounding matrix, or have no fill at all, they may react diagenetically in distinctive ways. In the following brief survey of the interrelationships among physical sedimentology, diagenesis, and trace fossils, I hope to indicate the inseparability of physical and biological processes.

2.1. Bioturbation in Recent Muds

Lamination in neritic sediments is rare because of the activities of infauna. Although the effects of burrowing infauna on marine sediments have been discussed theoretically for many years, only since the 1950s have they been widely recognized in the field.

It was known in the last century that deposit-feeding infauna rework marine muds (see Twenhofel, 1939). Twenhofel concluded that sediments will be reworked by infauna unless bottom conditions are hostile to life, and Dapples (1942) predicted that burrowing infauna such as annelids and holothurians would strongly alter marine sediments, the greatest change being found in areas of optimum living conditions. In 1950 Kuenen discussed laminated muds and concluded that they would be rare at shelf depth owing to both physical and biological reworking. He cited the clear-cut laminae of the Black Sea sediments as being very uncommon in normal shelf-depth seas. Pettijohn (1957) continued the theoretical discussion of lamination by pointing out that laminae could only be preserved below wave base in quiet water. He also postulated the role played by burrowing infauna: "if such borings are numerous the original stratification planes are largely destroyed and only vestiges remain. Complete eradication of bedding is possible and has probably occurred in some sediments" (Pettijohn, 1957, p. 193). This perspective was stratigraphic, an acceptance of bedding as the normal situation, with a marginal note that biological alteration of sediments was conceivable.

Moore and Scruton (1957) were the first to show that burrowing deposit feeders are a major factor in the final product of sedimentation of the Mississippi Delta. They found a decrease in the preservation of original sedimentary structures moving away from the Delta. Nearshore, where the sedimentation rate and physical reworking rates by waves and currents are high, the original structures in the sediment are preserved. Away from shore the physical sedimentary processes are slower owing to increasing distance from the sediment source and increasing depth. Here burrowing animals disrupt the regular bedding into irregular layers and lenses. Further offshore, biogenic activity produces sediments of mottled appearance, or even a homogeneous, structureless bioturbate. Moore and Scruton demonstrated that only in the very shallow nearshore region, where physical processes outstrip bioturbation, are regular layers preserved. At depths of 40 m or more the sediments are homogeneous muds, completely reworked by the infauna. Hence Moore and Scruton's conclusion: "the importance of animals in forming the final fabric of shallow-water marine sediments cannot be overstated. Were it not for the burrowing and crawling organisms, the appearance of many of our marine sediments would be entirely different."

Van Straaten's (1959) observations of the sediments of the Rhone Delta confirmed Moore and Scruton's conclusions. With increasing depth the effects of burrowing infauna overwhelm the processes of physical sedimentation. Van Straaten found that regular layering nearshore, a product of seasonal changes in discharge, was progressively destroyed away from the delta; below 80 m the muds were homogenized by infauna. Van Straaten thought such homogeneous muds to be the rule for outer shelf conditions worldwide, except for inclusions of relict Pleistocene sands and ice-rafted material on polar shelves.

Other workers noted the same gradient in sedimentary structures offshore from deltas. Allen (1965) described the sediments of the Niger Delta, where the platform (less than 20 m depth) sediments are wave- and current-winnowed and -layered, whereas the offshore prodelta and shelf clays are completely reworked by infaunal benthos.

Deltas are not unique in showing a decrease in physical structures away from shore. Moore and Scruton (1957) and Van Straaten (1959) described similar sedimentation patterns offshore from barrier bar sequences. Nearshore layering is a product of wave and tidal current reworking, rather than a rapid and fluctuating sediment supply. Reworking weakens with increasing depth, and bioturbation destroys the bedding. In 1967 Reineck discussed the patterns of sedimentary structures in the North Sea. He found decreasing physical influence and increasing bioturbation correlated with deeper water. Wave effects diminish greatly at depths of 20 m or more. Laminated sandy and silty deposits are confined to these very shallow depths, where the sediment can be wave-winnowed; bioturbation replaces physical structures completely in the seaward direction. By 1970, perspective had shifted: "open shelf sediments, particularly of the transitional and mud belts, are subject to extensive reworking by bottom dwelling and feeding animals, even to the extent that all traces of primary laminations are destroyed" (Allen, 1970).

Modern normal-marine sediments are dominated by biological processes except in areas of very rapid sedimentation influx or continual current activity. Hence the degree to which sediments have been bioturbated is an index of the physical energy of the environment. Environmental zonation of modern shelves based on the amount of bioturbation provides models for interpretation of the sedimentary record.

The absence of bioturbation in quiet-water, slow-deposition environments implies chemical barriers to benthic habitation. Very high salinity may restrict metazoans, but low oxygenation of the muds and/or bottom waters is a more common cause. The relationships among depth, oxygenation, metazoan life, and trace fossils have been reviewed by Rhoads and Morse (1971) and Byers (1977a).

2.2. Rate and Continuity of Sedimentation

In addition to indicating the overall energy of the depositional environment, trace fossils can provide information on the timing of sedimentary processes. The use of traces in this fashion has been pioneered by Goldring (1964) and especially by Howard (1975, 1978). Howard has identified patterns of sedimentation that can be distinguished on the basis of trace fossil distribution.

Slow deposition in the marine realm will give rise to a homogeneous sediment riddled with burrows; incipient stratification caused by changes in sediment grain type or size, or by current lamination, will be mostly obliterated. This is because the rate of bioturbation, especially by deposit-feeding infauna, is much more rapid than the rate of deposition. For example, Rhoads (1963) calculated that a single deposit-feeding species in Long Island Sound was able to rework completely the annual deposition twice in the course of a year. On average, each sedimentary grain that accumulates is disturbed (moved, eaten, pelleted) several times before it becomes part of the stratigraphic record. A thoroughly bioturbated texture in sediments or a sedimentary rock indicates continuity of deposition at a rate outstripped by bioturbation.

It is difficult to determine an actual rate, for bioturbation depends on population density. Abyssal sediments accumulate at rates of centimeters per thousand years, but benthic biomass is very low. Shallow estuary or marine shelf accumulation rates are two or three orders of magnitude higher, but population sizes are enormous because of the abundant food supply. Thus abyssal clays and shallow subtidal muds are both thoroughly bioturbated.

Rapid deposition will leave a record of physical stratification. Deposition of a turbidity current or a storm surge occurs much more rapidly than the processes of bioturbation—even large populations of active deposit feeders take at least days to churn a sedimentary layer (Howard and Elders, 1970; Elders, 1975). During the actual deposition of a graded bed or storm stratum, biological processes are simply overwhelmed. This influx of sediment will often destroy the existing benthic community, so that the stratified layer remains intact even after rapid deposition has ceased. If a period of slow deposition ensues, a new burrower population can be recruited and bioturbation may rework the upper part of the rapidly deposited layer. Burrowers typically forage within a few centimeters of the sediment–water interface, so a stratified layer thicker than this bioturbation zone should be preserved. A very thin (cm) stratum can be obliterated; conversely, a sequence of several rapid depositional events can form an amalgamated series of stratified units with no bioturbation. An isolated rapid depositional event in a prevailing slow-deposition environment should produce an asymmetric stratum. The base will sharply overlie bioturbated sediment; in fact, if the depositional event is preceded by erosional scour, as in a turbidity current or storm surge, the underlying burrows can be truncated. The top of the rapidly deposited stratum will commonly be less well-defined; it will grade into the overlying slow-deposition sediments.

These alternations in depositional rate are coming to be recognized as common in the stratigraphic record. They have long been known in turbidite basins, where the rapid depositional event produces graded beds and Bouma sequences, and the intervening slow pelagic sedimentation forms thin shale layers. Turbidites may have scoured bases and burrowed tops (Seilacher, 1978). More recently, the importance of storm deposition at shelf depth has been realized.

Howard (1972) noted that modern shoreface sediments consist of alternations of laminated and burrowed sediment. Individual laminated units are 10–50 cm thick and have basal scours. Burrows increase in density upward through the unit; the upper sediments are thoroughly bioturbated. A truncating erosional surface begins the next laminated stratum. Howard interpreted these sequences as alternating periods of storm and fairweather conditions in an environment at or just below fairweather wave base. He recognized similar cycles, termed parallel-laminated-to-burrowed, in the Cretaceous shoreline sediments of the Western Interior.

Bowen et al. (1974) showed that episodic scour and deposition were responsible for producing fossil coquinas in the Upper Devonian shelf sediments of New York. Coquinas form the bases of laminated strata, which grade upward into thoroughly bioturbated muds. The episodic depositional events were postulated to be storms that swept across the subaqueous delta platform west of the Catskill shoreline.

Bambach has been a strong advocate of the storm hypothesis to explain coquinas and laminated-to-burrowed strata. He and his co-workers (Bambach et al., 1978) have proposed that much of the Phanerozoic fossil record on the cratons consists of shell accumulations formed by storm reworking. According to this hypothesis, storm events are recorded by coquinas overlain by a hummocky-bedded stratum. The top of this stratum will be burrowed by a reestablished benthic infauna.

Probable storm events have also been recognized in unfossiliferous sandstones of the cratonic Cambrian (Porter and Byers, 1979; Byers and Dott, 1980). The very fine sandstones of the Norwalk Formation (Fig. 1)

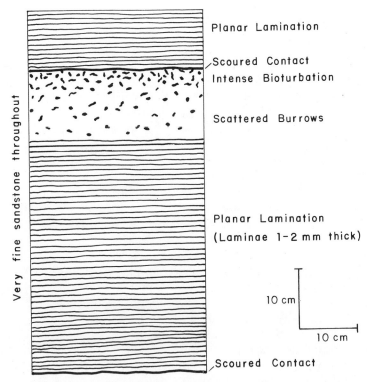

Figure 1. Field sketch of episodically burrowed sandstone in the Upper Cambrian Jordan Formation of Wisconsin. Laminated beds with scoured bases are interpreted to be storm surge deposits that were bioturbated by a reestablished benthic infauna.

display laminated-to-burrowed sequences similar in scale to those described by Howard. The Cambrian deposits cannot be related to a prograding clastic shoreline, unlike the Cretaceous or Recent examples Howard documented. Apparently, storm scour and deposition affected a broad subtidal shelf area, a low-gradient "clastic bank" that covered the central craton during the Late Cambrian.

In all the examples cited, the interpretation of episodicity of deposition comes from recognition of the asymmetry of beds with a sharp basal contact and gradational upper zone: "instants" of dominance by physical processes followed by longer periods of "normalcy," dominated by biological processes.

2.3. Substrate Consistency

Biogenic structures can provide direct evidence of the physical properties of the substrate. Rhoads (1970) showed that water content in fluid muds controlled the way in which the sediment deformed under stress by a burrower. The distinct burrows formed in a low-water, plastic mud are frequently seen in modern sediments and preserved in ancient rocks. High-water, thixotropic muds contain few distinct burrows; instead, both modern and ancient sediments appear swirled or mottled.

Thorough bioturbation in very liquid muds obliterates any preexisting stratification but does not impress a new biogenic architecture on the sediment; such muds may even appear homogeneous. Cryptic bioturbation can be inferred by the randomization of platy particles (organic carbon fragments, detrital micas) and by the lack of fissility in ancient mudrocks (Byers, 1974). There is also some evidence that bioturbation permanently randomizes clay particle fabrics. Strongly parallel clay plates are characteristic of black shales that have not undergone bioturbation; these fabrics are visible in high-magnification scanning electron micrographs taken perpendicular to stratification. My students and I have observed that this parallelism of microfabric is absent in burrow fills in black shale and in thoroughly bioturbated mudstones (Cluff, 1976; Larson, 1977). Apparently, disturbance of the mud microfabric is permanently impressed into fine sediments, and even compaction does not fully realign clay plates (cf. Moon, 1972; Bennett et al., 1977).

A subject in need of further study is the timing and method of burrow filling. A burrow that has been filled passively, by sedimentation, indicates that an actual hole existed on the seafloor, whereas a burrow backfilled by the organism was only "open" in small sequential segments, the volume occupied by the organism as it moved (Byers, 1973). Backfill is indicated where the burrow fill is meniscate (a series of nested crescent shapes) or swirled. Active filling is characteristic of deposit feeders mining the substrate. Backfill is derived from the matrix, with modifications resulting from the feeding process. The fill may show grain segregation (ingested versus rejected particles) (Fig. 2) or removal of the organic fraction to produce a fill lighter in color than the matrix. Backfilled burrows indicate that the sediment was plastic enough to record the passage of the animal (Rhoads, 1970); they do not necessarily mean that the sediment was solid enough to maintain an open tunnel.

Passive fill may be recognized by sediment obviously derived from an overlying stratum and distinct from the surrounding matrix (Fig. 3). An open burrow can be maintained in naturally strong (cohesive or semiindurated) sediment or by burrow linings constructed by the animal (Frey, 1978). Some linings are obvious and preservable, such as chitinous tubes, sand grain walls in mudstones (*Terebellina*), and pelleted sand walls (*Ophiomorpha*). Other linings are formed by mucus impregnation of the matrix to form a solid wall; these are not preserved in ancient burrows, but their existence is inferred where plainly passive fills or

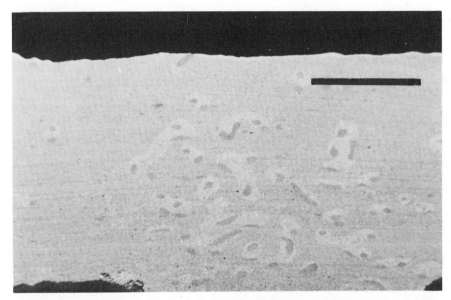

Figure 2. Backfilled burrows in the Fox Hills Formation of South Dakota. Burrow consists of a dark clay-rich fecal backfill surrounded by a halo of silt grains. The deposit-feeding animal apparently exercised size selection, ingesting clay and organic detritus and rejecting coarser particles. Scale bar: 1 cm.

actual open shafts are found in noncohesive sandy sediments (*Skolithos*). Also, such linings may be selectively cemented or diagenetically enhanced (see Section 2.4). The presence of linings indicates a substrate of little or no cohesion or induration.

Passive filling of burrows without linings would indicate sediments strong enough to hold open the burrow shaft or tunnel. Unfortunately, it is not always possible to distinguish between active and passive fill. A case in point is *Chondrites*, which has been interpreted by various authors as both actively filled and passively filled, and as lined and unlined (see Ekdale, 1977, for review). Osgood (1970) employed a model *Chondrites* burrow system to demonstrate experimentally that the burrow could be filled passively. Further research along these lines should be carried out.

True burrows are defined as excavations in unconsolidated sediments; if the substrate is lithified or consolidated, then the excavation is a *boring*. Borings are important in a sedimentologic context as agents of erosion and clast generation and as indicators of substrate hardness and its sequence of development (Warme and McHuron, 1978). Borings have sharp wall boundaries, which may transect sedimentary grains, crystals, or fossils.

Borings produced by macrofauna appeared by the Early Cambrian

Figure 3. Deposit feeder burrows in the New Albany Shale of Kentucky. Burrows extend downward from an intensely bioturbated greenish-gray shale layer into laminated black shale. Scale bar: 1 cm.

(James et al., 1977) and are abundant in hard substrates today. During Paleozoic time borings were mostly simple rod- or club-shaped holes in reefs or hardgrounds. During the Mesozoic and Cenozoic a great diversity and abundance of borings developed (Warme and McHuron, 1978).

Borings can be excavated in substrates of various degrees of consolidation, from stiff mud to igneous rock. Borings in calcareous substrata are especially significant for their abundance and the evidence they provide for submarine lithification. Breaks in sedimentation can lead to the development of an omission surface, dominated by a particular "omission suite" of trace fossils (Bromley, 1975). The omission suite traces typically extend down from a specific surface and may be filled by younger sediments after deposition resumes. Omission suite traces normally differ from traces produced during active sedimentation. An unconsolidated omission surface can become riddled with burrows that transect the older preomission burrows in the underlying sediment.

In carbonate environments, an omission surface is subject to submarine cementation (Purser, 1969); the omission suite on such a hardground surface will comprise borings (Bromley, 1975). Identification of borings is therefore crucial evidence in documenting hardgrounds in ancient carbonates (Byers and Stasko, 1978) (Fig. 4).

2.4. Diagenesis

Frey (1975) pointed out that the subject of diagenesis and mineralization of burrows is an "open field." This assessment remains true today.

The following short discussion focuses on a few observations that have recently been published and some hints of profitable research directions.

Differential diagenesis of burrows or burrow linings has been commented on by several authors. Frey (1970), Simpson (1957), and Byers and Stasko (1978) noted pyritization of burrow linings. Pyrite formation was probably a consequence of bacterial reduction in a local organic-rich concentration, the mucus lining of a burrow. Ekdale (1977) described two forms of "reduction burrows" that occur in deep-sea sediments. Rind burrows possess a whitened burrow wall; solid burrows are white throughout their entire filling. Ekdale proposed that these two forms are stages in a chemical reduction process that begins at the burrow margin and extends inward. Reducing conditions apparently convert Fe^{3+}, which is dominant in the oceanic red clay matrix, to Fe^{2+}. The soluble divalent ion is leached away, leaving a zone colored only by white siliceous or

Figure 4. Borings (*Trypanites*) in a hardground surface in the Platteville Formation of Wisconsin. Boring walls are sharp and mineralized by pyrite. Fill is derived from overlying unit. Scale bar: 1 cm.

calcareous microfossils. The reduction mechanism is not known, although Ekdale suggested that infaunal respiration might increase the CO_2 concentration in interstitial waters. The bacterial reduction of dissolved sulfate to form sulfide has been shown by Berner (1971) to be dependent on an adequate amount of metabolizable organic carbon in a nonoxidizing environment. Sulfide ions then react with detrital hematite to produce pyrite. In the deep-sea red clays, organic carbon content is very low, and the bottom waters and sediments are well-oxygenated. It is not clear how iron could be maintained in a divalent state and removed under these conditions.

Ekdale observed that the development of color contrasts between burrow fill and matrix may take millions of years. Burrow definition increases with age and is best in sediments of the Miocene or older. This slow diagenesis seems distinct from the bacterially mediated production of iron sulfides, which occurs just below the sediment–water interface.

Differential diagenesis of burrows has been studied in terms of how it affects burrow preservation. Shourd and Levin (1976), for example, figured beautifully preserved specimens of *Chondrites* that had been differentially dolomitized or chertified; the burrows were etched out by outcrop weathering of the limestone matrix. Byers and Stasko (1978) suggested that similar selective dolomitization of *Chondrites* was caused by localization of diagenetic fluid flow through burrow systems in impermeable limestone. Recently, Delgado (1980) demonstrated that large Ordovician *Thalassinoides* burrows were selectively dolomitized syndepositionally.

Finally, relating bioturbation to practical (financial) matters, it has been shown in a few published studies that biogenic structures may influence the migration of hydrocarbons. Dawson (1978) studied the mineralogy, grain size, and fabric of an Upper Cretaceous sandstone containing a *Cruziana* assemblage of trace fossils. Burrow fillings and matrix were analyzed separately for porosity. *Stipsellus*, *Planolites*, and small *Thalassinoides* were filled with fine, well-sorted sand; elongate grains paralleled the burrow walls. Initial porosity, determined by point counting cement, was 20%. The matrix was poorly sorted, clayey–micaceous sand with no apparent grain orientation. Matrix porosity was only 5–7%. In contrast, *Rhizocorallium*, *Pseudobilobites*, *Gyrolithes*, and large *Thalassinoides* exhibited no differences between fill and matrix. Dawson suggested that the increased porosity and probable increased permeability in some of the burrows could provide permeable conduits for hydrocarbons. An unbioturbated sandy unit of this sort would not be porous enough to form a reservoir; with the presence of the proper ichnogenera it would be. The important conclusion to be drawn from Dawson's study is that biogenic structures tend to contain their own sedimentary fabric, which may differ from that of the matrix. A formation's physical properties may depend on macroscopic internal structures (burrows) as well as on microscopic structures (average intergranular porosity).

Byers (1980) suggested that primary migration of petroleum from source beds might be facilitated by the presence of a laminated fabric, the laminae providing conduits for hydrocarbons. Conduits would be destroyed by bioturbation. Hence source rock potential might also depend on degree of bioturbation.

3. Paleobathymetry

3.1. Introduction

Sedimentary geologists have searched for years for a "magic wand," which, when knowledgeably waved, would conjure up the depositional environment of the strata under study. More especially, one would like to have a specific indicator of water depth in the ancient environment. When Seilacher, in a series of papers (1958, 1964, 1967, 1978), proposed that trace fossils might serve as such bathymetric indicators, the utility and elegance of his concept combined with the intensity of desire for the "magic wand" created a strong bandwagon effect. Strata were interpreted or reinterpreted in light of Seilacher's model, which quickly assumed the status of dogma. Seilacher's ichnofacies (Fig. 5) have been adopted universally. It would be superfluous to list here the large number of authors who have employed Seilacher's method; summaries have recently ap-

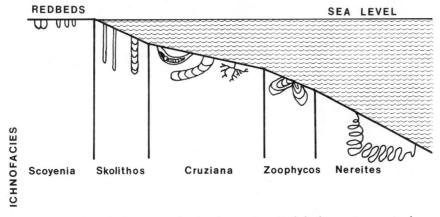

Figure 5. Seilacher's bathymetric ichnofacies zonation. Each bathymetric zone is characterized by a specific assemblage of trace fossils displaying similar behavior in response to depth. After Seilacher (1967).

peared in Frey (1975) and Howard (1978). Because the model has seemed to work, it has been accepted rather uncritically, especially considering its empirical basis. The Seilacher bathymetric model is not actually uniformitarian. It is a generalization taken from the stratigraphic record. As with all such geological models, we may know *that* they work, but not *why* they work, until we ground the model firmly in the Recent. Seilacher has always attempted to do this, looking at modern ocean sediments for traces similar to those in the record and attempting to rationalize the ichnofacies patterns as responses to physicochemical conditions in the ocean. However, nowhere does Seilacher clearly explain the dichotomy between his empirical method and his deductive explanation of it. Indeed, Seilacher's numerous papers on the topic are notable for the manner in which these two aspects are interwoven. As time has passed and Seilacher's model has become the paradigm, his theoretical explanation has also been accepted with little comment. Now, however, it seems time for a critical review of the basis of Seilacher's model. Evidence is mounting, from the study of modern oceans, that some of Seilacher's deductions are not correct. As the theoretical basis of the model has begun to be tested, contradictions have appeared. The new evidence and its implications will be discussed in Sections 3.2 and 3.3.

Before discussing the problems of Seilacher's model, it should be pointed out that when he first proposed it, in the early 1960s, the turbidite concept itself was new. Geologists were still adjusting to the very notion of deepwater sands, and the criteria used to interpret shallow versus deep were not yet codified. Note for instance that Seilacher's (1967) much-copied bathymetric diagram lists only *two* physical sedimentary features, "oscillation ripples" and "turbidites." Clearly this is a first-cut subdivision of the marine realm! Seilacher's ichnofacies are similar approximate categories—he was concerned with recognizing major depth zones. In recent years, resolution of physical criteria of depositional environments has improved (e.g., detailed subdivision of deep-sea fans), but equal progress has not been made in ichnofossil criteria. A major question is whether resolution *can* be improved. If enough effort is expended, will we be able to refine the depth zonation? Do ichnofacies display slight changes in response to bathymetry? Can we subdivide Seilacher's original depth regions of the sea? Or has the model reached the limits of its resolution? In order to answer these questions it is necessary to know why the bathymetric model works in the first place.

3.2. Problems with Seilacher's Model

The basis of Seilacher's bathymetric model was his observation (1958) that trace fossils do not occur in random mixtures. Rather, traces tend to occur in associations or suites. A suite of traces is commonly found with a particular assemblage of sedimentary structures as well; the inference is that a given sedimentary environment is recorded in the sediment in terms of the physical and biological processes acting therein. *Process* is an important word, for it is implicit in Seilacher's model that traces are to be taken mainly as records of behavior, and that behavior is dominated by the physical environment. This mechanistic view colors much trace fossil research (for obvious reasons—rarely do we have any idea what actual animal made a particular trace). Still, I think it is a critical point that the emphasis on behavior tends to relegate the organism to the status of perpetrator; the result is to focus on the distribution of traces and overlook the distribution of organisms. One has a sense that, given the proper stimulus (from the physical environment), it really does not matter much what kinds of animals are present, because surely at least one species will exhibit the requisite behavior. This is a view of organisms as readily moldable and programmable entities. Perhaps they are, but on the other hand a trace may be the work of only one species, and if the animal is not present for one of a variety of biological reasons, the trace will be absent as well. The distribution of traces is inextricably linked to the distribution of animals, and the latter is controlled by more than one factor (i.e., depth).

If we adopt Seilacher's viewpoint and assume that behavior is strongly controlled by physical conditions, then what is the evidence that depth is the major physical factor? It happens that, although this is a plausible assumption, Seilacher has simply asserted the point, not proved it. For example, in his 1967 paper, he says "The [trace fossil] communities are directly or indirectly related to depth no matter what factors caused the differences between them" (p. 414); this statement comes early in the paper, not as a conclusion, and is presented without qualification, evidence, or explanation. Again, a decade later (1978): "Water depth, by itself, is not a major ecological factor, but more important parameters such as turbulence, sedimentation roles and processes, diagenetic processes, productivity and evolutionary processes, are in one way or another depth-related. Therefore, we are justified to arrange different ichnocoenoses in a relative bathymetric sequence" (p. 194). This is not a very specific causality.

The argument could be sharpened a bit by focusing on physical energy level rather than depth *per se*. In Seilacher's original subdivision he was interested in separating very shallow (littoral) from very deep (flysch) environments. At this scale, depth and turbulence are inversely related. However, within depth zones, or along a depth gradient, the negative correlation may not hold. Several authors have pointed out that turbulence can be quite variable within a given depth zone, and that ichnofacies may change in response to turbulence variations involving minimal change in water depth.

Within a generalized shallow marine environment, the strict depth zonation of the *Skolithos* and *Cruziana* facies may not apply. *Skolithos* and other vertical burrows do seem to indicate more energetic conditions, but these do not everywhere correspond to shallower water. Likewise the horizontal burrows and trails of the *Cruziana* facies mean quieter but not necessarily deeper water. Ager and Wallace (1970), in a study of Jurassic traces of the Boulonnais, recognized an offshore-to-onshore gradient from horizontal to vertical burrows, but they also found horizontal *Rhizocorallium* and *Thalassinoides* in a nearshore setting protected by a barrier bar. Fursich (1975) further developed the concept of separating depth and energy in the shallow marine realm. In Upper Jurassic strata of England and Normandy he identified three trace fossil associations, primarily related to hydrodynamic conditions: a high-energy *Diplocraterion* association and low-energy *Rhizocorallium* and *Teichichnus* associations. The *Teichichnus* association was found in quiet offshore sediments and in shallow but protected lagoons. The reappearance of a quiet-water assemblage in a vertical sequence need not mean reversal of a transgressive or regressive cycle. Crimes (1970) observed that the Middle Cambrian strata of south Wales contain alternating beds of *Skolithos*-bearing sandstone and siltstone–mudstone with *Planolites* and *Chondrites*. He ascribed these alternations to temporary changes in environment (within a shallow marine setting) and not to paleogeographic change. Similar conclusions were advanced by Byers (1977b) and by Driese et al. (1981) for Cambrian sandstones in the Upper Mississippi Valley. In short, every *Skolithos–Planolites* couplet does not a transgression make.

Osgood (1970), in his monograph on Cincinnati trace fossils, reported that the prevalent traces belonged to Seilacher's *Cruziana* facies, but he found that the differences between the *Cruziana* and *Nereites* facies are not always clear-cut. He pointed out that Seilacher had not rigorously defined the depth zones, and he cited evidence that the facies might overlap. The *Zoophycos* facies seemed especially suspect to Osgood. *Zoophycos* is very common in shelf-depth rocks of the New York Devonian but had also been found in Recent deep-sea cores (cf. Seilacher, 1967). Osgood and Szmuc (1972) expanded the argument: they reported a widespread occurrence of *Zoophycos* in Lower Mississippian strata in association with oscillation ripples, large cross-laminae sets, and *Lingula*. The inferred shallow conditions prove that the range of *Zoophycos* extends from above wave base, through Seilacher's "intermediate" depths, to abyssal levels. Osgood and Szmuc concluded rather cautiously that "the validity of *Zoophycos* as a reliable depth indicator is thereby somewhat in doubt" (p. 1). Continuing study of DSDP cores has amply confirmed the presence of *Zoophycos* (also *Chondrites, Planolites,* and *Teichichnus*) at abyssal depths (Chamberlain, 1975; Ekdale, 1977, 1978; Ekdale and Berger, 1978). In fact these traces dominate cores of deep-sea sediment, even where the sediment surface is covered by traces of *Nereites*-facies. Ekdale concluded that the cores preserve the last imprint of burrowers, which are the infaunal deposit feeders rather than the surface grazers. Thus the *Nereites* facies may be present in the deep sea but go unpreserved; only in a turbiditic regime are the grazing traces occasionally buried by sandy layers and protected from destruction by bioturbation. The deep sea is actually a combination of the *Nereites* and *Zoophycos* zones, one on the surface and one infaunal, but under conditions of slow sedimentation only the *Zoophycos* assemblage is preserved.

Traces belonging to the *Cruziana* assemblage have also been reported in anomalously deep settings. Crimes (1977) described an Eocene deep-sea fan from northern Spain in which the middle fan facies contains typically a *Nereites* assemblage plus such "shallow-water" traces as *Arenicolites, Diplocraterion, Ophiomorpha, Pelycipodichnus, Rhizocorallium,* and *Thalassinoides.* Crimes interpreted this anomaly in terms of water turbulence; the middle fan environment is characterized by rapidly moving sand-laden currents, a sandy substrate, strong oxygenation, and low levels of organic detritus—similar conditions to those in intertidal and shoreface environments. The main controls on trace fossil distribution in this instance appear to be substrate and water movement, not depth.

Similar anomalous occurrences were reported by Kern (1978), who found ?*Arthrophycus* in the *Nereites* facies of an ancient Alpine deep-sea fan, and by Armentrout (1980), who described *Ophiomorpha* from a series of subdivisions of an Eocene fan. Armentrout estimated a paleobathymetry of 200–1200 m on the basis of associated foraminifera.

Both Seilacher (1967) and Rhoads (1975) have shown how turbulence may directly influence burrow morphology in the shallow marine realm. Deep vertical burrows are characteristic of the shifting substrate and rapid erosion or sedimentation events in a turbulent regime. Rhoads also demonstrated that deep burrowing allows an animal to escape the rigors of salinity or temperature changes, which may occur in a shallow environment.

It is in the switch from vertical to horizontal burrowing that turbulence loses direct control of burrow morphology. Seilacher postulated (1967) that horizontally directed burrows result from deposit feeding, that animals mine the sediment following nutritious seams, and that these seams form by settling of food particles in quiet water. Turbulence thus acts to control burrow form through the agency of feeding behavior. Seilacher expressed this postulate essentially as a conclusion (1967, p. 420), and, once the primacy of the feeding postulate is assumed, then it follows that the intricate patterns of the *Zoophycos* and *Nereites* facies could correspond to ever more careful mining techniques. This implies that food was ever more scarce as water depth increased. Thus the "food paradigm" was born. I review this didactically in order to point out that Seilacher did not use data on food in the modern ocean as the theoretical underpinning of his model.

3.3. Evidence from the Oceans

In a series of recent papers, Kitchell and her co-workers have attempted to test the food paradigm by comparing the distribution of food and trace types in the modern oceans (Kitchell, 1979; Kitchell et al., 1978a,b). The results of these studies cast doubt on the food paradigm and suggest that the distribution of traces is controlled by variables that are not depth-related.

Briefly stated, if the food paradigm were true, traces in the modern ocean should show a bathymetric gradient of increasing organization and complexity of form (Kitchell et al., 1978a). Because complexity of form can be expressed as complexity of commands needed to simulate the trace, it is possible to rank traces on a scale of complexity, from simple scribbles through spirals to meanders. Such a progression might be expected down a bathymetric gradient. In studying thousands of bottom photographs from the Antarctic and South Pacific, Kitchell et al. (1978a) found no depth difference between spirals and meanders, although both trace types occurred in a specific zone (3000–3500 m). This strongly suggests that trace distribution is not being controlled by a food availability gradient. Even more significant is the total lack of either spiral or meander traces in the Arctic Ocean at comparable depths; these trace types are thus shown to be species-specific and not simply depth-dependent.

Kitchell et al. (1978b) made a more extensive comparison between Arctic and Antarctic traces as a test of the hypothesis that productivity controls feeding behavior. The productivities of Arctic and Antarctic waters are approximately known; that of the Arctic is lower by a factor of 10^4. Food should be strongly limiting to benthic grazers in the Arctic sediments and much less so in the Antarctic. The food paradigm would predict intricate traces (*Nereites* facies) in the Arctic and scribble traces in the Antarctic. The actual situation is just opposite—a highly diverse assemblage of *Nereites* facies traces is present in the food-rich Antarctic, whereas only a few scribble traces are to be seen in the impoverished Arctic (Fig. 6). Kitchell concluded that the presence or absence of particular surface-grazing species exerts a greater influence on trace diversity than depth or nutrient supply.

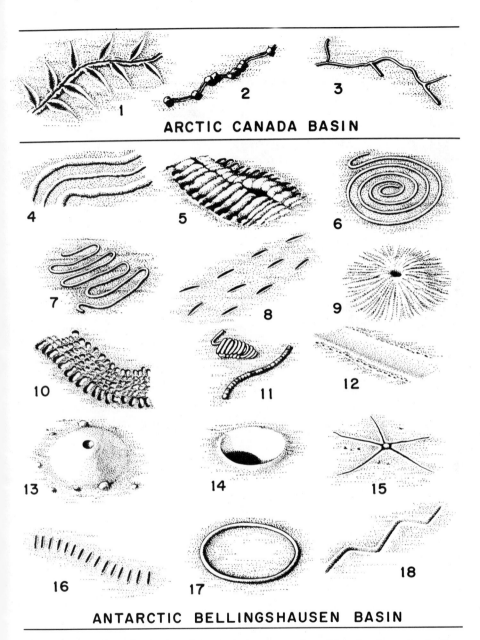

Figure 6. Traces from the Arctic and Antarctic deep-ocean floor. Diversity of complex feeding trails is much higher in the Antarctic in spite of the relatively larger food supply. 1, Pinnate; 2, groove–mound; 3, ridge; 4, plow; 5, crenulated plow; 6, spiral; 7, meander; 8, decapod; 9, rosette; 10, tread, 1,2; 11, fecal; 12, tread, 3; 13, cone; 14, crater; 15, star; 16, railroad; 17, circular ridge; 18, groove. Diagram and descriptive names from Kitchell *et al.* (1978b).

Kitchell's 1979 paper is a theoretical approach to the food paradigm; trace fossil morphology is discussed in terms of optimal foraging theory. The initial hypothesis reads thus: If trace morphology is controlled by feeding behavior and if food is assumed to decrease with increasing depth, then natural selection should produce optimal foraging. Feeding patterns should maximize coverage for energy expended. Optimum foraging implies nonrandom motion—down the bathymetric gradient, traces should show less and less random movement. Kitchell used various components of trace shape to quantify random motion in 302 traces (selected from 4806 deep-sea photos). Principal-components analysis extracted a single measure of randomness for each trace. This measure was plotted against depth (Fig. 7) and found to have a nonsignificant correlation (r = 0.26). In other words, at depths from about 600 m to 6000 m traces do not

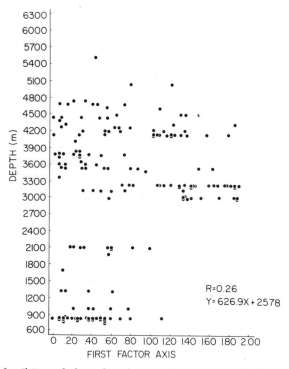

Figure 7. Trace fossil "morphological randomness" versus water depth. The horizontal axis is a principal-components analysis measure of randomness, which decreases to the right. Maximum randomness is exemplified by pinnate and plow traces (cf. Fig. 6), at values less than 0.8. Minimum randomness is shown by spirals and meanders, at values of 0.8–1.9. Multiple circles represent more than one score of the same value. No significant correlation is present between water depth and degree of randomness. From Kitchell (1979).

become less random (more efficient) with depth. Clearly, trace shape and the various ichnofacies are not controlled simply by depth.

If decreasing food supply is not the whole story, what is? Kitchell (1979) suggested two possibilities, control of trace morphology by patchy distribution of food and control by predation risk. The latter seems more plausible: Although spirals and meanders might be efficient ways to mine a rich food patch, Kitchell noted that the deep-sea food supply (for deposit feeders, at least) is probably homogeneously distributed. Given a uniform food distribution, an animal might just as well forage in a straight line; in terms of efficiency alone, a straight line and a spiral are equivalent. The spiral's advantage may be that it minimizes the risk of encountering a predator. The prevalence of spirals and meanders in the Antarctic, known to have a high population density, supports the hypothesis that trace-makers are attempting to maximize food encounter per unit length of travel and minimize total area covered. The overall import of Kitchell's work is that it casts doubt on the simple process–response aspect of the food paradigm and suggests that biological interactions are important in controlling trace morphology and distribution. In a word, the animal has been returned.

3.4. The Graphoglyptid Controversy

Another major problem in interpreting ancient bathymetry is that posed by the graphoglyptid traces (*Paleodictyon* and similar forms). These extremely regular burrow systems (usually networks of hexagons) are characteristic of many ancient flysch sequences and are a component of Seilacher's *Nereites* facies, but they are unknown (with one exception—see below) in the modern deep sea. Besides this enigma, the graphoglyptids do not fit the optimal foraging hypothesis implied in the food paradigm. Also, it is not even certain whether they were formed infaunally in muds and later filled by sand or formed by burrowing along a sand–mud interface. This latter point is actually most critical, for arguments regarding the food paradigm ultimately hinge on whether the graphoglyptids were pre- or postdepositional. The controversy is another instance of the tendency of trace fossil theorizing to outrace the evidence.

Postdepositional burrows along sand–mud interfaces imply deposit feeding, with the animal mining out a food-rich seam. Therefore graphoglyptid morphology should be controlled by the same factors as other *Nereites* traces, also thought to be feeding trails. Unfortunately, the open network of *Paleodictyon* and its allies is not very efficient. Large areas within the net are left unmined, so the burrow length/total area ratio is small, compared to that of spiral and meander burrows. Why should the

Paleodictyon animal go to the trouble of constructing such elaborate and perfect nets if they are an inefficient method of deposit feeding? Seilacher (1974, 1977a,b) recognized this paradox and sought to resolve it by postulating that graphoglyptids were not deposit feeding burrows at all, but open tunnels in mud, in which animals survived by microbial cultivation ("mushroom gardens"). The nets are not optimal expressions of foraging behavior because no foraging is occurring. Seilacher's hypothesis requires that the nets be open tunnels in muds, preserved as trace fossils by infillings of sand—in other words, the burrows must be predepositional, relative to the overlying sandy bed.

Are the actual burrows pre- or postdepositional? Ksiazkiewicz (1970, 1977) and Roniewicz and Pienkowski (1977), in their studies of the Carpathian flysch, concluded that *Paleodictyon* and some related graphoglyptids were produced by deposit feeders burrowing along sand–mud interfaces (postdepositional burrows). The evidence for this origin is the rare preservation of *Paleodictyon* nets covering flutes or other traces thought to be postdepositional and the fact that the basal parts of sand units are apparently reworked by burrows (Ksiazkiewicz, 1970, Plates 4h,q). Roniewicz and Pienkowski (1977) provided no photographic evidence, but simply asserted a postdepositional origin. Seilacher (1977a) took the opposite view and claimed that the burrows were predepositional tunnels in mud, which were scoured and filled by the overlying sandy bed. Seilacher provided little evidence except to cite instances in which burrow nets are cut by flutes. He published many drawings of reconstructed burrow tunnel systems; these and the accompanying text appear convincing, but fundamentally the burrow origin is simply not well-documented from the fossil record. One wishes for more photographs of traces showing their actual relationship to sole marks. Until the question of timing and preservation is settled, all the theoretical superstructure developed thereupon must be regarded cautiously.

For network burrows to be microbial farms, openings to the sediment–water interface would be essential for the introduction of organic detritus. Seilacher has found instances of vertical shafts leading upward from burrow networks, but it is impossible to prove whether these were truly openings. If the tunnels were predepositional then they can only be preserved where current scour cut down to the level of the network and filled the tunnels with sand, thus obliterating the tunnel openings to the original mud surface. Catch-22 is operating here: Without the scour-and-fill, the burrow is not preserved; with scour-and-fill, the crucial upper parts are destroyed. In his descriptions, Seilacher was forced to rely on plausibility and analogy in postulating shaft openings (1977a, e.g., uniramous burrows, p. 303; *Helicolithus*, p. 306; *Punctoraphe*, p. 307; *Uro-*

helminthoida, p. 307; *Hormosiroidea*, p. 309; *Oscillorhaphe*, p. 312; *Tuapseichnium*, p. 314).

The aforementioned absence of graphoglyptids from modern deep-sea bottom photos is not surprising if they were infaunally produced, whether pre- or postdepositionally. Ekdale (1980) recently reported the first instance of network burrows from the modern ocean: *Spiroraphe*, *Cosmoraphe*, and *Paleodictyon* were recovered in the tops of the cores. The burrows were open tunnels in soft sediment, a few millimeters below the original sediment–water interface. Ironically, the burrows were visible because the topmost millimeters of sediment had been washed away during core recovery; vertical shafts, if they existed, would have been eroded off. However, the presence of graphoglyptid tunnels in pelagic muds supports Seilacher's contention that the fossil nets are predepositional, and Ekdale cited the smooth walls and uniform tunnel diameters as evidence against a deposit-feeding origin.

Finally, it must be noted that Seilacher's (1977a) rather elaborate graphoglyptid taxonomy, burrow network reconstructions, and "mushroom garden" hypothesis all arise from the fact that network burrows fail to support the food paradigm. Perhaps the network burrows are simply additional evidence that the paradigm itself needs rethinking.

4. Precambrian Traces

Trace fossils have been utilized with some success in helping to unravel the history of life in the Precambrian. Most work has focused on two questions:

1. How do traces relate to the traditional Precambrian–Cambrian boundary based on body fossils? (Can we use traces to define the boundary in the absence of skeletal faunas?)
2. What is the oldest trace fossil, and what does it tell us about the evolution of the metazoans?

4.1. The Precambrian–Cambrian Boundary

4.1.1. Evolution of the Concept

As with so many important aspects of ichnology, the concept of using traces to define the basal Cambrian was enunciated by Seilacher (1956), who pointed out that a rapid increase in trace types and abundances took place across the Precambrian–Cambrian boundary. At the time Seilacher made this observation, Precambrian paleontology was steeped in mystery, and the basal Cambrian was regarded as the beginning of the fossil record of any consequence. Today, after the ensuing discoveries of Precambrian microbiota and the Ediacaran-age fossils, the Precambrian–Cambrian boundary is not regarded as such a fundamental break. Its definition is now important for orderly stratigraphic procedure, rather than as a marker of the origin of animal life.

The use of traces to define the basal Cambrian is made complex by several factors. First, the definition of the traditional (body fossil distribution) boundary has undergone revision during the last decade. In several regions of the world, but not in most of North America, the lowest trilobite zones are underlain by skeletal faunas dominated by archaeocyathids, sponges, brachiopods, gastropods, hyolithids, and problematic tubes. Soviet workers have established a new Lower Cambrian stage, the Tommotian, based on this fauna; this stage conformably overlies Upper Precambrian strata on the Siberian craton (Cowie and Glaessner, 1975).

Second, the discovery of the Ediacara fauna of soft-bodied metazoans fortuitiously preserved, and the subsequent recognition of similar fossils elsewhere, has extended the range of body fossils down below the traditional Cambrian. This raises the question of whether to define the Cambrian base at the lowest body fossil (the coelenterate-dominated Ediacara-type fauna) or at the lowest shells (the Tommotian fauna) or at the lowest trilobita (traditional in North America); see Stanley (1976) and Durham (1978) for reviews. The "Ediacaran" period (or "Eocambrian" or "Vendian") was approximately 100 m.y. [Basal Tommotian—570 m.y.; Basal Vendian—680 m.y. (Cowie and Glaessner, 1975)] in duration, about the same length as mid-Cretaceous to Recent; in trying to use trace fossils to define the basal Cambrian, it clearly makes a difference whether we push the Cambrian back to include the Ediacara faunas.

Young (1972) reported well-defined traces (e.g., *Cruziana, Rusophycus, Skolithos, Planolites*) occurring below the lowest trilobite zones in the Gog and Miette Groups of the Canadian Cordillera. The lowest trace occurrence, *Didymaulichnus*, perhaps a mollusc trail, lies 2000 m stratigraphically below the basal trilobite bed. Because the trace is rare and not associated with other ichnogenera, Young relegated it to the Late Precambrian. Similar situations were reported for other areas by Glaessner (1969), Banks (1970), Bergstrom (1970), Cowie and Spencer (1970), Webby (1970), and Crimes et al. (1977). The prevalent opinion has been that there is an upward increase in number and complexity of trace types from Precambrian into Cambrian strata (Crimes, 1975). In several of the cases cited, this ichnofaunal transformation takes place a few hundred to a few thousand meters below the overlying Cambrian body fossils, in apparently continuous sequence. The problem is where to draw the line. Because simple burrows and trails are known from rocks at least 100 m.y. older than Tommotian, it is hard to tell whether the so-called upward diversification is a synchronous, cosmopolitan progression (a 100-m.y. time envelope is hardly an "event"). As long as the biostratigraphic criterion is a sequential change or gradual diversification we will have little hope of improving the degree of precision. Alpert (1977) clearly recognized this problem and advanced a proposal to narrow the zone of confusion (see Section 4.1.2).

The third major difficulty is that of facies control. The best biostratigraphic guides are facies-independent. However, it is well understood that most traces are facies-linked—the whole bathymetric utility is built on that postulate. We should not be surprised to find that some burrows and trails persist far back into the Precambrian, ignoring the Cambrian threshold. Crimes et al. (1977) sounded an overdue note of caution in their analysis of Precambrian–Cambrian traces in Spain. Here again there is an upward increase in trace diversity, but it correlates with an upward facies change—a change only in detail, as all the rocks are cross-bedded quartzites and interbedded sand–shale sequences of shallow marine origin. The authors noted that the upward change through hundreds of meters of section could reflect changing environments rather than worldwide evolutionary change.

This latter point can be underlined by examining facies control in rocks known to be Cambrian. For example, the basal cratonic quartz sandstone in the northern midwestern U.S., the Mt. Simon Formation (~65 m thick), directly and conformably underlies the finer-grained Eau Claire Formation, which contains Upper Cambrian trilobites. There is essentially no doubt that the Mt. Simon is also Upper Cambrian, but it contains no diagnostic body fossils. *Cruziana* and allied traces are very common in the shaly Eau Claire, rarer in the upper Mt. Simon, and essentially absent in the very coarse sands of the basal Mt. Simon, where the only common burrow is *Skolithos* (Driese et al., 1981). Here is a situation in which trace diversity increases upward into a zone containing trilobite traces, surmounted by the lowest local body fossils. Because these lowest shells are Upper Cambrian, we may be sure that the upward diversification is simply facies-controlled, not evolutionary; for strata at the Precambrian–Cambrian boundary level that surety is lost. The strong, demonstrable facies linkage of traces makes them risky index fossils; the possibility of facies control needs to be carefully considered before evolutionary explanations for vertical changes are advanced. This is especially true for studies of local extent; if similar trends can be documented on a cosmopolitan basis, and in different sedimentologic settings, then traces may serve as guide fossils.

4.1.2. Alpert's Proposal

Alpert (1977) surveyed the recent literature on trace fossil definition of the Precambrian–Cambrian boundary and noted that the major problem is the existence of a long trace-containing time–stratigraphic interval (the 100-m.y. envelope referred to previously), which is commonly labeled "Cambrian or Precambrian?" This interval lies below beds containing shelly fossils, which are unequivocally Cambrian, and above beds lacking traces or containing rare, simple burrows; these latter strata may also contain tillites and are usually called "Precambrian." From his own work in California and from published reports of trace distributions in Canada, Greenland, Norway, Sweden, Africa, and Australia, Alpert compiled a range chart for trace fossils in the Upper Precambrian–Lower Cambrian (Table I). He found that the upward increase in trace abundance, diversity, and complexity is not gradual but occurs at specific horizons. Also, traces do not appear singly, in random sequence, but as groups. Many ichnogenera make their first appearance either at the levels of the lowest trilobite body fossils (undoubted Cambrian) or below, at the level of the lowest arthropod trace fossil. Alpert argued that trilobite traces (*Rusophycus*, for example) clearly indicate an anthropod grade of evolution; hence *Rusophycus* marks an evolutionary event. If the earliest Cambrian has traditionally (in North America) meant the beginning of trilobites, then the Cambrian should be extended downward to include the lowest evidence of trilobite activity. *Rusophycus, Cruziana, Diplichnites,* and other arthropod scratches would thus serve as index fossils, and non-arthropod traces that appear coincidentally (e.g., *Phycodes, Bergaueria*) could also serve, if arthropod traces were locally absent. In Alpert's scheme, the upper part of the vague "Cambrian or Precambrian?" interval would join the true Cambrian, on the basis of index ichnofossils.

Strata below the lowest Cambrian thus defined would be referred to the Upper Precambrian. These beds may contain trace fossils. Alpert specifically noted that *Skolithos, Planolites, Scolicia,* and various unnamed simple burrows are widespread in Upper Precambrian sediments.

Alpert's proposal has merit in that it does away with a cloudy interval in the time–stratigraphic column. The "Cambrian or Precambrian?" classification was based not on any particular trace type but on the general notion that metazoan life, as manifested by traces, was gradually advancing toward a "Cambrian grade." By keying his chronostratigraphy to the appearance of specific traces, Alpert is employing the same kind of zonation used for body fossils throughout the rest of the Phanerozoic. His basal Cambrian is thus objective and is not dependent on a gradualistic hypothesis regarding the diversification of life.

The major difficulty with Alpert's proposal is that it fails to address

Table I. Trace Fossil Zonation of the Late Precambrian–Early Cambrian[a]

	GROUP 1 Trace fossils indicative of early Cambrian age	GROUP 2 Trace fossils not useful in delineating the basal Cambrian boundary	GROUP 3 Trace fossils known only from the late Precambrian
Dwelling burrows	Diplocraterion Monocraterion Laevicyclus Dolopichnus Bergaueria	Skolithos	
Feeding burrows	Phycodes Rhizocorallium Teichichnus Arthrophycus Syringomorpha Zoophycos Dictyodora	Planolites Unbranched horizontal burrows Unbranched horizontal backfilled burrows	Archaeichnium
Trails or horizontal feeding burrows	Plagiogmus Psammichnites Cochlichnus Belorhaphe Helminthopsis Astropolithon Dactyloidites Oldhamia	Scolicia Curvolithus Didymaulichnus Torrowangea Helminthoidichnites Unbranched horizontal trails	Bunyerichnus Buchholzbrun- nichnus
Trilobite or arthropod traces	Rusophycus Cruziana Diplichnites Protichnites Dimorphichnus Monomorphichnus		

[a] After Alpert (1977).

the problem of the Tommotian. As noted previously, the Tommotian calcareous body fossils form a pretrilobite assemblage in some parts of the world and are recognized as being part of the Lower Cambrian. According to Alpert, the Cambrian commenced with the first trilobite traces. These commonly occur below the lowest trilobite body fossils, but Alpert is not specific on whether these traces antedate or are synchronous with the Tommotian. Apparently facies control returns to haunt us on this point. The best trace fossils are from terrigenous clastic facies, whereas the Tommotian faunas are found in carbonate facies. The basal Cambrian is thus unfortunately defined by two different fossil types, and as yet there is no tie point for correlation (Stanley, 1976). Almost certainly there is time overlap between Alpert's *Rusophycus*-defined basal Cambrian and the Tommotian basal Cambrian, but the exact temporal relationship is unknown. The problem is clearly exemplified by the Nama Group rocks in South West Africa (Germs, 1972, 1973), which contain sprigginid worms, probable Tommotian shelly fossils, and trace fossils that Alpert considers Precambrian! Perhaps careful trace fossil studies in Tommotian rocks will reveal traces common to both carbonate and clastic facies.

In summary, it appears that Seilacher was remarkably prescient in his 1956 paper: Trace fossil assemblages do display an abrupt increase in diversity, including the appearance of arthropod crawling traces, at or just before the beginning of the Cambrian. In fact, the exact placement in time of this diversification depends more on what is considered to mark the Cambrian base than on our knowledge of the trace record itself. Whether or not Alpert's proposal is accepted into formal stratigraphic nomenclature, a radical narrowing of the 100-m.y. time envelope has been accomplished.

A status report chronology:

1. Worldwide Late Precambrian glaciation at 680 m.y. BP (Cowie and Glaessner, 1975).
2. Ediacaran (Eocambrian, Vendian) strata containing rare soft-bodied fossil assemblages and simple burrows: 680–570 m.y. BP (Stanley, 1976).
3. First appearance of arthropod traces (*Rusophycus*) and calcareous faunas of Tommotian Stage (Alpert, 1977; Stanley, 1976): 570–560 m.y. BP. Trace fossils may be slightly older than earliest shelly fossils (see Cowie and Glaessner, 1975, Fig. 13, Australian correlation chart).
4. First trilobite body fossils (Soviet Atdabanian Stage, includes *Fallotaspis* Zone, traditional base of Cambrian in North America).

4.2. The Oldest Trace Fossil

It is a matter of considerable interest to locate the "oldest" body fossil or "oldest" trace; these Precambrian occurrences provide constraints on models of animal phylogeny and rate of diversification. As implied in the previous section, occurrences of both Ediacaran-type body fossils and simple burrows such as *Planolites* seem to be restricted to sedimentary rocks younger than about 680 m.y., postdating the Late Precambrian glaciation. This implies that metazoan animal life was a relatively late development, considering that microfossils and stromatolites are known from Archaean times. The late-development veiw has recently been summarized and championed by Stanley (1976), amplifying Cloud's (1968) critical review of Precambrian animal fossils. Cloud's recognition of many reported occurrences as pseudofossils cleared the decks for action; especially important was his rejection of numerous Proterozoic and Beltian "fossils." No one can doubt the biogenicity of the Ediacaran faunas, but supposed fossils or traces older than about 700 m.y. are regarded with healthy skepticism now that Proterozoic "ancestral" brachiopods, worms, or arthropods have been defrocked.

Traces from strata of Ediacaran age were divided by Alpert (1977) into two groups, those long-ranging forms that persist up into the Cambrian and a few traces that seem to be known only from the Ediacaran. All of these traces are rather simple burrows or surface trails. Vertical burrows have been lumped into the genus *Skolithos* and horizontal burrows into *Planolites*. It is extremely probable that many different biological species have produced such simple structures over geological time. These two ichnogenera are therefore even less useful as index fossils than the average trace.

Of the traces known only from the Ediacaran, perhaps the oldest is *Bunyerichnus*, a bilaterally symmetrical locomotion trace from the Brachina Formation in South Australia (Glaessner, 1969). The trace occurs as a bedding plane structure, apparently produced by an animal without discrete appendages; transverse ridging is reminiscent of traces made by the foot undulations of molluscs (cf. *Scolicia, Plagiogmus*). Webby (1970) questioned Glaessner's interpretation of the fossil and suggested that it might be a body fossil cast of an elongate organism, such as the Ediacaran annelid *Spriggina*. Germs (1973) reported a similar trace, which he named *Buchholzbrunnichnus*, from the Nama Group of South West Africa. A possible sprigginid body fossil (mold) was also found in the unit.

In Australia, *Bunyerichnus* occurs a short distance above the uppermost Late Precambrian tillite and 2000 m stratigraphically below the Pound Quartzite, the stratum containing the classical Ediacaran fauna. Glaessner's published photograph of *Bunyerichnus* shows a structure more than twice as long (12 cm) as the well-known specimens of *Spriggina*; in addition, *Bunyerichnus* has no hint of a head or tail, and the specimen is broken at both ends. I find Glaessner's interpretation plausible—*Bunyerichnus* does look like a trail. In terms of the origin of metazoans, the trace-versus-body-fossil question is moot; *Bunyerichnus* represents a primitive soft-bodied, macroscopic animal that lived nearly 700 m.y. ago.

Supposed traces older than *Bunyerichnus* are more controversial; their organic origin is questionable. Glaessner (1969) interpreted *Brooksella canyonensis* (1100–1300 m.y., from Grand Canyon Group) as a trace, related to *Asterosoma*. *Brooksella* had also been considered a medusoid impression (Bassler, 1941). Cloud (1968) rejected it as any sort of fossil and pronounced it a penecontemporaneous deformation structure. The published photographs of *Brooksella* (Glaessner, 1969) are not entirely convincing, and Glaessner's interpretation of this feature as a trace is both tenous and convoluted. The adjacent bedding surfaces have apparent raindrop imprints (Cloud, 1968), but Glaessner considers these to be algal colonies. Suffice to say that *Brooksella* does not instantly cause the viewer to exclaim, "That is a fossil!", whereas *Bunyerichnus* does.

Older so-called traces are even more problematic. Crimes et al. (1977) gave a range for *Planolites* as Upper Precambrian–Recent*; they cited Walcott's (1899) description of *Planolites* from the Belt Supergroup. Walcott described two traces from Belt rocks, *Helminthoidichnites* and *Planolites*. Cloud (1968) reexamined Walcott's specimens and described *Helminthoidichnites* as "featureless spiral films"; he decided it was not a trace but "probably algal." Walter et al. (1976) studied the microstructure of *Helminoidichnites* and concluded that it was probably a macroscopic eucaryotic algal filament, a body fossil preserved as a carbonized film on bedding planes. Cloud consigned Walcott's *Planolites* specimens to the category of "Algal?". Walter et al. rejected any organic origin for these specimens and considered them to be isolated ripple marks and concretions.

Byers (1976) examined Walcott's collecting localities in Montana and found no trace fossils, nor even any hint of bioturbation in thousands of meters of section. The Belt Supergroup spans an age of 1325–1100 m.y. BP (Obradovich and Peterman, 1968); Walcott's specimens are from the lower beds in the sequence, approximately twice as old as the Upper Precambrian tillites and *Bunyerichnus*. Although *Planolites* is present in rocks of Ediacaran age, the Belt occurrences can be discounted. The lack of bioturbation in Belt shales suggests that even very small infaunal metazoans, such as nematodes, had not yet evolved, much less a large coelomic burrower capable of forming *Planolites*.

A few other trace occurrences somewhat older than the Late Precambrian glaciation have been reported in the literature. Palij (1974) described *Rugoinfractus*, a bilobed trace 12 mm wide, from below the basal tillites in the Ukraine. The sedimentary rocks containing this trace overlie igneous rocks dated at 1100–1400 m.y. BP. This age would be a maximum; the traces are in a sedimentary sequence containing the tillites, suggesting an age closer to 700 m.y. than to the igneous date.

Faul (1950) described structures on bedding planes in the Ajibik

---

* *Planolites* is probably the biggest "wastebasket" ichnogenus extant, known from practically all environments throughout the Phanerozoic; this taxon is the convenient lodging place of any nondescript horizontal burrow to which no more specific name can be attached.

Quartzite of Michigan, which he interpreted as burrows. The published photographs show curving ropes of sand 3–5 mm wide and up to 15 cm long. Faul reported a maximum length of 60 cm. No branching or internal annulation was observable. The structures follow bedding planes but are separated from the bed top by "a thin film of ferruginous material." Faul gave a minimum age of 1200 m.y. for the age of the Ajibik Quartzite.

Similar curving marks and elongate sand tubes on bedding planes of various rippled Precambrian quartzites were interpreted as burrows by Frarey and McLaren (1963) and Hofmann (1967), but as mudcracks by Wheeler and Quinlan (1951) and Barnes and Smith (1964). In 1971, Hofmann published a detailed analysis of this structure, including specimens collected subsequent to his 1967 paper. One of the new specimens clearly indicates a shrinkage crack origin, modified by compaction deformation and removal (by dissolution?) of the clay layer. Although some specimens retain patches or thin films of the mud layer that originally cracked, other bedding planes contain only the curving sand tubes or spindles that filled the cracks. It is the lack of preserved mud that gives these inorganic structures their burrowlike appearance.

Clemmey (1976) reported burrows in sediments approximately 1000 m.y. old from Zambia. Cloud (1978) subsequently rejected Clemmey's interpretation. The structures figured by Clemmey are undoubtedly burrows, but their age is in question. Cloud suggested that the burrows are Recent, produced by termites digging preferentially down strongly weathered layers (in a steep-dipping section) from the lateritic soil zone.

Much older structures were described by Kauffman and Steidtmann (1976) in a talk entitled "Are These the Oldest Metazoan Trace Fossils?". The postulated traces are large, as big as Recent intertidal burrows. They occur in a cross-bedded metaquartzite, 2 billion years old, in Wyoming. Some of the various morphologies Kauffman and Steidtmann illustrated do resemble burrows; others could be soft-sediment deformation features. These structures deserve further study, but for the moment I would certainly retain the question mark in the title.

Durham (1978) recently reviewed the literature on Precambrian metazoans and cited several of the cases described previously to support his hypothesis of a very long history of metazoan evolution. Although geology often depends on one bit of evidence to support a theoretical superstructure (an unconformity visible at a single locality implying regional uplift and erosion; one unambiguous cross-cutting dike that gives the sequence of intrusion of big plutons), we must be careful in applying this method to the Precambrian fossil record. If one absolutely convincing occurrence of Middle Precambrian burrows were found, it would radically constrain phylogenetic models for the metazoa. I am convinced that we should follow Cloud's example and treat supposed traces of great age with skepticism, requiring careful documentation that the structures are (1) biogenic and (2) in situ, and that the rocks are (3) accurately dated.

The recurrent enigma inherent in the reports of trace fossils older than 700 m.y. is why they are so scattered and so few in number. We are dealing with fewer than ten possible occurrences over a time span of more than a billion years (cf. Durham, 1978).

It metazoan burrowers really did evolve in Middle Precambrian time, why don't we see scads of traces? A glance at normal marine Phanerozoic sediments will be rewarded with evidence of trace fossils or general bioturbation. Clemmey's (1976) comment is appropriate to the entire pre-Ediacaran record: "The question remains of why, after being at such an advanced stage of evolution . . . the animals did not continue to flourish. . . ." (p. 578).

As of this writing, it would appear that the oldest unambiguous trace fossil is approximately coincident with the Upper Precambrian glacial episode; metazoans began to leave their spoor about 700 m.y. ago.

References

Ager, D. V., and Wallace, P., 1970, The distribution and significance of trace fossils in the uppermost Jurassic rocks of the Boulonnais, northern France, in: *Trace Fossils* (T. P. Crimes and J. C. Harper, eds.), pp. 1–18, Seel House Press, Liverpool.

Allen, J. R. L., 1965, Late Quaternary Niger Delta, and adjacent areas: Sedimentary environments and lithofacies, *Am. Assoc. Pet. Geol. Bull.* **49**:547–600.

Allen, J. R. L., 1970, *Physical Processes of Sedimentation*, American Elsevier, New York, 248 pp.

Alpert, S. P., 1977, Trace fossils and the basal Cambrian boundary, in: *Trace Fossils 2* (T. P. Crimes and J. C. Harper, eds.), pp. 1–8, Seel House Press, Liverpool.

Armentrout, J. M., 1980, *Ophiomorpha* from upper bathyal Eocene subsea fan facies, northwestern Washington (Abstr.), *Am. Assoc. Pet. Geol. Bull.* **64**:670–671.

Bambach, R. K., Kreisa, R. D., and Whitehurst, H. F., 1978, Storm reworked but untransported faunal assemblages in Paleozoic age shelf environments, *Geol. Soc. Am. Abstr.* **10**:362.

Banks, N. L., 1970, Trace fossils from the late Precambrian and Lower Cambrian of Finnmark, Norway, in: *Trace Fossils* (T. P. Crimes and J. C. Harper, eds.), pp. 19–32, Seel House Press, Liverpool.

Barnes, A. G., and Smith, A. G., 1964, Some markings associated with ripple-marks from the Proterozoic of North America, *Nature* **201**:1018–1019.

Bassler, R. S., 1941, A supposed jellyfish from the Pre-Cambrian of the Grand Canyon, *Proc. U. S. Nat. Mus.* **89**:519–522.

Bennett, R. H., Bryant, W. R., and Keller, G. H., 1977, Clay fabric and geotechnical properties of selected submarine sediment cores from the Mississippi Delta, *Natl. Oceanic Atmos. Adm. U. S. Prof. Pap.* **9**, 86 pp.

Bergstrom, J., 1970, *Rusophycus* as an indicator of early Cambrian age, in: *Trace Fossils* (T. P. Crimes and J. C. Harper, eds.), pp. 35–42, Seel House Press, Liverpool.

Berner, R. A., 1971, *Principles of Chemical Sedimentology*, McGraw-Hill, New York, 240 pp.

Bowen, Z. P., Rhoads, D. C., and McAlester, A. L., 1974, Marine benthic communities in the Upper Devonian of New York, *Lethaia* **7**:93–120.
Bromley, R. G., 1975, Trace fossils at omission surfaces, in: *The Study of Trace Fossils* (R. W. Frey, ed.), pp. 399–428, Springer-Verlag, New York.
Byers, C. W., 1973, Biogenic structures of black shale paleoenvironments, Ph.D. dissertation, Yale University, New Haven, Connecticut, 134 pp.
Byers, C. W., 1974, Shale fissility: Relation to bioturbation, *Sedimentology* **21**:479–484.
Byers, C. W., 1976, Bioturbation and the origin of the metazoans: Evidence from the Belt Supergroup, Montana, *Geology* **4**:565–567.
Byers, C. W., 1977a, Biofacies patterns in euxinic basins: A general model, in: *Deep-water Carbonate Environments* (H. E. Cook and P. Enos, eds.), *Soc. Econ. Paleontol. Mineral Spec. Publ.* **25**:5–17.
Byers, C. W., 1977b, Ichnofacies and paleoenvironments in the Upper Cambrian sandstones of Wisconsin (Abstr.), *J. Paleontol.* **51**(Suppl. to No. 2):5.
Byers, C. W., 1980, Bioturbation as factor in hydrocarbon generation—Example from Mowry Shale (Abstr.), *Am. Assoc. Pet. Geol. Bull.* **64**:685.
Byers, C. W., and Dott, R. H., Jr., 1980, Depositional environments in Upper Cambrian Jordan Sandstone in Wisconsin (Abstr.), *Am. Assoc. Pet. Geol. Bull.* **64**:685.
Byers, C. W., and Stasko, L. E., 1978, Trace fossils and sedimentologic interpretation—McGregor Member of Platteville Formation (Ordovician) of Wisconsin, *J. Sediment Petrol.* **48**:1303–1310.
Chamberlain, C. K., 1975, Trace fossils in DSDP cores of the Pacific, *J. Paleontol.* **49**:1074–1096.
Clemmey, H., 1976, World's oldest animal traces, *Nature* **261**:576–578.
Cloud, P. E., Jr., 1968, Pre-metazoan evolution and the origins of the metazoa, in: *Evolution and Environment* (E. T. Drake, ed.), pp. 1–72, Yale University Press, New Haven.
Cloud, P. E., 1978, World's oldest animal traces, *Nature* **275**:344.
Cluff, R. M., 1976, Paleoecology and depositional environment of the Mowry Shale (Albian), Black Hills region, M.S. thesis, University of Wisconsin, Madison, 104 pp.
Cowie, J. W., and Glaessner, M. F., 1975, The Precambrian–Cambrian boundary: A symposium, *Earth Sci. Rev.* **11**:209–251.
Cowie, J. W., and Spencer, A. M., 1970, Trace fossils from the Late Precambrian/Lower Cambrian of East Greenland, in: *Trace Fossils* (T. P. Crimes and J. C. Harper, eds.), pp. 91–100, Seel House Press, Liverpool.
Crimes, T. P., 1970, A facies analysis of the Cambrian of Wales, *Palaeogeogr. Palaeoclimatol. Palaeoecol.* **7**:113–170.
Crimes, T. P., 1975, The stratigraphical significance of trace fossils, in: *The Study of Trace Fossils* (R. W. Frey, ed.), pp. 109–130, Springer-Verlag, New York.
Crimes, T. P., 1977, Trace fossils of an Eocene deep-sea sand fan, northern Spain, in: *Trace Fossils 2* (T. P. Crimes and J. C. Harper, eds.), pp. 71–90, Seel House Press, Liverpool.
Crimes, T. P., Legg, I., Marcos, A., and Arboleya, M., 1977, ?Late Precambrian–low Lower Cambrian trace fossils from Spain, in: *Trace Fossils 2* (T. P. Crimes and J. C. Harper, eds.), pp. 91–138, Seel House Press, Liverpool.
Dapples, E. C., 1942, The effect of macro-organisms upon near-shore marine sediments, *J. Sediment. Petrol.* **12**:118–126.
Dawson, W. C., 1978, Improvement of sandstone porosity during bioturbation (Abstr.), *Am. Assoc. Pet. Geol. Bull.* **62**:508–509.
Delgado, D. J., 1980, Submarine diagenesis (aragonite dissolution, cementation by calcite, and dolomitization in Ordovician Galena Group, Upper Mississippi Valley (Abstr.), *Am. Assoc. Pet. Geol. Bull.* **64**:697.
Driese, S. G., Byers, C. W., and Dott, R. H., Jr., 1981, Tidal deposition in the basal Upper Cambrian Mt. Simon Formation in Wisconsin, *J. Sediment. Petrol.* **51**:367–381.

Durham, J. W., 1978, The probable metazoan biota of the Precambrian as indicated by the subsequent record, *Annu. Rev. Earth Planet. Sci.* **6**:21–42.

Ekdale, A. A., 1977, Abyssal trace fossils in worldwide Deep Sea Drilling Project cores, in: *Trace Fossils 2* (T. P. Crimes and J. C Harper, eds.), pp. 163–182, Seel House Press, Liverpool.

Ekdale, A. A., 1978, Trace fossils in Leg 42A cores, *Init. Rept. Deep Sea Drilling Project* **XLII** (Part 1):821–827.

Ekdale, A. A., 1980, Graphoglyptid burrows in modern deep-sea sediment, *Science* **207**:304–306.

Ekdale, A. A., and Berger, W. H., 1978, Deep-sea ichnofacies: Modern organism traces on and in pelagic carbonates of the western equatorial Pacific, *Palaeogeogr. Palaeoclimatol. Palaeoecol.* **23**:263–278.

Elders, C. A., 1975, Experimental approaches in neoichnology, in: *The Study of Trace Fossils* (R. W. Frey, ed.), pp. 513–536, Springer-Verlag, New York.

Faul, H., 1950, Fossil burrows from the Precambrian Ajibik Quartzite of Michigan, *J. Paleontol.* **24**:102–106.

Frarey, M. J., and McLaren, D. J., 1963, Possible metazoans from the Early Proterozoic of the Canadian Shield, *Nature* **200**:461–462.

Frey, R. W., 1970, Trace fossils of the Fort Hays limestone member of the Niobrara Chalk (Upper Cretaceous), west-central Kansas *Univ. Kans. Paleontol. Contrib.* **53**, 41 pp.

Frey, R. W. (ed.), 1975, *The Study of Trace Fossils*, Springer-Verlag, New York, 562 pp.

Frey, R. W., 1978, Behavioral and ecological implications of trace fossils, in: *Trace Fossil Concepts* (P. B. Basan, ed.), *Soc. Econ. Paleontol. Mineral. Short Course* **5**:49–75.

Fursich, F. T., 1975, Trace fossils as environmental indicators in the Corallian of England and Normandy, *Lethaia* **8**:151–172.

Germs, G. J. B., 1972, New shelly fossils from Nama Group, South West Africa, *Am. J. Sci.* **272**:752–761.

Germs, G. J. B., 1973, Possible sprigginid worm and a new trace fossil from the Nama Group, South West Africa, *Geology* **1**:69–70.

Glaessner, M. F., 1969, Trace fossils from the Precambrian and basal Cambrian, *Lethaia* **2**:369–393.

Goldring, R., 1964, Trace-fossils and the sedimentary surface in shallow-water marine sediments, in: *Deltaic and Shallow Marine Deposits* (L. M. J. U. van Straaten, ed.), pp. 136–143, Elsevier, New York.

Gould, S. J., 1980, The promise of paleobiology as a nomothetic, evolutionary discipline, *Paleobiology* **6**:96–118.

Häntzschel, W., 1975, *Trace Fossils and Problematica*, Treatise on Invertebrate Paleontology (C. Teichert, ed.), Volume W, Miscellanea (Supplement 1), Kansas University Press, Lawrence, 269 pp.

Hofmann, H. J., 1967, Precambrian fossils(?) near Elliot Lake, Ontario, *Science* **156**:500–504.

Hofmann, H. J., 1971, Precambrian fossils, pseudofossils and problematica in Canada, *Geol. Surv. Can. Bull.* **189**, 146 p.

Howard, J. D., 1972, Trace fossils as criteria for recognizing shorelines in stratigraphic record, in: *Recognition of Ancient Sedimentary Environments* (J. K. Rigby and W. K. Hamblin, eds.), *Soc. Econ. Paleontol. Mineral. Spec. Publ.* **16**:215–225.

Howard, J. D., 1975, The sedimentological significance of trace fossils, in: *The Study of Trace Fossils* (R. W. Frey, ed.), pp. 131–146, Springer-Verlag, New York.

Howard, J. D., 1978, Sedimentology and trace fossils, in: *Trace Fossil Concepts* (P. B. Basan, ed.), *Soc. Econ. Paleontol. Mineral. Short Course* **5**:13–47.

Howard, J. D., and Elders, C. A., 1970, Burrowing patterns of haustoriid amphipods from Sapelo Island, Georgia, in: *Trace Fossils* (T. P. Crimes and J. C. Harper, eds.), pp. 243–262, Seel House Press, Liverpool.

James, N. P., Kobluk, D. R., and Pemberton, S. G., 1977, The oldest macro borers: Lower Cambrian of Labrador, *Science* **197**:980–983.

Kauffman, E. G., and Steidtmann, J. R., 1976, Are these the oldest metazoan trace fossils? *Geol. Soc. Am. Abstr.* **8**:947–948.

Kern, J. P., 1978, Trails from the Vienna Woods, *Palaeogeogr. Palaeoclimatol. Palaeoecol.* **23**:231–262.

Kitchell, J. A., 1979, Deep-sea foraging pathways: An analysis of randomness and resource exploitation, *Paleobiology* **5**:107–125.

Kitchell, J. A., Kitchell, J. K., Clark, D. L., and Dangeard, L., 1978a, Deep-sea foraging behavior: Its bathymetric potential in the fossil record, *Science* **200**:1289–1291.

Kitchell, J. A., Kitchell, J. K., Johnson, G. L., and Hunkins, K. L., 1978b, Abyssal traces and megafauna: Comparison of productivity, diversity and density in the Artic and Antarctic, *Paleobiology* **4**:171–180.

Ksiazkiewicz, M., 1970, Observations on the ichnofauna of the Polish Carpathians, in: *Trace Fossils* (T. P. Crimes and J. C. Harper, eds.), pp. 283–322, Seel House Press, Liverpool.

Ksiazkiewicz, M., 1977, Trace fossils in the flysch of the Polish Carpathians, *Paleontol. Pol.* **36**:1–208.

Kuenen, Ph. H., 1950, *Marine Geology*, John Wiley and Sons, New York, 568 pp.

Larson, D. W., 1977, Paleoenvironment of the Mowry Shale Lower Cretaceous), western and central Wyoming, as determined from biogenic structures, M.S. thesis, University of Wisconsin, Madison, 121 pp.

Moon, C. F., 1972, The microstructure of clay sediments, *Earth Sci. Rev.* **8**:303–321.

Moore, D. G., and Scruton, P. C., 1957, Minor internal structures of some Recent unconsolidated sediments, *Am. Assoc. Pet. Geol. Bull.* **41**:2723–2751.

Obradovich, J. D., and Peterman, Z. E., 1968, Geochronology of the Belt Series, Montana, *Can. J. Earth Sci.* **5**:737–747.

Osgood, R. G., Jr., 1970, Trace fossils of the Cincinnati area, *Palaeontogr. Am.* **VI**(41):281–438.

Osgood, R. G., Jr., and Szmuc, E. J., 1972, The trace fossil *Zoophycos* as an indicator of water depth, *Bull. Am. Paleontol.* **62**:1–22.

Palij, V. M., 1974, On finding of the trace fossil in the Riphean deposits of the Ovruch Ridge, *Rep. Acad. Sci. Ukr. SSR Ser. B. Geol. Geophys. Chem. Biol. Jahrg.* **36**:34–37 [in Ukrainian].

Pettijohn, F. J., 1957, *Sedimentary Rocks*, Harper and Brothers, New York, 718 pp.

Porter, J., and Byers, C. W., 1979, Depositional environments of the Norwalk Member of the Jordan Formation (Upper Cambrian), Southwest Wisconsin, *Geol. Soc. Am. Abstr.* **11**:253

Purser, B. H., 1969, Syn-sedimentary marine lithification of Middle Jurassic limestones of the Paris Basin, *Sedimentology* **12**:205–230.

Reineck, H. E., 1967, Layered sediments of tidal flats, beaches, and shelf bottoms of the North Sea, in: *Estuaries* (G. H. Lauff, ed.), pp. 191–206, American Association for the Advancement of Science, Washington, D. C.

Rhoads, D. C., 1963, Rates of sediment reworking by *Yoldia limatula* in Buzzard's Bay, Massachusetts and Long Island Sound, *J. Sediment. Petrol.* **33**:723–727.

Rhoads, D. C., 1970, Mass properties, stability, and ecology of marine muds related to burrowing activity, in: *Trace Fossils* (T. P. Crimes and J. C. Harper, eds.), pp. 391–406, Seel House Press, Liverpool.

Rhoads, D. C., 1975, The paleoecologic and environmental significance of trace fossils, in: *The Study of Trace Fossils* (R. W. Frey, ed.), pp. 147–160, Springer-Verlag, New York.

Rhoads, D. C., and Morse, J. W., 1971, Evolutionary and ecologic significance of oxygen-deficient marine basins, *Lethaia* **4**:413–428.

Roniewicz, P., and Pienkowski, G., 1977, Trace fossils of the Podhale flysch basin, in: *Trace Fossils 2* (T. P. Crimes and J. C. Harper, eds.), pp. 273–288, Seel House Press, Liverpool.

Seilacher, A., 1956, Der Beginn des Kambriums als biologische Wende, *Neues Jahrb. Geol. Palaeontol. Abh.* **103**:155–180.
Seilacher, A., 1958, Zur okologischen Charakteristik von Flysch und Molasse, *Eclogae Geol. Helv.* **51**:1062–1078.
Seilacher, A., 1964, Biogenic sedimentary structures, in: *Approaches to Paleoecology* (J. Imbrie and N. Newell, eds.), pp. 296–316, John Wiley and Sons, New York.
Seilacher, A., 1967, Bathymetry of trace fossils, *Mar. Geol.* **5**:413–428.
Seilacher, A., 1974, Flysch trace fossils: Evolution of behavioural diversity in the deep-sea, *Neues Jahrb. Geol. Palaeontol. Monatsh.* **4**:233–245.
Seilacher, A., 1977a, Pattern analysis of *Paleodictyon* and related trace fossils, in: *Trace Fossils 2* (T. P. Crimes and J. C. Harper, eds.), pp. 289–334, Seel House Press, Liverpool.
Seilacher, A., 1977b, Evolution of trace fossil communities, in: *Patterns of Evolution* (A. Hallam, ed.), pp. 359–376, Elsevier, New York.
Seilacher, A., 1978, Use of trace fossil assemblages for recognizing depositional environments, in: *Trace Fossil Concepts* (P. B. Basan, ed.), *Soc. Econ. Paleontol. Mineral. Short Course* **5**:185–201.
Shourd, M. L., and Levin, H. L., 1976, *Chondrites* in the upper Plattin Subgroup (Middle Ordovician) of eastern Missouri, *J. Paleontol.* **50**:260–268.
Simpson, S., 1957, On the trace fossil *Chondrites*, *Q. J. Geol. Soc. London* **112**:475–499.
Stanley, S. M., 1976, Fossil data and the Precambrian–Cambrian evolutionary transition, *Am. J. Sci.* **276**:56–76.
Twenhofel, W. H., 1939, Environments of origin of black shales, *Am. Assoc. Pet. Geol. Bull.* **23**:1178–1198.
Van Straaten, L. M. J. U., 1959, Minor structures of some Recent littoral and neritic sediments, *Geol. Mijnbouw* **21**:197–216.
Walcott, C. D., 1899, Pre-Cambrian fossiliferous formations, *Geol. Soc. Am. Bull.* **10**:199–244.
Walter, M. R., Oehler, J. H., and Oehler, D. Z., 1976, Megascopic algae 1,300 million years old from the Belt Supergroup, Montana: A reinterpretation of Walcott's *Helminthoidichnites*, *J. Paleontol.* **50**:872–881.
Warme, J. E., and McHuron, E. J., 1978, Marine borers: Trace fossils and geological significance, in: *Trace Fossil Concepts* (P. B. Basan, ed.), *Soc. Econ. Paleontol. Mineral. Short Course* **5**:77–131.
Webby, B. D., 1970, Late Precambrian trace fossils from New South Wales, *Lethaia* **3**:79–109.
Wheeler, H. E., and Quinlan, J. J., 1951, Pre-cambrian sinuous mud cracks from Idaho and Montana, *J. Sediment. Petrol.* **21**:141–146.
Young, F. G., 1972, Early Cambrian and older trace fossils from the Southern Cordillera of Canada, *Can. J. Earth Sci.* **9**:1–17.

Chapter 6
Geological Significance of Aquatic Nonmarine Trace Fossils

MICHAEL J. S. TEVESZ and PETER L. McCALL

1. Introduction 257
2. Identification of Traces 259
3. Trophic Level Reconstruction 270
4. Paleoenvironmental Reconstruction 271
   4.1. Environment of Deposition 271
   4.2. Vectorial Features 273
5. Discussion 277
   5.1. Usefulness of Systematics Information and Suggestions for Future Work 277
   5.2. Usefulness of Paleoenvironmental Information and Suggestions for Future Work 278
6. Concluding Remarks 280
   References 281

1. Introduction

In Chapters 3 and 4 it was shown that benthic invertebrates interact with modern fluvial and lacustrine sediments and, through these interactions, alter the physical, chemical, and biological properties of sediments. In this chapter, the literature documenting invertebrate–sediment interactions in ancient fluvial, lacustrine, and associated terrestrial environments is reviewed in order to show that invertebrate activities influenced benthic processes and properties in these environments in the past. The literature concerning invertebrate traces of nonmarine origin is scant and scattered and has not been the subject of recent review. Curran (1980) recently emphasized the need for a review of this kind. We will attempt to show that aquatic nonmarine trace fossils are abundant and widely distributed (Table I) and are useful for a variety of geological purposes.

MICHAEL J. S. TEVESZ • Department of Geological Sciences, Cleveland State University, Cleveland, Ohio 44115. PETER L. McCALL • Department of Geological Sciences, Case Western Reserve University, Cleveland, Ohio 44106.

Further research on these traces will probably improve the resolution of paleoecological and paleoenvironmental reconstructions involving nonmarine rocks and increase the understanding of the origin and history of freshwater life.

Table I. Examples of the Stratigraphic and Geographic Extent of Aquatic Nonmarine and Associated Trace Fossils

Geologic Age	North America	Europe	Other
Pleistocene	Tarr (1935) Banerjee (1973) Ashley (1975)	Gibbard and Stuart (1974)	
Pliocene	Link and Osborne (1978)		
Miocene	Toots (1967) K. O. Stanley and Fagerstrom (1974)	Riding (1979)	
Oligocene	Edwards (1975)	Daley (1968)	
Eocene	Picard and High (1972a) Ryder et al. (1976) (Early Tertiary) Moussa (1966, 1968, 1970) Brown (1934) Surdam and Wolfbauer (1975) Eugster and Hardie (1975) Peterson (1976)	Truc (1978)	
Paleocene	Gilliland and LaRocque (1952)		
Cretaceous	Hubert et al. (1972) Siemers (1970, 1971)		
Jurassic			Goldberg and Friedman (1974) (Israel)
Triassic	Reinemund (1955) Sanders (1968) Van Houten (1964) Klein (1962a)	Klein (1962b)	Turner (1978) (South Africa) Bromley and Asgaard (1972, 1979) (Greenland) Webby (1970) (Australia)

(Continued)

Table I. (Continued)

Geologic Age	North America	Europe	Other
Permian	Hanley et al. (1971) Olson and Bolles (1975) White (1929)		Barrett (1965) (Antarctica) Van Dijk et al. (1978) (Permo-Triassic, South Africa) Savage (1971) (Late Carboniferous–Early Permian, South Africa)
Carboniferous	Mutch (1968) Hesse and Reading (1978) D. J. Stanley (1968) Belt (1968)	R. E. Elliott (1968) T. Elliott (1976) Seilacher (1963) Eagar (1974) Chisholm (1970)	Rattigan (1967) (Australia) Glaessner (1957) (Australia)
Devonian	Allen (1970) Cant and Walker (1976) Allen and Friend (1968) Berg (1977) Miller (1979)	Allen (1964, 1970) Rayner (1963) Read and Johnson (1967) Trewin (1976)	Conolly (1965) (Australia) Friend (1965) (Spitzbergen) Webby (1968) (Antarctica)
Cambro-Ordovician			Selley (1972) (Jordan) Selley (1970) ("Lower Paleozoic, Jordan")

2. Identification of Traces

One major goal of the study of aquatic nonmarine trace fossils is to classify a trace in its proper ichnotaxon and then determine the identity of the trace-making organism. This exercise is important because it often provides information on the taxonomic compostion and richness of the fauna, both of which may be useful tools for inferring benthic processes and properties.

Table II shows that nonmarine trace fossils made by invertebrates are taxonomically diverse and environmentally widespread. Common tracemakers in these environments include molluscs, insect larvae and adults, annelids, and nematodes. The trace-makers were identified by different means. One commonly employed method was for workers to use direct observations of the trace-making activities of modern organisms. For ex-

Table II. Examples of Taxonomic, Lithologic, and Paleoenvironmental Diversity of Aquatic Nonmarine and Associated Trace Fossils

Author(s)	Ichnogenus	Trace-maker	Lithology	Paleoenvironment
Allen and Friend (1968)	cf. *Chondrites*	"Root marks" (Barrell, 1913, p. 461), "worm burrows" (Dyson, 1963, p. 25)	Sandstone	Deltaic
	—	Arthropod tracks	—	Deltaic
Belt (1968)	cf. *Cruziana*	Arthropods larger than ostracods and conchostracans	—	—
Berg (1977)	—	*Archanodon* (Mollusca: Bivalvia)	Sandstone	Fluvial
Bromley and Asgaard (1972)	See Bromley and Asgaard (1979) for discussion of *Cruziana* taxonomy.	Notostracan branchiopods (Arthropoda)	Siltstone	—
	Merostomichnites triassicus Linck	Notostracan branchiopods (Arthropoda)	Siltstone?	—
	Planolites regulosus Reineck	Insect larvae?	Siltstone?	—
Bromley and Asgaard (1979)	*Arenicolites* sp.	Annelida	Sandstone intercalations	Lacustrine
	Pelecypodichnus amygdaloides Seilacher	Bivalves (Mollusca)	Sandstone	Lacustrine
	Skolithos spp.	Terrestrial insect burrows	Sandstone	Dessicated fluvial (environment)
	?*Margaritichnus*	Insects?	Sandstone	—
	Scoyenia gracilis White	Annelids?	Sandstone	Very shallow lacustrine
	Steinichnus carlsbergi Bromley and Asgaard	Terrestrial insect	Sandstone	"Terrestrial suite"

Aquatic Nonmarine Trace Fossils

Reference	Trace fossil	Tracemaker	Lithology	Environment
	Fuersichnus communis Bromley and Asgaard	—	Sandstone	Lacustrine
	Diplichnites triassicus Linck	Arthropod	Sandstone	Aquatic?
	Rusophycus eutendorfensis (Linck)	Notostracan crustaceans (Arthropoda)	Sandstone?	"Aquatic suite"
	Cruziana problematica (Schindewolf)	Arthropod	Sandstone?	"Aquatic suite"
	Cruziana spp.	Arthropod	Sandstone?	"Aquatic suite"
Brown (1934)	Celliforma spirifer Brown	Larval chamber of mining bees	Preserved as calcite with or without greenish clay	Lacustrine
Chisholm (1970)	Planolites	—	Siltstones and sandstones	Nonmarine
	Monocraterion	—	Argillites	Nonmarine
Eagar (1974)	Pelecypodichnus	Carbonicola (Mollusca: Bivalvia)	Sandstone, siltstone, flagstone, shaley flagstones	Nonmarine
	Planolites	—	Shaley flagstones	Nonmarine
	Arenicolites	—	Crossbedded, ripple-marked flagstones	Nonmarine
Eugster and Hardie (1975)	—	Insect larvae burrows	Oil shale facies	Lacustrine
Gibbard and Stuart (1974)	—	Isopod	Laminated silty clays	Proglacial lake
	—	Arthropod	Laminated silty clays	Proglacial lake
	—	Crustacean	Laminated silty clays	Proglacial lake
	—	Gastropod	Laminated silty clays	Proglacial lake
	—	Wormlike animal	Laminated silty clays	Proglacial lake
	—	Insect larvae	Laminated silty clays	Proglacial lake
Gilliland and LaRocque (1952)	Xenohelix	—	Limestone	Lacustrine?

(Continued)

Table II. (Continued)

Author(s)	Ichnogenus	Trace-maker	Lithology	Paleoenvironment
Glaessner (1957)	*Isopodichnus osbornei* Glaessner [but refer to the taxonomic revision of Bromley and Asgaard (1979)]	Arthropod	Shales	Periglacial lake
Hanley et al. (1971)	—	Mole beetles? [but referred to as "mole crickets" by Ratcliffe and Fagerstrom (1980)]	Sandstone	Subenvironment of coastal region? Fluvial subenvironment?
Hesse and Reading (1978)	—	Limulid arthropods	Shale	Lacustrine
Miller (1979)	?*Chondrites*	—	—	Alluvial
	Planolites sp.	—	—	Alluvial
	Skolithos sp.	—	—	Alluvial
	Spirophyton	—	—	Alluvial
Moussa (1970)	—	Nematodes (see also Tarr, 1935)	Limestone	Lacustrine
	—	Diptera larvae	Limestone	Lacustrine
Olson and Bolles (1975)	—	Large crustaceans	Sandstone?	Lacustrine
Peterson (1976)	cf. *Arenicolites*	—	Sandstone	Freshwater delta
	cf. *Thalassinoides*	—	Sandstone	Freshwater delta
	cf. *Bifungites*	—	Sandstone	Freshwater delta
Rattigan (1967)	*Isopodichnus osbornei* Glaessner [but refer to the taxonomic revision of Bromley and Asgaard (1979)]	Arthropods	Silts	Aqueoglacial

Aquatic Nonmarine Trace Fossils

Author	Trace fossil	Producer	Lithology	Environment
Riding (1979)	—	?Hydrobia (gastropod) fecal pellets	Algal cones, in micritic crust	Lacustrine
Savage (1971)	Umfolozia sinuosa Savage	Syncarid or peracarid crustaceans	Very fine-grained varvite rocks	Periglacial lakes
	Diplichnites govenderi Savage	Syncarid or peracarid crustaceans	Very fine-grained varvite rocks	Periglacial lakes
	Diplichnites sp.	Syncarid or peracarid crustaceans	Very fine-grained varvite rocks	Periglacial lakes
	Protichnites spp.	Syncarid or peracarid crustaceans	Very fine-grained varvite rocks	Periglacial lakes
	Gyrochorte	Syncarid or peracarid crustaceans	Very fine-grained varvite rocks	Periglacial lakes
	Isopodichnus	Syncarid or peracarid crustaceans	Very fine-grained varvite rocks	Periglacial lakes
	Kingella natalensis Savage	Syncarid or peracarid crustaceans	Very fine-grained varvite rocks	Periglacial lakes
	Gluckstadtella cooperi Savage	Syncarid or peracarid crustaceans	Very fine-grained varvite rocks	Periglacial lakes
Selley (1970)	Cruziana furcifera d'Orbigny[a]	—	Shale	Fluvial; channel infillings
	Cruziana goldfussi (Roualt)[a]	—	Shale	Fluvial; channel infillings
	Two distinct types of Cruziana	—	Silts, fine sands	Fluvial; channel floor
	cf. Diplichnites[b]	—	Silts, fine sands	Fluvial; channel floor
	cf. Ichnyspica[b]	—	Silts, fine sands	Fluvial; channel floor
	cf. Incisifex[b]	—	Silts, fine sands	Fluvial; channel floor
	cf. Tasmanadia[b]	—	Silts, fine sands	Fluvial; channel floor
	cf. Merostomichnites[b]	—	Silts, fine sands	Fluvial; channel floor
Selley (1972)	Cruziana	—	Sandstone	Abandoned channel

(Continued)

Table II. (Continued)

Author(s)	Ichnogenus	Trace-maker	Lithology	Paleoenvironment
Siemers (1970)	*Planolites*	—	Mudstone	—
D. J. Stanley and Fagerstrom (1974)	—	Beetles	Sand, ash lenses	Braided fluvial
Toots (1967)	*Taenidium*	—	Sandstone	Shallow lacustrine or paludal
Trewin (1976)	*Isopodichnus stromnessi* Trewin	Arthropod	Various positions with respect to sand–mud interface	Temporary lacustrine
Turner (1978)	*Scolicia*	Gastropods or bivalves	Sandstone	Fluvial
Van Dijk *et al.* (1978)	*Scolicia*	Paired arthropod tracks	Siltstones, mudstones	Lacustrine shelf facies
	—	—	Silty sands	Fluvio-lacustrine offlap facies
	—	Arthropod tracks	Silty sands	Fluvio-lacustrine offlap facies
	—	Arthropod tracks	Siltstones	Bay facies
	—	Gastropod trails	—	Playa lake facies
	—	Arthropod tracks	—	Playa lake facies
Webby (1968)	*Skolithos*	—	Sandstone	Shallow marine to littoral, or fluviatile to littoral
	Tigillites	—	Sandstone	Shallow marine to littoral, or fluviatile to littoral
	Cylindricum	—	Sandstone	Shallow marine to littoral, or fluviatile to littoral

Aquatic Nonmarine Trace Fossils

"Scolicia"	—	Sandstone	Shallow marine to littoral, or fluviatile to littoral
Beaconites	—	Sandstone	Shallow marine to littoral, or fluviatile to littoral
Brookvalichnus obliquus Webby	Wormlike organisms or insect larvae	Silty shale	Lacustrine
Scoyenia gracilis White	Worms?	—	Fluviatile?
Walpia hermitensis White	cf. Worms or crustaceans	—	Fluviatile?

Webby (1970)

White (1929)

[a] See Bender (1963, 1968).
[b] See Häntzschel (1962), per instructions of Selley (1970).

ample, in order to determine the kind of organisms that produced the *Cruziana problematica* in Triassic freshwater sediments from Greenland (Fig. 1), Bromley and Asgaard (1972) made *in situ* observations on the trace-making activity of the Recent freshwater notostracan branchiopod *Lepidurus arcticus*. Because of the similarities of the ancient and modern traces, they concluded that the *Cruziana* were made by a notostracan like modern *Lepidurus*.

Moussa (1966, 1968, 1970) published a series of papers in an attempt to identify wavelike trails from Eocene lacustrine limestones from Utah. He was finally able to identify the trace-maker when he made *in situ* observations on flood-deposited sediments still covered by a thin film of water and discovered modern nematodes making traces similar to the fossil trails (Fig. 2).

Brown (1934) provided a detailed description of *Celliforma spirifer*, a cylindrical mold capped by a low detral spiral (average length and width 2.7 × 1.2 cm) from Eocene lacustrine deposits of Wyoming (Fig. 3). After dismissing a few possible explanations of its origin (e.g., bivalve burrows, insect eggs), he demonstrated that the structure was similar to the larval

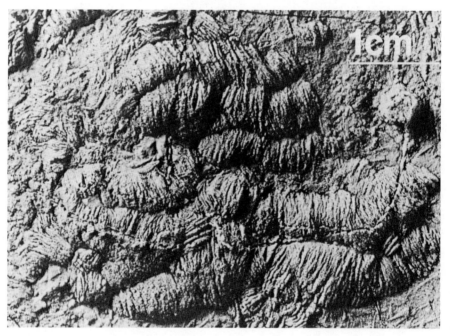

Figure 1. *Cruziana problematica* (Schindewolf, 1921). From Bromley and Asgaard (1979). Reprinted with the permission of Elsevier Scientific Publishing Co. (Amsterdam), R. Bromley, and U. Asgaard.

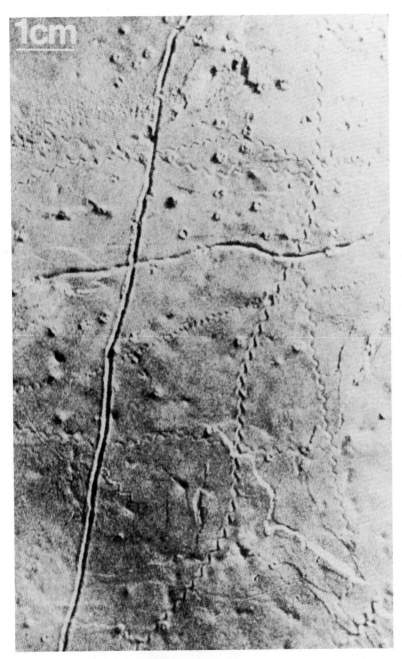

Figure 2. Trails from the Green River Formation. Wavelike trails are nematode trails; insect larvae may have made the irregular-line trails. From Moussa (1970). Reprinted with the permission of the Society of Economic Paleontologists and Mineralogists.

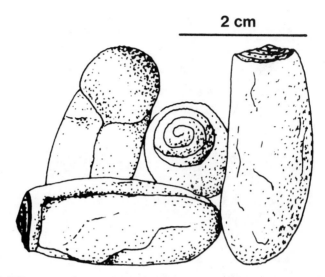

Figure 3. *Celliforma spirifer* Brown: fossil larval chambers of mining bees. Redrawn from Brown (1934).

chamber of modern mining bees. Although mining bees are terrestrial, he cited several possible mechanisms (p. 535: "... faulting, land slipping, subsidence, climatic change, collapse of a sink....") that could have brought the chambers to their site of entombment in freshwater sediments. Hanley et al. (1971), K. O. Stanley and Fagerstrom (1974), and Turner (1978) also made extensive use of *in situ* observations of living analogues to identify trace-makers.

Other writers such as Glaessner (1957), Rattigan (1967), Trewin (1976), and Bromley and Asgaard (1979) made their identifications of trace-making organisms on the basis of available literature. Except for Bromley and Asgaard (1979), the cited literature largely concerned marine forms. Use of this marine-based literature provided identifications of nonmarine organisms only at very high taxonomic levels.

There are several examples of ichnotaxa whose origin is unknown. For example, Gilliland and LaRocque (1952) described a helical "burrow" from the flagstaff Limestone (Paleocene) of Utah that they tentatively assigned to the new genus *Xenohelix* (Fig. 4). Their problem in un-being able to identify the trace-maker was typical: they could not find a modern analogue of the trace (cf. Olson and Bolles, 1975).

A few (mostly old) papers discuss at length the taxonomy of certain nonmarine trace fossils. For example, Schindewolf (1928) and Linck (1942) go into the interpretation of "*Isopodichnus*" in detail. Reineck (1955) provides a sound discussion of *Planolites rugulosus* (= *Scoyenia*

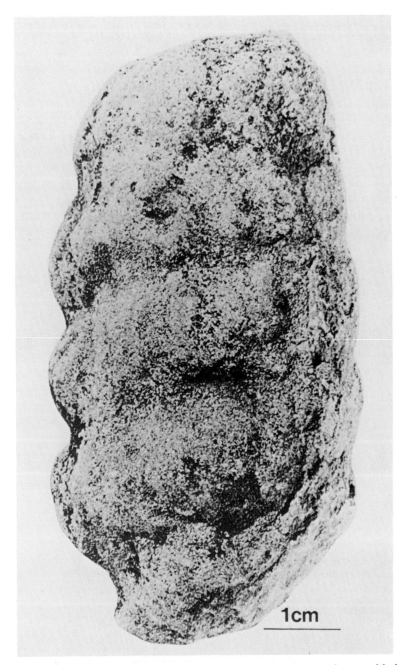

Figure 4. *Xenohelix utahensis* Gilliland and LaRocque 1952 (external view of holotype). From Gilliland and LaRocque (1952). Reprinted with the permission of the Society of Economic Paleontologists and Mineralogists.

gracilis). Examples of more current discussions of the taxonomy of nonmarine traces include Trewin (1976) and Bromley and Asgaard (1979).

3. Trophic Level Reconstruction

Toots (1967) described cylindrical burrows, 11–13 mm in diameter with meniscate internal structure, from the Miocene Sheep Creek formation of Wyoming (Fig. 5). The burrows were associated with fluviatile sediments and were referred to as "*Taenidium*." Toots suggested that the meniscate internal structure of the burrows was the result of backfilling of the burrows by sediment as a result of the deposit-feeding activities of the organism. Several other workers, including Daley (1968), Hanley *et*

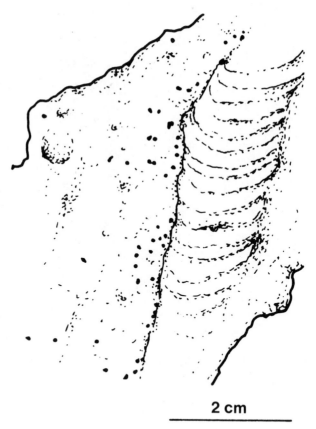

Figure 5. *Taenidium*: burrow of a deposit-feeding organism as indicated by meniscate infilling. Redrawn from Toots (1967).

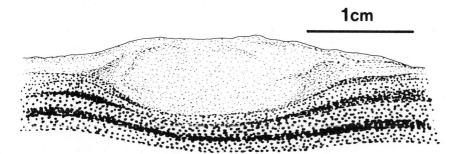

Figure 6. Cross-section of a burrow made by a deposit feeder. Redrawn from Rayner (1963).

al. (1971), K. O. Stanley and Fagerstrom (1974), Edwards (1975), and Bromley and Asgaard (1979), also used meniscate burrow structure as evidence of deposit feeding.

Rayner (1963) employed another aspect of burrow structure for inferring former deposit-feeding activity in freshwater sediments. By analyzing burrow cross-sections in Devonian lacustrine deposits from Scotland, she concluded from grain-size differences (coarse exterior, fine interior) (Fig. 6) that the burrows were filled by sediment that had passed through the gut of the organism.

4. Paleoenvironmental Reconstruction

4.1. Environment of Deposition

Seilacher was one of the first researchers to use trace fossil assemblages as major clues to the nature of the depositional environment. In a 1963 paper, he showed that a trace fossil assemblage, called the *Scoyenia* association, was useful in recognizing nonmarine environments in general. In a 1978 paper he summarized some of his thoughts on this association (p. 188):

> The real problem, however, is the recognition of *non-marine* aquatic environments. In red-bed sequences of different ages we meet a lowly diverse association of rather small trace fossils, for which the name *Scoyenia* has been proposed. Unfortunately, none of the elements of this ichnofacies is by itself a reliable indicator (see Chamberlain, 1975, for examples of modern fresh water traces).
>
> *Scoyenia* is a backstuffed, cylindrical burrow probably produced by insect larvae, but similar burrows (e.g., *Planolites*) may be produced by other organisms, including marine ones. The coffeebean trace *Isopodichnus* and associated biserial tracks can be assigned to non-marine phyllopod shrimp, but they are difficult to distinguish from the burrows and tracks of small trilobites

(Trewin, 1976), which reduces their indicator value except in post-Paleozoic sediments.

The third element is tracks of small limulid and merostome arthropods. These animals are primarily marine, but tolerate low-salinity conditions. The fact that their tracks are mainly found in non-marine deposits is a preservational bias. Varved, silty sediments, which favor the formation and preservation of recognizable undertracks, are deposited more frequently in smaller non-marine basins than in the marine realm.

Nevertheless, the *Scoyenia* association can be used in conjunction with other sedimentological criteria to define environments in the terrestrial realm.

Seilacher's *Scoyenia* association is illustrated here in Fig. 7. Another more general way of recognizing nonmarine deposits was proposed by Frey (1978), who suggested using trace fossil diversity gradients.

Bromley and Asgaard (1979) used trace fossils to help distinguish between aquatic and terrestrial nonmarine deposits. They found that trace fossils formed in sediments exposed to air have three major sets of characteristics that are absent in subaqueously produced forms. In the terrestrial forms, the burrow walls are sharply defined and striated as a result of the organisms' burrowing into a stiff substratum. Second, the burrow filling has a vuggy fabric that would collapse underwater. While this vuggy material may often block the burrow mouth, the lower end of the burrow may contain no sediment. Voids within the burrow may be filled with calcitic cement, and sometimes the burrows are deformed by compaction. Finally, mudcracks are always associated with these traces, and

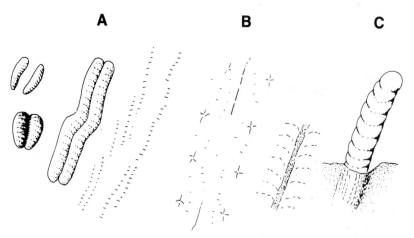

Figure 7. *Scoyenia* ichnofacies. (A) Phyllopod tracks and burrows; (B) Merostome tracks; (C) *Scoyenia*. Redrawn from Seilacher (1963).

burrows commonly cross or follow the fill of the cracks. Where the crack was empty when the animal entered, both the burrow and the crack will contain similar fill. On the same general topic, Fagerstrom and Ratcliffe (1975) observed that the density of Recent insect-produced sediment traces appears to be considerably higher in moist floodplain substrates than in uplands.

Trace fossils have also been used to distinguish among different aquatic nonmarine environments. Some of the most successful of these studies matched up trace fossil assemblages with analogous modern freshwater traces and used the degree of similarity to determine the environment of deposition. Two of the best examples of studies that used this approach are now summarized.

Hanley et al. (1971) used fluviatile trace fossils to determine the environment of deposition of cross-stratified units from the Permian Casper Sandstone, southern Laramie Basin, Wyoming and Colorado. Before this study, most workers had supposed that the cross-stratified beds originated in a nonaquatic environment. The absence of body fossils was an important reason for the uncertainty. Further searching by Hanley's group uncovered trace fossils that had modern analogues on a beach of the Seminole River, Wyoming (Fig. 8). The modern traces were made by mole beetles when low water exposed the sand but still left it moist. Based on this observation and the reasoning that the ridgelike trace fossils probably would not have been preserved in loose sand, Hanley et al. (1971) concluded that the Casper Sandstone had formed in a similarly moist environment.

K. O. Stanley and Fagerstrom (1974) did a study involving nonmarine trace fossils in braided river deposits from the Miocene of Nebraska. They described a trace fossil assemblage that they interpreted as consisting of vertical shelter burrows, horizontal deposit-feeding burrows, bioturbated layers, and vertical passageways between bioturbated layers. Then they compared these Miocene burrows with burrow populations formed by modern beetles and their larvae in Platte River, Nebraska, sediments. They found that these modern burrows were only formed where the river sediment periodically became subaerially exposed. Finally, they used this information from the Platte River as a basis for inferring a similar genesis for the Miocene burrows.

4.2. Vectorial Features

Bromley and Asgaard (1972) observed that the modern notostracan branchiopod *Lepidurus* is rheotactic, and on the basis of this observation

Figure 8. Ribbonlike burrows (A) on beach at Seminole Reservoir, Wyoming; possible modern counterparts for trace fossils pictured in B. From Hanley et al. (1971). Reprinted with the permission of the Society of Economic Paleontologists and Mineralogists.

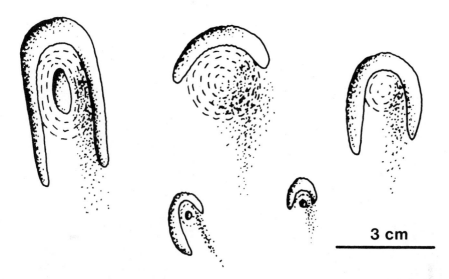

Figure 9. Current crescents developed around the infillings of vertical burrows. Redrawn from Allen and Friend (1968).

suggested that the orientation of notostracan trace fossils (certain *Cruziana*) could serve as clues to paleocurrent direction. Another way of determining paleocurrent direction from nonmarine trace fossils is suggested from observations by Allen and Friend (1968). These authors illustrated current crescents that formed around the resistant infillings of vertical burrows from the Devonian Catskill facies (Fig. 9). Thus paleocurrent direction could be determined by plotting the orientation of the convex portion of the crescent.

Thoms and Berg (1974) inferred that burrow structures from a deltaic paleoenvironment from the Devonian of New York were made by upward escape from burial by the nonmarine bivalve *Archanodon*. They needed to find and study a Recent analogue of the burrow-producing organism in order to use the traces in paleoenvironmental reconstruction, and, after field and laboratory observations, decided that studies of modern *Margaritifera margaritifera* would provide the needed information. Earlier, Berg (1973) had used the bivalve *Siliqua* as an analogue. Because *Siliqua* burrows in an oceanward direction, Berg concluded that burrow curvature and hinge impressions found in the burrows could be useful vectorial features in paleoenvironmental analysis. Based on more studies of Recent analogues, Berg (1977) further advanced the idea of the usefulness of *Archanodon* in determining paleocurrent direction. Eagar (1974) provides good illustrations of Paleozoic nonmarine bivalve burrows (Fig. 10).

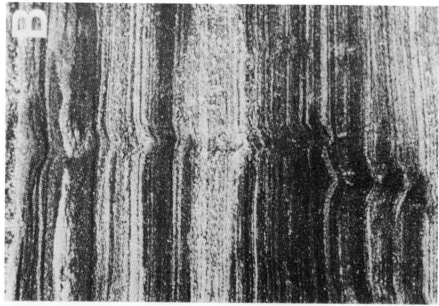

Figure 10. Sectional views of *Pelecypodichnus* burrows. From Eagar (1974). Reprinted with the permission of the Editor, *Lethaia*.

5. Discussion

5.1. Usefulness of Systematics Information and Suggestions for Future Work

Diverse ichnotaxa occur in nonmarine rocks; these fossils represent a record of several major invertebrate phyla. It is evident that the available information on the systematics of both nonmarine trace fossils and trace-makers is limited, and that there is much room for more research. We think that this research needs to be done on two complementary subjects: the trace fossils themselves and modern traces in aquatic nonmarine sediments.

Taxonomic resolution is poor when marine analogues are used to identify nonmarine trace-makers, and identification of trace-makers is greatly impeded when modern analogues of an ichnotaxon are unknown. This all suggests that the ichnology of modern sediments should be much more extensively studied. Such studies have recently begun to appear. The most important include Chamberlain (1975), Wells (1977), and Ratcliffe and Fagerstrom (1980). Chamberlain provides a descriptive catalogue of traces from aquatic nonmarine environments. One important aspect of the article is that it emphasizes the wealth of nonmarine traces that are potentially preservable, and therefore serves as a stimulus for paleoecologists to look for them. The Wells thesis describes freshwater invertebrate traces from the Mississippi alluvial valley and defines relationships between trace fossil assemblages and subenvironments, suggesting their potential usefulness in paleoenvironmental reconstructions. Like Chamberlain, Wells demonstrates that diverse invertebrate traces are produced in modern freshwater sediments. Ratcliffe and Fagerstrom (1980) also emphasize this wealth of potentially fossilizable information, but point out some problems in interpreting it: for example, although the taxonomic diversity of sediment-disturbing spiders and insects on floodplains is great, they found convergence in burrow form among taxonomically dissimilar groups (cf. Wells, 1977). Also, within the same species, ontogenetic stage and environmental factors such as substrate conditioning and weather may further affect burrow form. Therefore, it is difficult to infer taxonomic diversity from trace fossil diversity.

By comparing the literature on modern trace-makers with the literature on their trace fossil counterparts, it is obvious that the great majority of traces identified in modern sediments have yet to be described from the fossil record. Thus there is clearly a need for more careful ichnological study of nonmarine rocks. The trace fossils are there to study. We agree with the sentiments of Howard (1978, p. 15) that aquatic nonmarine ichnology has suffered more from lack of interest than from lack of fossils:

"Traces are not always abundant in non-marine units. However, as more and more geologists include trace fossils on their check-off lists, there are more and more reports of fluvial and lacustrine traces."

Aquatic nonmarine environments are inhabited today by diverse and abundant invertebrates. For example, Chamberlain (1975, p. 433, Table 19.1) lists approximately 30 groups (phyla, classes, or orders) that inhabit freshwater environments. Nevertheless, the body fossil record of nonmarine invertebrates is not very impressive. The only invertebrate body fossils that are usually referred to as being "abundant" or "well preserved" in nonmarine deposits are bivalves, gastropods, and ostracods (see also Picard and High, 1972b, Table 2). But several papers that mentioned the paucity of body fossils also mentioned an abundance of trace fossils (K. O. Stanley and Fagerstrom, 1974; also D. J. Stanley, 1968; Reinemund, 1955; Conolly, 1965; Bromley and Asgaard, 1979). Thus, in several cases, trace fossils are both better preserved and more abundant than body fossils. It would therefore appear that further studies of trace fossils will add much new information concerning the taxonomic nature and distribution of organisms inhabiting aquatic nonmarine environments.

5.2. Usefulness of Paleoenvironmental Information and Suggestions for Future Work

It has been shown previously that some nonmarine trace fossil assemblages are facies-specific (see also Toots, 1975). This facies specificity makes them a very useful tool for geologists because nonmarine, particularly freshwater, depositional environments often exhibit rapid lateral facies changes. Correlation across facies boundaries may be difficult, particularly if the fossils are not diagnostic of environment (Picard and High, 1972a). Because most workers have relied on body fossils for these purposes, it has often been extremely difficult to define particular paleoenvironments within nonmarine deposits because body fossils are frequently rare or poorly preserved. But since trace fossils may be abundant where body fossils are not, further study of these trace fossils may provide an additional useful tool for more rapidly and accurately defining paleoenvironments in nonmarine deposits. While this would obviously help stratigraphers and paleoecologists with their work, there would be benefits for other groups of scientists as well. For example, paleoclimates and climatic changes are often reconstructed based on the extent of ancient lake deposits (Grove et al., 1975). Since certain trace fossil assemblages are well represented in ancient lakes (Table II), these fossils could be used by paleoclimatologists to help them map the extent of these lakes.

The limits of resolution of aquatic nonmarine trace fossils in delimiting depositional environments are suggested by Wells (1977). He found that his Recent Mississippi Valley assemblages could not be used to delimit subenvironments on the scale of point bar, ridge and swale, levee or backswamp. But he felt that the assemblages could at least be used to distinguish marine from nonmarine environments on a regional scale. Fine zonations were not preserved because of the limited temporal and spatial extent of the distribution-controlling factors plus the fluctuation of shorelines. This would lead to a fossil assemblage that would be time-averaged throughout the whole of the alluvial deposits.

Most of the papers that have used aquatic nonmarine trace fossils in environmental reconstructions employ them to define the general environment of deposition, as indicated previously. Far less attention has been paid to them as tools for reconstructing specific environmental parameters, such as mass properties of the bottom, sediment organic content, and pore water chemistry, within a general environment. Judging from studies based in the marine realm (see Chapter 5), it is clear that trace fossils are useful for such purposes. Nevertheless, we found only one paper (Moussa, 1970) that used nonmarine trace fossils in this context. In it, Moussa showed that modern sinusoidal nematode trails form under an extremely limited set of environmental conditions: they form only in moist sediments covered by a film of water that is thinner than the animal. Consequently, the sinusoidal trails he found in Eocene lacustrine deposits precisely delimited these aspects of a portion of the depositional environment. But Moussa is practically alone in his detailed work, and these high-resolution reconstructions therefore are another promising subject for aquatic nonmarine ichnological research.

But do distinctive trace fossil assemblages characterize a sufficiently broad range of environments to be generally useful for the several purposes discussed in this section? Not everyone thinks so. Turner (1978), for one, believes that trace fossils are scarce in fluvial sediments of Pre-Cretaceous age. Nevertheless, we found several references to trace fossils being associated with Paleozoic fluvial deposits (e.g., D. J. Stanley, 1968; Friend, 1955; Cant and Walker, 1976; Allen and Friend, 1968; Allen, 1970). Moreover, Turner says that trace fossils tend to be more common in finer-grained sediments. While this is generally true, it should be pointed out that they also occur in coarser deposits such as sands (e.g., Toots, 1967) and even conglomerates (e.g., Dineley and Williams, 1968) (see also Table II). In addition, while workers like K. O. Stanley and Fagerstrom (1974) and Hanley et al. (1971) showed that some trace fossil assemblages in nonmarine environments formed when the sediments became subaerially exposed, works such as those by Chamberlain (1975), McCall et al. (1979), and Tevesz et al. (1980) showed that extensive trace-making activity by freshwater invertebrates occurs under subaqueous conditions in a variety of environments. These latter observations suggest that the new paleontological information that trace fossils could provide is potentially broad in its environmental scope.

6. Concluding Remarks

It is apparent that trace fossils produced by aquatic nonmarine invertebrates are important sources of information that are still not widely used by geologists. There is room for much further research on trace-fossil and trace-maker systematics and the paleoecological significance of these fossils. It is hoped that this review has served to organize a large part of the interpretive literature for prospective researchers.

Even cursory notice of aquatic nonmarine trace fossils can be extremely useful to geologists. Truc (1978), for instance, studied evaporitic sequences from an ancient saline lake from the Paleogene of the Mormoiron Basin, France, and noted that

> Over the whole of the evaporitic area, well exposed bedding surfaces of both the gypsum and green marls show mud-cracks and are covered with footprints of birds and the trails of animals. The marls and gypsum beds are also extensively burrowed. This can explain the abundance of bird footprints since the birds (probably waders) would have been feeding on the infaunal annelids and insect grubs responsible for the bioturbation. The presence of an infauna might indicate that the formation of gypsum was not the result of excessive salinities.

In this brief statement, Truc employs these trace fossils to infer the taxonomic identity of the organisms making them, reconstruct predator–prey relationships, explain trace fossil abundance, and infer chemical conditions that lead to the formation of gypsum.

Finally, it is our opinion that geologists effecting this research should establish nonmarine ichnological studies on a basis that is separate and distinct from marine ichnology. It is tempting to do otherwise, because there is a great variety of interpretive literature concerning marine ichnology to serve as a guide, and, moreover, there are similar-appearing traces common to both realms.

But major differences between the taxonomic composition and functional associations in marine and nonmarine communities indicate that nonmarine environments merit separate study. For example, insects are perhaps the most common producers of nonmarine traces and trace fossils, but are vastly different from marine arthropods and are virtually absent in the marine realm. Tubificid oligochaetes are another group that commonly produce traces in freshwater sediment but are uncommon tracemakers in the oceans, probably because marine tubificids are as a rule smaller than their freshwater relatives. Thus similar-appearing traces in nonmarine and marine settings may have vastly different "meanings" because of the different life habits and adaptive needs of the organisms producing them.

There may also be differences in the functional associations of organisms that inhabit the two realms which would produce distinctive sedimentary effects in each. For instance, filter-feeding and deposit-feeding organisms have an important impact on sediments. Moreover, the net effect of the activities of deposit-feeding organisms in muddy substrata in the marine realm has been to exclude large suspension-feeding organisms. This concept is termed *trophic group amensalism* (e.g., Rhoads and Young, 1971; Aller and Dodge, 1974). But in nonmarine aquatic environments, the net effect of deposit feeders on suspension feeders is less clear. Oligochaetes, for example, co-occur with large suspension-feeding bivalves in various environments (McCall et al., 1979; Tevesz et al., 1980), including those with muddy substrata. This unexpected association may be caused by freshwater suspension feeders having an expanded niche owing to lower competition/predation pressure than that to which their counterparts in the ocean are subjected. Thus this kind of amensalistic interaction may not be as important in structuring communities in aquatic nonmarine environments (Tevesz and McCall, 1978, 1979). Since deposit feeders and large suspension feeders may regularly co-occur in nonmarine sediments, the combined effects of the benthos on sediments are likely to be qualitatively different from those in marine communities. Using trace fossils, it may be possible to determine the origin and longevity of this association.

ACKNOWLEDGMENTS. We thank J. A. Fagerstrom and R. Bromley for their critical reviews of an earlier version of this chapter.

References

Allen, J. R. L., 1964, Studies in fluviatile sedimentation: Six cyclothems from the Lower Old Red Sandstone, Anglo-Welsh Basin, *Sedimentology* **3**:163–198.

Allen, J. R. L., 1970, Studies in fluviatile sedimentation: A comparison of fining-upwards cyclothems, with special reference to coarse-member composition and interpretation, *J. Sediment. Petrol.* **40**:298–323.

Allen, J. R. L., and Friend, P. F., 1968, Deposition of the Catskill facies, Appalachian region: With notes on some other Old Red Sandstone basins, in: *Late Paleozoic and Mesozoic Continental Sedimentation, Northeastern North America* (G. deV. Klein, ed.), *Geol. Soc. Am. Spec. Pap.* **106**:21–74.

Aller, R. C., and Dodge, R. E., 1974, Animal–sediment relations in a tropical lagoon—Discovery Bay, Jamaica, *J. Mar. Res.* **32**:209–232.

Ashley, G. M., 1975, Rhythmic sedimentation in glacial Lake Hitchcock,

Massachusetts–Connecticut, in: *Glaciofluvial and Glaciolacustrine Sedimentation* (A. V. Jopling and B. C. McDonald, eds.), *Soc. Econ. Paleontol. Mineral. Spec. Publ.* **23**:304–332.

Banerjee, I., 1973, Sedimentology of Pleistocene glacial varves in Ontario, Canada, *Can. Geol. Surv. Bull.* **226,A**:1–44.

Barrett, P. J., 1965, Geology of the area between Axel Heiberg and Shackleton glaciers, Queen Maud Range, Antarctica. Part 2: Beacon Group, *N. Z. J. Geol. Geophys.* **8**:344–370.

Belt, E. S., 1968, Carboniferous continental sedimentation, Atlantic provinces, Canada, in: *Late Paleozoic and Mesozoic Continental Sedimentation, Northeastern North America* (G. deV. Klein, ed.), *Geol. Soc. Am. Spec. Pap.* **106**:127–176.

Bender, F., 1963, Stratigraphie der "Nubischen Sandsteine" in Sud-Jordanien, *Geol. Jahrb.* **81**:237–276.

Bender, F., 1968, *Geologie von Jordanien*, Beiträge zur regionalen Geologie der Erde 7, Borntraeger, Berlin, 230 pp.

Berg, T. M., 1973, Pelecypod burrows in the basal sandstone member of the Catskill Formation, northeastern Pennsylvania, *Geol. Soc. Am. Abstr. Progr.* **5**:137.

Berg, T. M., 1977, Bivalve burrow structures in the Bellvale Sandstone, New Jersey and New York, *Bull. N. J. Acad. Sci.* **22**:1–5.

Bromley, R. B., and Asgaard, U., 1972, Notes on Greenland trace fossils. I. Freshwater *Cruziana* ffrom the Upper Triassic of Jameson Land, East Greenland, *Bull. Grønlands Geol. Unders.* **49**:7–13.

Bromley, R., and Asgaard, U., 1979, Triassic freshwater ichnocoenoses from Carsberg Fjord, East Greenland, *Palaeogeogr. Palaeoclimatol. Palaeoecol.* **28**:39–80.

Brown, R. W., 1934, *Celliforma spirifer*, the fossil larval chamber of mining bees, *Wash. Acad. Sci. J.* **24**:532–539.

Cant, D. J., and Walker, R. G., 1976, Development of a braided-fluvial facies model for the Devonian Battery Point Sandstone, Quebec, *Can. J. Earth Sci.* **13**:102–119.

Chamberlain, C. K., 1975, Recent lebensspuren in nonmarine aquatic environments, in: *The Study of Trace Fossils* (R. W. Frey, ed.), pp. 431–458, Springer-Verlag, Berlin.

Chisholm, J. I., 1970, Lower Carboniferous trace fossils from the Geological Survey boreholes in west Fife (1965–66), *Geol. Surv. G. B. Bull.* **31**:19–35.

Conolly, J. R., 1965, Petrology and origin of the Hervey Group, Upper Devonian, central New South Wales, *J. Geol. Soc. Aust.* **12**:123–166.

Curran, H. A., 1980, Ichnology, *J. Sediment. Petrol.* **50**:312–314.

Daley, B., 1968, Sedimentary structures from a non-marine horizon in the Bembridge marls (Oligocene) of the Isle of Wight, Hampshire, England, *J. Sediment. Petrol.* **38**:114–127.

Dinely, D. L., and Williams, B. P. J., 1968, Sedimentation and paleoecology of the Devonian Escuminac Formation and related strata, Escuminac Bay, Quebec, in: *Late Paleozoic and Mesozoic Continental Sedimentation, Northeastern North America* (G. deV. Klein, ed.), *Geol. Soc. Am. Spec. Pap.* **106**:241–264.

Eagar, R. M. C., 1974, Shell shape of *Carbonicola* in relation to burrowing, *Lethaia* **7**:219–238.

Edwards, P., 1975, Invertebrate burrows in an Oligocene freshwater limestone, *Univ. Wyo. Contrib. Geol.* **14**:7–8.

Elliott, R. E., 1968, Facies, sedimentation successions and cyclothems in productive coal measures in the East Midlands, Great Britain, *Mercian Geol.* **2**:351–372.

Elliott, T., 1976, Upper Carboniferous sedimentary cycles produced by river-dominated, elongate deltas, *J. Geol. Soc. London* **132**:199–208.

Eugster, H. P., and Hardie, L. A., 1975, Sedimentation in an ancient playa–lake complex: The Wilkins Peak Member of the Green River Formation of Wyoming, *Geol. Soc. Am. Bull.* **86**:319–334.

Fagerstrom, J. A., and Ratcliffe, B. C., 1975, Synopsis of Recent burrowing insects as analogues of non-marine trace fossils, *Geol. Soc. Am. Abstr. Progr.* **7**:1070–1071.
Frey, R. W., 1978, Behavioral and ecological implications of trace fossils, in: *Trace Fossil Concepts* (P. B. Basan, ed.), *Soc. Econ. Paleontol. Mineral. Short Course* **5**:49–75.
Friend, P. F., 1965, Fluviatile sedimentary structures in the Wood Bay Series (Devonian) of Spitzbergen, *Sedimentology* **5**:39–68.
Gibbard, P. L., and Stuart, A. J., 1974, Trace fossils from proglacial Lake sediments, *Boreas* **3**(2):69–74.
Gilliland, W. N., and LaRocque, A., 1952, A new *Xenohelix?* from the Paleocene of Utah, *J. Paleontol.* **26**:501–504.
Glaessner, M. F., 1957, Paleozoic arthropod trails from Australia, *Palaeontol. Z.* **31**:103–109.
Goldberg, M., and Friedman, G. M., 1974, Paleoenvironments and paleogeographic evolution of the Jurassic system in southern Israel, *Bull. Geol. Surv. Isr.* **61**, 44 pp.
Grove, A. T., Street, F. A., and Goudie, A. S., 1975, Former lake levels and climatic change on the rift valley of southern Ethiopia, *Geogr. J.* **141**(2):177–202.
Hanley, J. H., Steidtman, J. R., and Toots, H., 1971, Trace fossils from the Casper Sandstone (Permian) southern Laramie Basin, Wyoming and Colorado, *J. Sediment. Petrol.* **41**:1065–1068.
Häntzschel, W., 1962, Trace fossils and problematica, in: *Treatise on Invertebrate Paleontology* (R. C. Moore, ed.), Part W, pp. 177–245, Geological Society of America and University of Kansas Press, Lawrence.
Hesse, R., and Reading, H. G., 1978, Subaqueous clastic fissure eruptions and other examples of sedimentary transposition in the lacustrine Horton Bluff Formation (Mississippian), Nova Scotia, Canada, in: *Modern and Ancient Lake Sediments* (A. Matter and M. E. Tucker, eds.), *Spec. Publ. Int. Assoc. Sedimentol.* **2**:241–257.
Howard, J. D., 1978, Sedimentology and trace fossils, in: *Trace Fossil Concepts* (P. B. Basan, ed.), *Soc. Econ. Paleontol. Mineral. Short Course* **5**:13–47.
Hubert, J. F., Butera, J. G., and Rice, R. F., 1972, Sedimentology of Upper Cretaceous Cody Parkman Delta, southwestern Powder River Basin, Wyoming, *Geol. Soc. Am. Bull.* **83**:1649–1670.
Klein, G. deV., 1962a, Triassic sedimentation, Maritime Provinces, Canada, *Geol. Soc. Am. Bull.* **73**:1127–1146.
Klein, G. deV., 1962b, Sedimentary structures in the Keuper Marl (Upper Triassic), *Geol. Mag.* **99**:137–144.
Linck, O., 1942, Die Spur *Isopodichnus*, *Senckenbergiana* **25**:232–255.
Link, M. H., and Osborne, R. H., 1978, Lacustrine facies in the Pliocene Ridge Basin Group: Ridge Basin, California, in: *Modern and Ancient Lake Sediments* (A. Matter and M. E. Tucker, eds.), *Spec. Publ. Int. Assoc. Sedimentol.* **2**:169–187.
McCall, P. L., Tevesz, M. J. S., and Schwelgien, S. F., 1979, Sediment mixing by *Lampsilis radiata siliquoidea* (Mollusca) from western Lake Erie, *J. Great Lakes Res.* **5**:105–111.
Miller, M. F., 1979, Paleoenvironmental distribution of trace fossils in the Catskill deltaic complex, New York state, *Palaeogeogr. Palaeoclimatol. Palaeoecol.* **28**:117–141.
Moussa, M. T., 1966, Insect tracks? *Entomol. Soc. Am. Bull.* **12**:377.
Moussa, M. T., 1968, Fossil tracks from the Green River Formation (Eocene) near Soldier Summit, Utah, *J. Paleontol.* **42**:1433–1438.
Moussa, M. T., 1970, Nematode fossil trails from the Green River Formation (Eocene) in the Uinta Basin, Utah, *J. Paleontol.* **44**:304–307.
Mutch, T. A., 1968, Pennsylvanian nonmarine sediments of the Narragansett Basin, Massachusetts–Rhode Island, in: *Late Paleozoic and Mesozoic Continental Sedimentation, Northeastern North America* (G. deV. Klein, ed.), *Geol. Soc. Am. Spec. Pap.* **106**:177–209.

Olson, E. C., and Bolles, K., 1975, Permo-Carboniferous freshwater burrows, *Fieldiana Geol.* **33:**271–290.

Peterson, A. R., 1976, Paleoenvironments of the Colton Formation, Colton, Utah, *Brigham Young Univ. Geol. Stud.* **23:**3–35.

Picard, M. D., and High, L. R., Jr., 1972a, Paleoenvironmental reconstructions in an area of rapid facies change, Parachute Creek Member of Green River Formation (Eocene), Uinta Basin, Utah, *Geol. Soc. Am. Bull.* **83:**2689–2708.

Picard, M. D., and High, L. R., Jr., 1972b, Criteria for recognizing lacustrine rocks, in: *Recognition of Ancient Sedimentary Environments* (J. K. Rigby and W. K. Hamblin, eds.), *Soc. Econ. Paleontol. Mineral. Spec. Publ.* **16:**108–145.

Ratcliffe, B. C., and Fagerstrom, J. A., 1980, Invertebrate lebensspuren of Holocene floodplains: Their morphology, origin and paleoecological significance, *J. Paleontol.* **54:**614–630.

Rattigan, J. H., 1967, Deposition, soft sediment and post-consolidation structures in Palaeozoic aqueoglacial sequence, *J. Geol. Soc. Aust.* **14:**5–18.

Rayner, D. H., 1963, The Achanarras Limestone of the Middle Old Red Sandstone, Caithness, Scotland, *Geol. Soc. Yorkshire Proc.* **34:**117–138.

Read, W. A., and Johnson, S. R. H., 1967, The sedimentology of sandstone formations within the Upper Old Red Sandstone and lowest Calciferous Sandstone Measures west of Stirling Scotland, *Scott. J. Geol.* **3:**242–267.

Reineck, H.-E., 1955, Marken, Spuren, und Fahrten in den Waderner Schichten (ro) bei Martinstein/Nahe, *Neues Jahrb. Geol. Palaeontol. (Abh.)* **101:**75–101.

Reinemund, J. A., 1955, Geology of the Deep River coal field, North Carolina, *U. S. Geol. Surv. Prof. Pap.* **246**, 159 pp.

Rhoads, D. C., and Young, D. K., 1971, Animal–sediment relations in Cape Cod Bay, Massachusetts. Part II: Reworking by *Molpadia oolitica* (Holothuroidea), *Mar. Biol.* **11:**255–261.

Riding, R., 1979, Origin and diagenesis of lacustrine algal bioherms at the margin of the Ries crater, Upper Miocene, southern Germany, *Sedimentology* **29:**645–680.

Ryder, R. T., Fouch, T. D., and Elison, J. H., 1976, Early Tertiary sedimentation in the western Uinta Basin, Utah, *Geol. Soc. Am. Bull.* **86:**496–512.

Sanders, J. E., 1968, Stratigraphy and primary sedimentary structures of fine-grained, well-bedded strata, inferred lake deposits, Upper Traissic, central and southern Connecticut, in: *Late Paleozoic and Mesozoic Continental Sedimentation, Northeastern North America* (G. deV. Klein, ed.), *Geol. Soc. Am. Spec. Pap.* **106:**265–305.

Savage, N. M., 1971, A varvite ichnocoenosis from the Syka Series of Natal, *Lethaia* **4:**217–233.

Schindewolf, O. H., 1928, Studien aus dem Marburger Buntsandstein III–VII; IV. *Isopodichnus problematicus* (Schdwf.) im Unteren und Mittleren Buntsandstein, *Senckenbergiana* **10:**27–37.

Seilacher, A., 1963, Lebensspuren und Salinitats-Fazies, *Fortschr. Geol. Rheinland. Westf.* **10:**81–94.

Seilacher, A., 1978, Use of trace fossil assemblages for recognizing depositional environments, in: *Trace Fossil Concepts* (P. B. Basan, ed.), *Soc. Econ. Paleontol. Mineral. Short Course* **5:**185–201.

Selley, R. C., 1970, Ichnology of Paleozoic sandstones in the Southern Desert of Jordan: A study of trace fossils in their sedimentologic context, in: *Trace Fossils* (T. P. Crimes and J. C. Harper, eds.), pp. 477–488, Seel House Press, Liverpool.

Selley, R. C., 1972, Diagnosis of marine and non-marine environments from the Cambro-Ordovician Sandstones of Jordan, *J. Geol. Soc. Lond.* **128:**135–150.

Siemers, C. T., 1970, Facies distribution of trace fossils in a deltaic environmental complex: Upper part of Dakota formation (Upper Cretaceous), central Kansas, *Geol. Soc. Am. Abstr. Progr.* **2:**683–684.

Siemers, C. T., 1971, Deltaic deposits of upper part of Dakota formation (Upper Cretaceous), central Kansas, *Am. Assoc. Petrol. Geol. Bull.* **55**:364.

Stanley, D. J., 1968, Graded bedding–sole marking–greywacke assemblage and related sedimentary structures in some Carboniferous flood deposits, eastern Massachusetts, in: *Late Paleozoic and Mesozoic Continental Sedimentation, Northeastern North America* (G. deV. Klein, ed.), *Geol. Soc. Am. Spec. Pap.* **106**:211–239.

Stanley, K. O., and Fagerstrom, J. A., 1974, Miocene invertebrate trace fossils from a braided river environment, western Nebraska, U.S.A., *Palaeogeogr. Palaeoclimatol. Palaeoecol.* **15**:63–82.

Surdam, R. C., and Wolfbauer, C. A., 1975, Green River Formation, Wyoming: A playa–lake complex, *Bull. Geol. Soc. Am.* **86**:335–345.

Tarr, W. A., 1935, Concretions in the Champlain Formation of the Connecticut River Valley, *Geol. Soc. Am. Bull.* **46**:1493–1533.

Tevesz, M. J. S., and McCall, P. L., 1978, Niche width in freshwater bivalves and its paleoecological implications, *Geol. Soc. Am. Abstr. Progr.* **10**:504.

Tevesz, M. J. S., and McCall, P. L., 1979, Evolution of substratum preference in bivalves, *J. Paleontol.* **53**:112–120.

Tevesz, M. J. S., Soster, F. M., and McCall, P. L., 1980, The effects of size-selective feeding by oligochaetes in the physical properties of river sediments, *J. Sediment. Petrol.* **50**:561–568.

Thoms, R. E., and Berg, T. M., 1974, Comparison of the burrowing habits of a Devonian pelecypod with those of a Recent analogue, *Geol. Soc. Am. Abstr. Progr.* **6**:267.

Toots, H., 1967, Invertebrate burrows in the non-marine Miocene of Wyoming, *Univ. Wyo. Contrib. Geol.* **6**:93–96.

Toots, H., 1975, Distribution of meniscate burrows in non-marine Tertiary sediments of the western U.S., *Univ. Wyo. Contrib. Geol.* **14**:9–10.

Trewin, N. K., 1976, *Isopodichnus* in a trace fossil assemblage from the Old Red Sandstone, *Lethaia* **9**:29–37.

Truc, G., 1978, Lacustrine sedimentation in an evaporitic environment: The Ludian (Palaeogene) of the Mormoiron basin, southeastern France, in: *Modern and Ancient Lake Sediments* (A. Matter and M. E. Tucker, eds.), *Spec. Publ. Int. Assoc. Sedimentol.* **2**:189–203.

Turner, B. R., 1978, Trace fossils from the Upper Triassic fluviatile Molteno Formation of the Karoo (Gondwana) supergroup, Lesotho, *J. Paleontol.* **52**:959–963.

Van Dijk, D. E., Hobday, D. K., and Tankard, A. J., 1978, Permo-Triassic lacustrine deposits in the eastern Karoo Basin, Natal, South Africa, in: *Modern and Ancient Lake Sediments* (A. Matter and M. E. Tucker, eds.), *Spec. Publ. Int. Assoc. Sedimentol.* **2**:225–239.

Van Houten, F. B., 1964, Cyclic lacustrine sedimentation, Upper Triassic Lockatong Formation, central New Jersey and Adjacent Pennsylvania, *Kans. Geol. Surv. Bull.* **169**:497–531.

Webby, B. D., 1968, Devonian trace fossils from the Beacon Group of Antarctica, *N. Z. J. Geol. Geophys.* **11**:1001–1008.

Webby, B. D., 1970, *Brookvalichnus*, a new trace fossil from the Triassic of the Sydney Basin, Australia, in: *Trace Fossils* (T. P. Crimes and J. C. Harper, eds.), pp. 527–530, Seel House Press, Liverpool.

Wells, R. F., 1977, Freshwater invertebrate living traces of the Mississippi alluvial valley near Baton Rouge, Louisiana, M.S. thesis, Louisiana State University, Baton Rouge, 253 pp.

White, C. D., 1929, Flora of the Hermit Shale, Grand Canyon, Arizona, *Carnegie Inst. Washington Publ.* **405**, 221 pp.

IV
Models

Chapter 7
Mathematical Models of Bioturbation

GERALD MATISOFF

1. Introduction 289
   1.1. Processes to be Modeled 289
   1.2. Kinds of Models and Their Objectives 291
2. Particle Transport Models 293
   2.1. Diffusion Models 293
   2.2. Box Models 305
   2.3. Signal Processing Models 309
   2.4. Markov Models 310
3. Fluid Transport Models 313
   3.1. Diffusion–Reaction Models 314
   3.2. Advection Models 322
4. Conclusions 325
   References 327

1. Introduction

The purpose of this chapter is to examine existing mathematical models of important chemical, physical, and biological effects of organisms on sediments. The objectives and nature of the models will be discussed, the mathematical solution techniques will be identified, and the advantages and disadvantages of each type of model will highlighted.

1.1. Processes to be Modeled

Let us consider first a broad and practical subdivision of the effects of organisms on sediments. Sedimentary materials exist in three states—

GERALD MATISOFF • Department of Geological Sciences, Case Western Reserve University, Cleveland, Ohio 44106.

solid particles, pore fluids, and gases—but the vast bulk is comprised of particulates and interstitial water. It is the movement of these latter two constituents by biological agents that most models of animal–sediment relations address, and this is how we will divide our discussion. There are good reasons for this division: particulates and fluids can move in different ways at different rates and in different directions, and are acted on by different organisms, as previous chapters have demonstrated.

The redistribution of sediment particulates by organisms can obscure primary stratigraphic features and create other secondary structures (Rhoads, 1974). Graded bedding, for example, may be either a primary depositional feature or caused by size-selective feeding activities of benthos. Mixing invariably leads to poorer resolution of microfossil dating of deep-sea sediments by lengthening the sediment section in which index fossils are found and by overlapping the horizons of fossils indicative of distinct time intervals (Berger and Heath, 1968). Biological mixing of particles may also affect the chemical diagenesis of sediments (Aller, 1977). For example, in the typical case, fine-grained sediments in marine and freshwater environments are chemically layered, with an oxidizing zone extending a few millimeters or centimeters from the sediment–water interface overlying a much thicker reduced zone. Iron is present in reducing sediments in the solids as ferrous carbonate, phosphate, and sulfides. Organisms may transport particles from a reduced zone to the sediment–water interface and cause the iron in the solid phases to oxidize and precipitate as iron oxides and hydroxides. This refluxing of reduced sediment to the sediment–water interface enhances the sediment oxygen demand and affects the oxygen budget of the environment. An understanding of the effects of bioturbation on particle redistribution is essential to a proper interpretation of these problems, and mathematical models that assist in deciphering the information are highly desirable.

Being able to model approximately the fluid motion (or its effects) is also essential to the understanding of the sediment's chemical and depositional dynamics. The exchanges of interstitial and overlying water and solutes are important processes in many chemical balances (e.g., phosphorus in Lake Erie) and in sediment diagenesis (Aller, 1980b). For example, in the Three-Basin Phosphorus Budget Model for Lake Erie (United States Army Corps of Engineers, 1975) the flux of phosphorus from sediments to the lakewater is slightly greater than the entire external loading of phosphorus to the lake. Thus the postdepositional transfer of phosphorus is thought to be an important part of the Lake Erie phosphorus cycle. Benthos often alter sediment fabric (for example, by creating a fecal pellet layer) and indirectly modify the exchange across the interface owing to diffusion through the pelletal layer (McCall and Fisher, 1980). Some organisms construct rigid tubes in the sediment that allow free exchange of water and solutes from several centimeters deep to the overlying water (irrigation) (Aller, 1980a). Other organisms inject water directly into the sediment or pump water through their habitats during feeding, burrowing, and locomotory activities (McCall et al., 1979; McCaffrey et al., 1980). These biological activities result in an increased exchange of overlying water and solutes with those in the sediments, and mathematical models are essential to interpret properly the varied effects of bioturbation on fluid and solute exchange.

1.2. Kinds of Models and Their Objectives

The term *model* is a very general one and can be applied to any process or mathematical expression capable of giving information about the approximate behavior of a real system. In this way an experiment can be a good model. Similarly, a good understanding of one physical phenomenon and the theory developed for it could be used as a model to simulate other physical processes as a first approximation. For example, heat transfer in fluids may be used to model dispersion in fluids. Defined here, a *mathematical model* will be a mathematical expression or group of equations that can simulate the processes that occur in the real world and that may be used to predict approximately the reaction of the physical system to perturbations or simply to predict its state at future times.

Mathematical models may be classified as either deterministic or probabilistic. A model is *deterministic* if, for each given input to the model, there is a unique output. *Stochastic* or *probabilistic* models assume that changes in the physical state from one condition to another occur instantaneously. They can then be used to predict the current or future state of the system or the likelihood or probability of the existence of different states at some other time. Inputs to a model may be prescribed either as *known functions* or as *random functions*, which can be described only in a statistical or probabilistic sense. In the formulation of complex probabilistic models deterministic models are often used and the randomness is introduced either in the input or somewhere in the system itself, depending upon the nature of the physical problem (Kowal, 1971).

The construction of a mathematical model to aid in the deciphering of mixed sediments is dependent upon the nature of the results desired. In deterministic models, the physical process is represented mathematically and the results of the model compared to field and experimental data to determine if the model output is not inconsistent with the effects of the organism's activities and can approximate them adequately. Diffusion models are of this type. The solid particles are clearly not diffusing, but the random transport of the particulates on a macroscopic scale caused by the actions of organisms may be described by such a mechanism in the appropriate spatial and temporal context. On the other hand, it may be advantageous to know the likelihood of a particle infilling a burrow hole or becoming incorporated into a fecal pellet. Such models usually have an output of "yes or no or a degree of maybe" and are stochastic or probabilistic by their nature. In practice, then, the selection of one type of model over another is dependent upon the type of information available as input data and the type of output desired.

Many models can be classified as either deterministic or probabilistic, depending upon the scale of observation. For this reason, the bioturbation models discussed in this chapter have not been formally labeled as either "deterministic" or "probabilistic." Rather, the classification presented here is a convenient framework for discussing models of similar objectives and formulation. We will consider four classes of particle transport models: diffusion models, box models, signal processing models, and Markov models; and two classes of fluid transport models: diffusion models and advection models.

Diffusion models are by far the most popular and make up the bulk of the literature. These models have great appeal because of the degree of sophistication that can be achieved with them. All of these models assume that the transport may be mathematically described as a diffusional process. This permits the differential equations describing diffusion to be used, to which other effects can then be easily incorporated as additional terms in the equations. In addition to the mixing described by diffusion, processes that have been examined include sedimentation, mixing depth, radioactive decay, directed vertical transport, adsorption, and chemical reactions. These models will be discussed in more detail in Sections 2.1 and 3.1.

Box models, first introduced by Berger and Heath (1968), assume that the upper layer of sediment is mixed rapidly with respect to sedimentation. This is a good assumption in many environments, so that box models remain popular as a first approximation for a bioturbation model. These models are discussed in Section 2.2.

Another model that deserves mention because of its originality and simplicity was proposed by Goreau (1977). His *signal-theory-based model* calculates the output (i.e., ultimate distribution in the sediment) for any given input (distribution in the sediment in the absence of mixing) by a convolution integral once the mixing function is specified. The mixing function mathematically represents the sediment redistribution processes. The model may also be applied to solute transport where the mixing function represents pore water advection caused by sediment redistribution and faunal water pumping. This model is discussed in detail in Sections 2.3 and 3.2.

Recently, Jumars et al. (1981) presented a *Markov model* of particle redistribution. Their probabilistic model is based on different assumptions and can have quite different structures from the other types of models. In general, it assumes no memory of previous events. The likelihood of the occurrence of another event depends only upon the probability distribution within the system immediately before the event occurs. Thus the history of preceding events is contained *only* in the present state of the system, and exact calculations back in time are not possible. On the other hand, the model can generate the probability of all possible future states of the system. This model is discussed in Section 2.4.

2. Particle Transport Models

Redistribution of sediment particles by benthos can occur by a variety of processes. They may be ingested and defecated, advected by flow, cemented together by egested materials, or simply pushed aside.

The mixing may occur on a variety of time scales. For example, stirring in the horizontal or planar dimensions may be very slow, while vertical turbation of particles by advection and dispersion by smearing may be quite rapid (Piper and Marshall, 1969). In this section we will examine existing models of particle redistribution.

2.1. Diffusion Models

The diffusion equation has been frequently applied to the redistribution of solid particles. Clearly, the particles themselves are not diffusing, and neither the actions nor distribution of the organisms in sediments is random. However, the redistribution of solid particles by large numbers of organisms and over a large number of individual transport events may be mathematically described by the diffusion analogy. This model cannot be used for strongly directional motion or transient conditions without the addition of terms to describe the other effects. The ability to add an additional term to account for another process is a highly desirable feature of any model and is probably the reason that diffusional models are the most popular mathematical models of bioturbation.

Goldberg and Koide (1962) developed the first diffusional model to explain observed homogeneity in the ionium-to-thorium ratio in the upper portion of a pelagic sediment column. Since the publication of their model, a large number of diffusion models have been utilized to explain observed distributions of many radionuclides in sediments modified by biological mixing. The models have most frequently been employed to "unravel" the effects of particle redistribution in a determination of sedimentation rates in mixed sediments. The model has been applied to ^{210}Pb distributions by Bruland (1974), Spencer (1975), Shokes (1976), Turekian et al. (1978), and Santchi et al. (1980) for estuarine and coastal marine sediments; by Nozaki et al. (1977), Turekian et al. (1978), and Peng et al. (1979) for deep-sea sediments; and by Robbins et al. (1977) and Robbins (1978) for lake sediments. Robbins (1978) discusses in detail the application and interpretation of ^{210}Pb data in mixed sediments, and Imboden and Stiller (1982) consider the effects of radon diffusion on the ^{210}Pb distribution. The model has also been applied to the distributions of ^{239,240}Pu (Schink et al., 1975; Schink and Guinasso, 1982) for deep-sea sediments and (Benninger et al., 1979; Santchi et al., 1980; Olsen et al., 1981) for estuarine and coastal marine sediments; to ^{234}Th (Turekian et al., 1978; Aller et al., 1980; Cochran and Aller, 1980; Demaster et al., 1980; Santchi et al., 1980) for estuarine and coastal marine sediments and (Kadko, 1980a) for deep-sea sediments; to ^{7}Be (Krishnaswami et al., 1980) for estuarine and lake sediments; and to ^{137}Cs (Robbins et al., 1977, 1979; Robbins, 1978; Fisher et al., 1980) for lake sediments and laboratory experiments and (Olsen et al., 1981) for estuarine sediments.

The distribution of a particle-bound radionuclide whose source is constant in time and at the sediment–water interface is pictured in Fig. 1. The radionuclide reaches the sediment surface sorbed onto a sediment particle. Stirring of particles occurs to a depth m in the sediment, and, if the mixing is fast with respect to the decay constant and sedimentation, then the activity is uniform in the mixed layer. It is important to note that

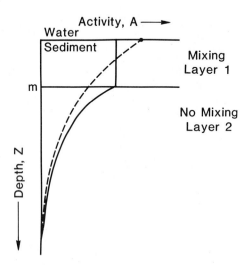

Figure 1. Distribution of a particle-bound radionuclide whose source is constant in time and at the sediment–water interface. If mixing is fast with respect to radioactive decay and sedimentation (solid line) then the activity is uniform in the mixed layer and decreases exponentially with deeper burial. The distribution with no mixing is given by the dashed line. The area under the two curves is the same.

mixing *decreases* the activity at the sediment–water interface and *increases* the activity everywhere below the mixed zone. Slower rates of mixing, greater sedimentation rates, or faster rates of radioactive decay will generate profiles that show a decrease in the mixed zone.

The general equation for the distribution of the radionuclide depicted in Fig. 1 is

$$\frac{\partial A}{\partial t} = \frac{\partial}{\partial z}\left[D_B\left(\frac{\partial A}{\partial z}\right)\right] - \omega\left(\frac{\partial A}{\partial z}\right) - \lambda A \qquad (1)$$

where A is the activity of the radionuclide, t is time, D_B is the particle biodiffusion coefficient, z is the depth in the sediment (positive downward), ω is the sedimentation rate, and λ is the decay constant of the radionuclide. The first term on the right-hand side of equation (1) represents the rate of change of activity at depth z owing to the effects of mixing. D_B, the particle biodiffusion coefficient, is treated mathematically like an eddy diffusion coefficient and represents the rate of biological mixing. If $D_B = 0$, then there is no mixing. D_B can be assumed constant within the mixed layer (most models) or can be assigned a functional form such as a linear decrease with depth (McCall and Fisher, 1980); an exponential decrease with depth (Schink and Guinasso, 1977; Peng et al., 1979; Santchi et al., 1980; Olsen et al., 1981); or a gaussian distribution within the mixed layer (Christensen, 1982). The second term describes the change of activity as a function of depth resulting from the accumulation of sediment, and the third term accounts for radioactive decay. Under steady-state conditions, $\partial A/\partial t = 0$. If it is assumed that there is no compaction and D_B is constant down to $z = m$ and 0 below, i.e., the organisms uniformly mix the sediment to a depth m, then equation (1) may be expressed as

$$D_B\left(\frac{\partial^2 A_1}{\partial z^2}\right) - \frac{\partial A_1}{\partial z} - \lambda A_1 = 0 \qquad z \leq m$$
$$\omega\left(\frac{\partial A_2}{\partial z}\right) - \lambda A_2 = 0 \qquad z > m \qquad (2)$$

The boundary conditions for these equations specify the depositional flux of the radionuclide, require continuity of concentrations and material fluxes between the layers, and provide for the eventual decay of the isotope. A_1 refers to the activity in the upper layer and A_2 to the activity below the zone of mixing. If the particle-bound substance were not radioactive, then $\lambda = 0$, and the general solution for the activity in the mixed layer is

$$A_1 = ae^{-(\omega/D_B)z_0} + b \tag{3}$$

The solution is an exponential like those derived from the homogeneous mixing models in a subsequent section of this chapter. Thus, under these restricted sets of conditions, the different models yield the same results. The power of the diffusion model lies in the ease of examining the independent effects of sedimentation, mixing, and mixing depth, and of including such factors as chemical reaction, adsorption, and radioactive decay as necessary. A few examples of previous applications of this model follow.

Guinasso and Schink (1975) adopted a model similar to that of Goldberg and Koide (1962), but with the concentration dependent upon time as well as depth. They modeled the time-dependent redistribution of a source layer deposited at the sediment surface and concluded that the appearance of the historical record depends upon the relative rates of biological mixing and sedimentation and the depth of the mixed layer. They quantified this relationship in the form of a mixing parameter, G:

$$G = D_B/m\omega \tag{4}$$

where m is the thickness of the mixed layer. With little or no mixing (D_B small) and/or a high sedimentation rate (ω large), the impulse layer is buried below the zone of mixing before much reworking takes place (G less than ~ 0.05) so that the final distribution is gaussian. On the other hand, a high mixing rate or a low sedimentation rate will result in a homogenized mixed layer in a time short compared to the residence time in the mixed layer, and the final distribution will be an exponential (G greater than ~ 1.0) like those in Berger and Heath (1968) (Fig. 4). Figure 2 illustrates this result. They also show that dimensional analysis may be used to estimate the particle biodiffusion mixing coefficient D_B:

$$D_B = mv_c \tag{5}$$

Since v_c has units of length per time, it may be thought of as the apparent sedimentation rate or apparent reworking rate. Thus, estimates of the thickness of the mixed layer and the reworking rate may be used to evaluate the particle biodiffusion coefficients.

Benninger et al. (1979) developed a three-layer mixing model in which the top 2–3 cm were mixed about 50 times faster than the middle layer (3–10 cm). No mixing occurred in the bottom layer (>10 cm). They interpreted discontinuous profiles of excess ^{210}Pb and ^{234}Th as caused

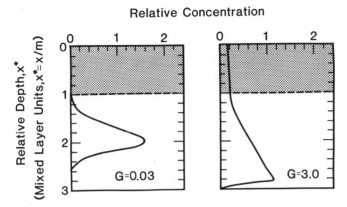

Figure 2. Concentration profiles of identical impulse sources for different values of the mixing parameter $G = D_B/m\omega$. The elapsed time after deposition of the impulse source is that required for sediments of two mixed layer thicknesses to accumulate ($t^* = \omega t/m = 2$). The mixed layer is indicated by the shaded area. Note the similarity with the homogeneous mixed layer case (Fig. 4) for $G > 1$. After Guinasso and Schink (1975).

by deep burrows filled by accumulating suspended solids rather than the collapse of burrow walls. They found that, in an area of slow sediment accumulation, a low rate of bioturbation below the surficial zone of rapid mixing causes an increase of at least a factor of two in apparent accumulation rate. This demonstrates the necessity of characterizing D_B as a function of depth.

Peng et al. (1979) used a numerical extension of the mixing model of Guinasso and Schink (1975) to calculate sedimentation rates from ^{14}C and ^{210}Pb data. They quantified how a thicker mixed layer results in older core-top ages and younger burial ages. They also observed and modeled a marked decrease in sedimentation rate apparently associated with the climatic transition from glacial to recent interglacial.

Robbins et al. (1979) performed a laboratory experiment in which ^{137}Cs-labeled surface sediment was redistributed by the oligochaete Tubifex tubifex and the amphipod Pontoporeia hoyi. They found that a simple diffusional model accurately described the results of the P. hoyi experiments, but could not account for the mixing processes of the oligochaetes. They concluded that particle redistribution caused by the life activities of benthos that move particles in a more or less random order should be adequately described by the particle diffusion model. This model, however, cannot be used to describe mixing processes that exhibit preferential directionality, such as feeding by the tubificids ("conveyor belt" transport—see previous chapters). Some types of bioturbation are distinctly advective on short time scales (Amiard-Triquet, 1974; Aller and Dodge,

1974). Fisher et al. (1980) modeled experimental data for the tubificid *T. tubifex* by including advective transport:

$$\frac{\partial A}{\partial t} = D_B \left(\frac{\partial^2 A}{\partial z^2}\right) - \frac{\partial(\omega A)}{\partial z} + S(z) \qquad (6)$$

where $\omega = \omega(z)$ is the vertical velocity of a sediment layer at depth z and S(z) is the rate of radioactive sediment loss from a layer owing to feeding. In their model, they assumed a depth-dependent advective velocity to account for the depth-dependent feeding activities. The feeding function was assumed to be gaussian about the zone of maximum feeding and to decrease linearly to 0 at the surface. For a population density of 10^5 organisms/m^2, the advective velocity was constant at 0.12 cm/day from the surface to just above the zone of maximum feeding, decreased by about a factor of two at the depth of maximum feeding, and equal to 0 when the feeding function was equal to 0. In order to evaluate the relative importance of tubificid particle mixing in Lake Erie, the authors calculated the ratio of particle mixing rates ($\bar{V}_t$) (based upon population densities of tubificids) to sedimentation rates ($\bar{V}_s$). The ratios ranged from 0.3 to 48.0. This indicates that in most locations the biological turnover of the sediment is faster than the rate of burial, and the authors expected the sediment to be significantly mixed.

The experimental results of Fisher et al. (1980) may be substituted into equation (4) in order to quantify G, the mixing parameter of Guinasso and Schink (1975). In the Fisher et al. (1980) model, D_B was taken to be 7.5×10^{-3} cm^2/day, and the base of the mixed layer was set at z = 8 cm. Robbins (personal communication) has identified the mixed layer in a field core to be 4.5 cm in a Lake Erie central basin sediment that is undergoing sedimentation at a rate of 0.168 cm/yr (4.60×10^{-4} cm/day). Substituting values, G = 2.04 for an 8-cm mixed layer and 3.62 for a 4.5-cm mixed layer. According to Guinasso and Schink's classification, this sediment would not exhibit a completely homogeneous upper layer because G < 10, but it is sufficiently mixed so that an impulse source deposited at the sediment surface would appear with an exponential distribution upon burial in the historical layers.

There have been several models that couple the particulate and pore water phases (Schink et al., 1975; Schink and Guinasso, 1977, 1978, 1982; Wong and Grosch, 1978; Cochran and Krishnaswami, 1980). Since these models primarily address the effects of soluble and/or adsorbed particulates on the pore water concentrations, they will be discussed with the pore water models in Section 3.

All the diffusional models require the evaluation of D_B, the particle biodiffusion coefficient. A tabulation of reported values of D_B is given in

Table I. The values of D_B vary about six orders of magnitude, from a low of 1.6×10^{-10} cm²/sec to a high of 1.5×10^{-4} cm²/sec. There does appear to be a decrease in biological reworking from shallow water ($D_B \sim 10^{-6}$ cm²/sec) to the deep sea ($D_B \sim 10^{-8}$ cm²/sec). The range in values is probably attributable in part to real differences in the magnitude of mixing of different organisms, to differences in biomass, and to temperature dependencies. Some of the variation is probably also due to the fact that not all of the diffusion models are the same, and hence other terms in those models may account for some of the mixing. This would modify the calculated values of D_B. For example, the two entries from Fisher et al. (1980) are in significant disagreement. The first value, 8.68×10^{-8}, comes from their model, which accounts for the majority of the sediment reworking by an advective term. The second value, 1.4×10^{-6}, was calculated using equation (5). This approximation was presented by Guinasso and Schink (1975) as a means to estimate D_B in a model that considers diffusion to be the only means of redistributing sediment particles. In this case, then, it is not unreasonable to expect the diffusion coefficient to be larger in the model that accounts for mixing only by diffusion.

The reported biodiffusivities should be thought of as a net biodiffusivity, D_B, which is the product of the biomass density biodiffusivity, $\bar{D}_B$, and the biomass density, b (g dry weight organism/cm²):

$$D_B = \bar{D}_B b \qquad (7)$$

In this manner, a comparison of the relative magnitudes of $\bar{D}_B$ between organisms (and with different organism densities) becomes more meaningful. Unfortunately, much of the data in Table I were collected without accompanying community or population density data. In shallow-water environments, the community at the time of sampling may be significantly different in numbers and species from that which has caused the previous mixing history (McCall, 1977). Laboratory studies provide a better control on community distributions and population densities, but there are only a few of these studies. Table II contains those studies from which a comparison of $\bar{D}_B$ can be made. The organism biomass density, b, has been calculated from the product of the population density of the benthos and the dry weight per individual. For *Yoldia limatula*, the dry weight per individual has been estimated as 100 mg and for *Pectinaria gouldii*, as 30 mg. Fisher (1978) gives a dry weight of 4 mg per tubificid, and Marzolf (1965) gives 1 mg for the weight of the amphipod *Pontoporeia hoyi*.

Benthos biodiffusion coefficients are very strongly temperature-dependent. Although the data for *Y. limatula* are significantly improved when corrected for organism densities, the data also reflect the fact that Buzzards Bay is cooler than Long Island Sound (Rhoads, 1963). The high-

Table I. Reported Values of D_B, the Particle Biodiffusion Coefficient, and m, the Depth of the Mixed Layer

D_B (cm²/sec)	m (cm)	Method	Location	Organism(s)	Reference
5.4×10^{-6}		Impulse source[a]	Chesapeake Bay		Duursma and Gross (1971)
2.5×10^{-8}		^{210}Pb[b]	San Diego Basin		Bruland (1974)
3.5×10^{-9}		^{210}Pb[b]	San Clemente Basin		Bruland (1974)
1.5×10^{-8}	2	Rhoads (1963)[c]	Buzzards Bay	Yoldia limatula	Guinasso and Schink (1975)
3.2×10^{-7}	2	Rhoads (1963)[c]	Long Island Sound	Yoldia limatula	Guinasso and Schink (1975)
7.6×10^{-7}	2	Rhoads (1963)[c]	Long Island Sound	Yoldia limatula	Guinasso and Schink (1975)
4.1×10^{-8}	6	Gordon (1966)[c]	Barnstable Harbor	Pectinaria gouldii	Guinasso and Schink (1975)
7.6×10^{-5}	38	Davison (1891)[c]	Holy Island sands	Arenicola	Guinasso and Schink (1975)
1.5×10^{-6}	38	Davison (1891)[c]	Caves Haven	Arenicola	Guinasso and Schink (1975)
4.4×10^{-4}	30	Fox et al. (1948)[c]	Intertidal sand	Thoracophelia mucronata	Guinasso and Schink (1975)
1.0×10^{-8}	6	Davis (1974)[c]	Freshwater lake	Tubiflex	Guinasso and Schink (1975)
1.8×10^{-9}	23	Microtektite data[a]	Indian Ocean		Schink et al. (1975)
3×10^{-10}	34	Microtektite data[a]	Eastern Indian Ocean		Schink et al. (1975)
7×10^{-8}	20	Microtektite data[a]	Gulf of Mexico		Schink et al. (1975)
1.6×10^{-10}	83	Microtektite data[c]	Indian Ocean		Schink et al. (1975)
4.1×10^{-10}	48	Microtektite data[a]	Southern Indian Ocean		Schink et al. (1975)
1.6×10^{-8}	17	Microtektite data[a]	Equatorial Atlantic Ocean		Schink et al. (1975)
2.4×10^{-10}	42	Microtektite data[a]	Northern equatorial Pacific Ocean		Schink et al. (1975)
1.2×10^{-10}	22	Microtektite data[a]	Indian Ocean		Schink et al. (1975)
7.0×10^{-7}	12	^{239,240}Pu[a]	Mediterranean Sea		Schink et al. (1975)
8.0×10^{-8}	8	^{239,240}Pu[a]	North Atlantic		Schink et al. (1975)
4.4×10^{-8}	9	^{239,240}Pu[a]	Northern equatorial Atlantic		Schink et al. (1975)
3.2×10^{-8}	4	^{239,240}Pu[a]	South Atlantic		Schink et al. (1975)
6.3×10^{-8}	6	^{239,240}Pu[a]	Southern equatorial Atlantic		Schink et al. (1975)
1.2×10^{-6}		^{210}Pb[b]	Buzzards Bay		Spencer (1975)
	4	c	Long Island Sound	Yoldia limatula and Nucula annulata	Aller and Cochran (1976)

3.5×10^{-6}	4	c	Long Island Sound	Yoldia limatula and Nucula annulata	Aller and Cochran (1976)
1.3×10^{-6}	4	c	Long Island Sound	Yoldia limatula and Nucula annulata	Aller and Cochran (1976)
6×10^{-9}	8	$^{210}Pb^b$	North Atlantic		Nozaki et al. (1977)
1.0×10^{-7}	3	$^{210}Pb^b$	Lake Huron	Tubificids and Pontoporeia affinis	Robbins et al. (1977)
1.8×10^{-7}	6	$^{210}Pb^b$	Lake Huron	Tubificids and Pontoporeia affinis	Robbins et al. (1977)
6.3×10^{-9}	4	$^{210}Pb^b$	Lake Ohrid		Robbins (1978)
9.5×10^{-6}		Assumed	Gulf of Mexico		Schink and Guinasso (1978)
0.2×10^{-6}		$^{234}Th^b$	Long Island Sound		Turekian et al. (1978)
1.2×10^{-6}		$^{234}Th^b$	Long Island Sound		Turekian et al. (1978)
0.1×10^{-6}		$^{234}Th^b$	Long Island Sound		Turekian et al. (1978)
0.5×10^{-6}		$^{210}Pb^b$	New York Bight		Turekian et al. (1978)
6×10^{-9}		$^{210}Pb^b$	North Atlantic		Turekian et al. (1978)
2×10^{-9}		$^{210}Pb^b$	South Atlantic		Turekian et al. (1978)
14×10^{-9}		$^{210}Pb^b$	Northern equatorial Pacific		Turekian et al. (1978)
8×10^{-9}		$^{210}Pb^b$	Northern equatorial Atlantic		Turekian et al. (1978)
10×10^{-9}		$^{210}Pb^b$	Northern equatorial Pacific		Turekian et al. (1978)
1×10^{-9}		$^{210}Pb^b$	Antarctic		Turekian et al. (1978)
7×10^{-9}		$^{210}Pb^b$	Antarctic		Turekian et al. (1978)
8×10^{-9}		$^{210}Pb^b$	Antarctic		Turekian et al. (1978)
2×10^{-9}		$^{210}Pb^b$	Antarctic		Turekian et al. (1978)
2×10^{-8}	3	$^{239,240}Pu^b$	Long Island Sound	Squilla sp.	Benninger et al. (1979)
3×10^{-8}	2	$^{239,240}Pu^b$	Long Island Sound	Squilla sp.	Benninger et al. (1979)
3.8×10^{-9}	8	$^{210}Pb\ ^{14}C^b$	Western equatorial Pacific		Peng et al. (1979)
3.8×10^{-9}	3	$^{210}Pb\ ^{14}C^b$	Western equatorial Pacific		Peng et al. (1979)
1.6×10^{-7}	1.5	$^{137}Cs^b$	Laboratory	Pontoporeia hoyi	Robbins et al. (1979)
5.4×10^{-6}	9	Robbins et al. (1979)^c	Laboratory	Tubifex tubifex	This work
1.3×10^{-6}	5	$^{234}Th^b$	Long Island Sound		Aller et al. (1980)

(Continued)

Table I. (Continued)

D_B (cm²/sec)	m (cm)	Method	Location	Organism(s)	Reference
0.01×10^{-6}	5	^{234}Th[b]	Long Island Sound		Aller et al. (1980)
1.6×10^{-6}	5	^{234}Th[b]	Long Island Sound		Aller et al. (1980)
1.3×10^{-6}	5	^{234}Th[b]	Long Island Sound		Aller et al. (1980)
0.59×10^{-6}	5	^{234}Th[b]	Long Island Sound		Aller et al. (1980)
0.99×10^{-6}	5	^{234}Th[b]	Long Island Sound		Aller et al. (1980)
0.47×10^{-6}	5	^{234}Th[b]	Long Island Sound		Aller et al. (1980)
0.33×10^{-6}	5	^{234}Th[b]	Long Island Sound		Aller et al. (1980)
0.21×10^{-6}	5	^{234}Th[b]	Long Island Sound		Aller et al. (1980)
0.25×10^{-6}	5	^{234}Th[b]	Long Island Sound		Aller et al. (1980)
0.24×10^{-6}	5	^{234}Th[b]	Long Island Sound		Aller et al. (1980)
0.43×10^{-6}	5	^{234}Th[b]	Long Island Sound		Aller et al. (1980)
0.15×10^{-6}	5	^{234}Th[b]	Long Island Sound		Aller et al. (1980)
9.5×10^{-7}		^{234}Th	Amazon Shelf		Demaster et al. (1980)
8.7×10^{-8}	8	^{137}Cs[b]	Laboratory	Tubifex tubifex	Fisher et al. (1980)
1.4×10^{-6}	8	Fisher et al. (1980)[c]	Laboratory	Tubifex tubifex	This work
1×10^{-6}	15	Assumed ^{234}Th	Central Pacific		Kadko (1980a,b)
4.6×10^{-6}	12	^{210}Pb[b]	Central Pacific		Kadko (1980a)
2.9×10^{-7}	3	^{7}Be[b]	Long Island Sound		Krishnaswami et al. (1980)
1.2×10^{-7}	2	^{7}Be[b]	Long Island Sound		Krishnaswami et al. (1980)
0.47×10^{-7}	3	^{7}Be[b]	Lake Whitney		Krishnaswami et al. (1980)
2.4×10^{-7}	5	^{234}Th, ^{210}Pb, ^{239,240}Pu[b]	Narragansett Bay		Santchi et al. (1980)
$1.6–10 \times 10^{-7}$	9	^{234}Th, ^{210}Pb, ^{239,240}Pu[b]	Narragansett Bay		Santchi et al. (1980)
3.4×10^{-7}	7.5	^{234}Th, ^{210}Pb, ^{239,240}Pu[b]	New York Bight		Santchi et al. (1980)
1.2×10^{-7}	7.5	^{234}Th, ^{210}Pb, ^{239,240}Pu[b]	New York Bight		Santchi et al. (1980)
4.1×10^{-8}		^{234}Th, ^{210}Pb, ^{239,240}Pu[b]	New York Bight		Santchi et al. (1980)

3.2×10^{-9}		^{234}Th, ^{210}Pb, ^{239,240}Pu^b	New York Bight	Santchi et al. (1980)
8.2×10^{-9}	8	^{239,240}Pu^a	Western Atlantic	Schink and Guinasso (1982)
1.3×10^{-8}	15	^{239,240}Pu^a	Western Atlantic	Schink and Guinasso (1982)
1.6×10^{-8}	11	^{239,240}Pu^a	Western Atlantic	Schink and Guinasso (1982)
9.8×10^{-9}	10	^{239,240}Pu^a	Western Atlantic	Schink and Guinasso (1982)
6.3×10^{-9}	10	^{239,240}Pu^a	Western Atlantic	Schink and Guinasso (1982)
1.9×10^{-8}	8	^{239,240}Pu^a	Puerto Rican Trench	Schink and Guinasso (1982)
1.6×10^{-8}	10	^{239,240}Pu^a	Eastern Atlantic	Schink and Guinasso (1982)
1.0×10^{-8}	9	^{239,240}Pu^a	Eastern Atlantic	Schink and Guinasso (1982)
3.2×10^{-8}	10	^{239,240}Pu^a	Eastern equatorial Atlantic	Schink and Guinasso (1982)

^{a} Standard deviation technique as employed by Guinasso and Schink (1975).
^{b} Fit parameter in diffusion model.
^{c} Dimensional analysis [equation (5) in text].

Table II. Comparison of Selected Values of D_B from Table I with $\bar{D}_B$, the Values Corrected for b, the Biomass Density

D_B (cm²/sec)	b (g dry weight/m²)	$\bar{D}_B$ (cm⁴/g dry weight per sec)	Temperature (°C)	Organism(s)	Location
3.5×10^{-8}	2.2	1.6×10^{-4}		Yoldia limatula	Buzzards Bay
1.5×10^{-7}	6.1	2.5×10^{-4}		Yoldia limatula	Long Island Sound
3.2×10^{-7}	13.7	2.3×10^{-4}		Yoldia limatula	Long Island Sound
7.6×10^{-8}	0.3	2.5×10^{-3}	19	Pectinaria gouldii	Barnstable Harbor
1.3×10^{-8}	0.3	2.5×10^{-4}	13	Pectinaria gouldii	Barnstable Harbor
4.4×10^{-8}	0.89	4.9×10^{-4}	10	Limnodrilus and Tubifex tibifex	Messalonskee Lake
1.6×10^{-7}	16	1.0×10^{-4}	7	Pontoporeia hoyi	Laboratory
5.4×10^{-6}	200	2.7×10^{-4}	20	Tubifex tubifex	Laboratory
8.7×10^{-8}	260	3.3×10^{-6}	15	Tubifex tubifex	Laboratory
1.4×10^{-6}	260	5.4×10^{-5}	15	Tubifex tubifex	Laboratory

est value of $\bar{D}_B$ reported, 2.5×10^{-3} cm²/sec for *P. gouldii*, was calculated at 19°C, far warmer than the mean annual temperature at any of the field sites. The 13°C value for the polychaete is comparable to those of the other organisms.

Except for the low value of $\bar{D}_B$ from the advection-containing model of Fisher et al. (1980), the variation between studies is improved by a factor of about 8 when corrected for benthic biomass. The range in values of D_B is more than a factor of 400, and the range of $\bar{D}_B$ is about a factor of 50. The remaining variation in D_B is probably due to temperature effects, effects of organisms not reported, sampling and experimental error, and processes besides biodiffusion that cause mixing.

The studies listed in Table II do not include any estimates of D_B from the deep sea. Schink et al. (1975) estimate oceanic values of D_B to range from about 10^{-8} to 10^{-10} cm²/sec. Although they did not determine the benthic community and population densities at these sites, an estimate of the biomass may be obtained from the work of Rowe et al. (1974). They derived a log-linear relationship between biomass density and water depth for the Atlantic:

$$\log_{10} \text{biomass [mg/m}^2\text{]} = 3.03 - 0.0031(\text{water depth in meters} - 2104 \text{ m}) \quad (8)$$

Deep-sea sediment could be expected to contain a biomass of about 150 mg/m² ($b = 0.15$ g/m²). Substituting into equation (7) gives a range for

$\tilde{D}_B$ of 6.7×10^{-4} to 6.7×10^{-6} cm⁴/g dry weight per sec. These values are comparable to those in Table II and indicate that correcting the diffusivities for biomass density can account for the majority of the enormous variation in reported values of D_B. This finding also suggests that equation (7) may be readily utilized to estimate D_B quantitatively.

Diffusion models provide a mechanistic interpretation of the bioturbation process. Techniques for quantitatively determining D_B are well established. In cases where the mixing processes are controlled by mechanisms other than just diffusion, the mathematics of diffusion models readily permit the incorporation of these other processes. This feature has made the diffusion model the most popular approach to describing bioturbation.

2.2. Box Models

Figure 3 represents the simplest mixing model. This approach has been classified as a "box model" (Berner, 1980). The model assumes continuous sedimentation and incorporates mixing in a layer immediately beneath the sediment–water interface. This layer is treated as being completely homogenized to a fixed depth (Berger and Heath, 1968; Ruddiman

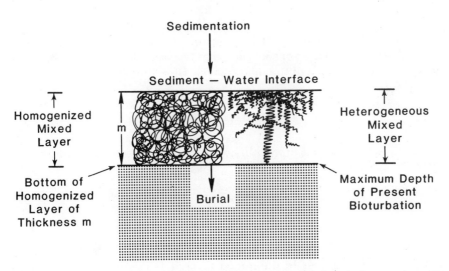

Figure 3. Schematic diagram of box mixing models. Particles deposited upon the sediment–water interface by sedimentation are rapidly and uniformly incorporated into the mixed layer in the homogenous model and are redistributed with depth and spatial constraints in the heterogeneous model. The rate of deposition determines the rate of removal of sediment from the mixed layer into buried sediment.

and Glover, 1972; Davis, 1974; Peng et al., 1977; Sundquist et al., 1977; Benoit et al., 1979; Cochran, 1980) or as being heterogeneously mixed (Piper and Marshall, 1969; Hanor and Marshall, 1971; Håkanson and Källström, 1978). Homogeneous mixing implies complete disordering of the particles within the mixed layer and hence is analogous to very rapid "particle diffusion." The homogeneous models may be used to study only steady-state mixing, because only two conditions are permitted, i.e., unmixed and totally homogenized. Heterogeneous mixing models account for partial mixing in the mixed layer corresponding to burrows by having small volumes of homogeneous mixing, or allowing the mixing to decrease with depth in the sediment.

A box model is a mass balance model. When a particle lands on the sediment surface, it is incorporated into the mixed layer. In the homogeneous model, the rate of mixing is large compared with the sedimentation rate, while in the heterogeneous model it is not. A small increment of sediment deposition results in a small increment of sediment removed from the mixed zone and buried. Thus, there is a certain fraction of the newly deposited sediment that is buried below the mixed zone. A few examples of this model follow.

Berger and Heath (1968) expressed the mass balance as

$$\frac{dP}{P} = \frac{-dL}{m} \tag{9}$$

where P is the probability of finding the particle in the homogeneous layer, dL is the thickness of sediment that has been deposited on the layer, and m is the thickness of the mixed layer. Direct integration of this equation yields

$$P = P_0 \exp(-L/m) \tag{10}$$

where P_0 is the original probability. The mass balance of Sundquist et al. (1977) may be expressed as (ignoring dissolution)

$$\frac{dg}{dt} = \frac{F}{m\rho}(f - g) \tag{11}$$

where F is the total sediment mass flux to the mixed layer, g is the mass fraction of the particles in the mixed layer, f is the mass fraction of the particles in the sediment supply, m is the thickness of the mixed layer, ρ is the dry density of sediment in the mixed layer, and t is time. The solution to equation (11) if all parameters except g are assumed to be constant is

$$g = g_0 \exp\left[\frac{-F}{m\rho}(t - t_0)\right] + f\left\{1 - \exp\left[\frac{-F}{m\rho}(t - t_0)\right]\right\} \quad (12)$$

where g_0 and t_0 refer to initial conditions. If sedimentation occurs over a time interval dt, then a sediment layer of thickness dL will have been deposited. Hence, $(F/\rho)(dt) = dL$ or $(F/\rho)(t - t_0) = L$. The first term in equation (12) thus represents the fraction of the new particles in the homogeneous layer and may be expressed as $g_0\exp(-L/m)$, which is identical to the right-hand side of equation (10). This demonstrates that the mass balance formulations of Berger and Heath (1968) and Sundquist et al. (1977) are the same. Furthermore, note that the exponential form of the distribution in the mixed layer in these homogeneous box models is the same as that derived from the Guinasso and Schink (1975) model for $G > 1$ [equation (4); compare Figs. 2 and 4].

Berger and Heath (1968) developed the mixing model for the interpretation of microfossil distributions in deep-sea sediments. Figure 4 shows a typical result of their homogeneous model. Application of this model to a particle-bound radionuclide whose source is at the sediment–water interface would generate a profile similar to that of the solid line in Fig. 1. Berger and Heath (1968) concluded that marked stratigraphic errors can appear if the layers representing the time ranges are similar in thickness to the mixed layer.

Davis (1974) reached a similar conclusion by calculating age frequency distribution curves of pollen in sediments reworked by tubificids. He measured pollen displacement rates in 1-cm depth intervals in laboratory experiments. His mass balance model was expressed as

$$Z_i = PC_0(F) + PC_1(Y_1) + PC_2(Y_2) + \cdots \\ + PC_n(Y_n) + \cdots + PC_N(Y_N) \quad (13)$$

where Z_i is the number of pollen grains present in the nth depth interval at the beginning of the current year, Y_n is the number of pollen grains in the nth depth interval at the beginning of the previous year, N is the number of depth intervals from which pollen grains in the ith interval originated during the previous year, PC is percent, and F is the number of pollen grains added from the water column during the previous year. His model predicted that 36% of surface pollen would be older than 30 years and 5% older than 90. His age frequency distribution curves give results that are similar to Fig. 4, but are plotted in a different manner.

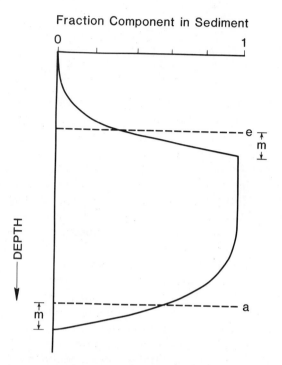

Figure 4. Vertical distribution of a component in the sediment resulting from homogeneous mixing and sedimentation. a and e mark time-equivalent levels in the sediment of the appearance and disappearance (extinction) of the component. Note that the component is found to a depth of one mixed layer thickness below the time horizon of the first arrival of the component. Similarly, note that the concentration of the component begins to decrease at a depth of one mixed layer thickness deeper than the time associated with the extinction. This is because of dilution by downward mixing of fresh sediment absent in this component. Finally, note the similarity with the rapid particle diffusion case (Fig. 2). After Berger and Heath (1968).

Peng et al. (1977), Sundquist et al. (1977), and Benoit et al. (1979) used the box model to show that bioturbation causes systematic deviations of apparent ^{14}C ages from true ages of carbonate and humic-rich sediments. Sundquist et al. (1977) included a term for chemical reaction (dissolution of $CaCO_3$) and concluded that sediment mixing may account for discrepancies in correlating the high-carbonate top layer of sediments with the last glacial period.

The heterogeneous models based on the same ideas were developed by Piper and Marshall (1969) and Hanor and Marshall (1971) to examine progressive mixing of a given volume of sediment. Piper and Marshall (1969) examined the mixing of a simple two-component mud-over-sand system. They estimated the amount of reworking at the junction of sand

and mud beds by comparing observed distributions of sand-size materials within the mud. They assumed that the sediment affected by a hypothetical burrow is completely homogenized. They let the number of vertical burrows decrease linearly with depth, while the number of horizontal burrows remained the same at all depths. Otherwise a random distribution of burrows was assumed. Their calculated redistribution curves are similar to the fraction distribution curves of Berger and Heath (1968) (Fig. 4) and the age frequency distribution curves of Davis (1974), but are plotted in a different manner. They found that the relationship between the amount of disturbance of a primary sedimentary structure and the total amount of reworking is dependent on the spatial distribution of burrows, and concluded that only where there is a very low degree of disturbance can an estimate be made directly of the amount of reworking without making assumptions as to the distribution of burrows. Thus, a high rate of reworking of sediments makes a heterogeneous model reduce to a homogeneous model, depending upon the scale of observation (Hanor and Marshall, 1971).

The last model of this type mentioned here is that of Håkanson and Källström (1978), who presented a "biotransport" model of the porosity in the top 20 cm of lake sediments, where the porosity is determined by gravitational compaction of sediment and a net upward advection of sediment by organisms. The model permits the calculation of age frequency distribution for defined sediment layers and the results agree well with those of Davis (1974).

2.3. Signal Processing Models

Goreau (1977) presented a signal-theory-based model that allows the calculation of the effect of any series of time-varying mixing events on any time-varying input to the sediment. The input function, $I(x)$, is the profile that would result if some time-varying input were delivered to the sediments in the absence of mixing. The output of the model, $O(x)$, is the final mixed profile that results from the action of the mixing function, $F(x, x')$:

$$O(x) = \int_{\infty}^{0} I(x') F(x - x') \, dx' \qquad (14)$$

Figure 5 schematically demonstrates the input–output nature of the model.

In order to utilize the model, it is necessary that the mixing function $F(x)$, be known. $F(x)$ can be determined from measurements of $O(x)$ when

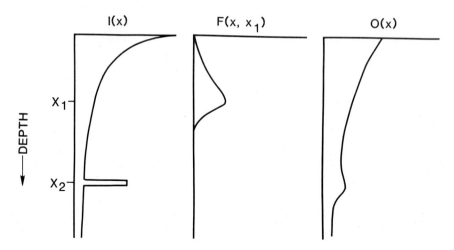

Figure 5. Signal processing model of Goreau (1977). $I(x)$ represents the profile that would be observed in the complete absence of mixing of a sedimentary input that is increasing exponentially with time and with a single source layer superimposed. $F(x,x_1)$ is the mixing function, identified here at $x = x_1$, for a discrete input. This function is convolved with $I(x)$ at all depths to obtain the observed historical record, $O(x)$. Note that the mixing function $F(x,x_1)$ is similar to the *profiles* obtained with an intermediate value of G (Fig. 2). After Goreau (1977).

the input to the sediment is known. However, the calculation of the mixing function does not necessarily describe the actual mixing processes of the organisms, only their net effect on the redistribution of the sediment particles. Goreau suggests that when $F(x)$ is known it can be used in conjunction with measured mixed profiles, $O(x)$, to deconvolve the effects of mixing by inversion of equation (14) to derive the original input history, $I(x)$. The potential importance of the deconvolution technique to interpreting the historical record for detailed stratigraphic and pollution information is significant. Successful deconvolution of equation (14), however, remains to be demonstrated.

Application of the signal theory model appears to be straightforward in the presence of a discrete input signal as seen in Fig. 5. The model does not give a time history of the sediment redistribution, but rather the net result of the mixing as defined by the input–output nature of the model. Much work needs to be performed with this model to demonstrate its full capabilities.

2.4. Markov Models

Recently, Jumars et al. (1981) presented the formulation of a Markov chain model of flow–sediment–organism interactions. A Markov process is a stochastic process in which the probability of a change in state of a parameter depends only on the present state of the parameter and not on the entire past evolution of the system. A Markov chain is a Markov process with a finite number of states (Blanc-Lapierre and Fortet, 1965).

Figure 6 graphically illustrates the Jumars et al. (1981) model of selective feeding and fecal pellet production coupled with sediment burial. An individual sediment grain may be selected stochastically for ingestion by a deposit feeder and move from the free state to the pelletal state. Conversely, a pellet may break down into its constituent particles. Buried sediment is considered removed from the activities of benthos. The required input data are the probabilities of transition from one compartment to another. The probability of buried sediment becoming pelletized is zero, by definition of buried sediment, and, for the sake of simplicity, the probability of burial of an intact pellet was set equal to zero. Thus, the model is built upon the assumption that the transition probabilities depend only on the present state and not on the history of the sediment particles. As Jumars et al. (1981) point out, the model in absence of sedimentation is homologous with the infinite series model of deposit feeding developed by Levinton and Lopez (1977).

The solution procedure of the Markov model is described by Kemeny and Snell (1960) and Blanc-Lapierre and Fortet (1965). A matrix of tran-

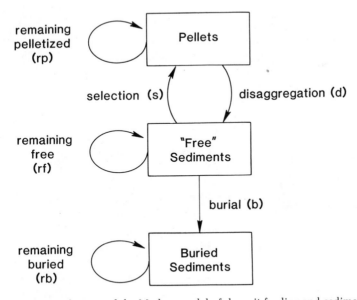

Figure 6. Schematic diagram of the Markov model of deposit feeding and sediment burial. A particle may exist in one of three compartments: (1) in a pellet, (2) as "free" sediment, or (3) as buried sediment. Transfer between compartments is accomplished by assigning probabilities to the transitions in particle state. After Jumars et al. (1981).

sition probabilities is devised and constrained by the known transition probabilities. For the model in Fig. 6, the transition matrix is

$$\text{From} \begin{cases} \text{Buried} \\ \text{Pellet} \\ \text{Free} \end{cases} \overbrace{\begin{matrix} \text{Buried} & \text{Pellet} & \text{Free} \end{matrix}}^{\text{To}} \\ \begin{bmatrix} 1 & 0 & 0 \\ 0.00 & rp & d \\ 0.01 & s & rf \end{bmatrix} \quad (15)$$

In this case, as noted earlier, the probability of a buried particle remaining buried is equal to 1, and that of a pellet becoming buried is equal to 0. Similarly, buried particles cannot become pellets or free particles. The other "known" information is that the probability of a particle input to the system as a free particle becoming buried is 0.01. Mass balance constraints are that all rows in the matrix add to 1:

$$rb = 1.00$$
$$rp + d = 1.00 \quad (16)$$
$$s + rf = 0.99$$

In the above equations the dependent variables rp and rf can be calculated as a function of the transitional probabilities, d, s, for breakdown and selection, which will in general be variable- and problem-dependent (i.e., they depend on the type of particle, feeder, physical environments, etc.) and should be determined experimentally.

The model predicts that particles that are more strongly selected will have greater residence times in surficial sediments before they become part of the stratigraphic record and will therefore achieve higher concentrations in surficial sediments than will less-preferred particles. Selective feeding thus increases the average age of surficial sediments. The implications for deep-sea stratigraphy are enormous. The model perhaps provides a more biologically realistic alternative explanation of the frequently occuring anomalously great ages of surficial sediments than core-top loss or deep mixing. The authors extend the capabilities of the model to include lateral transport of surficial sediments. Since the transport is size-dependent, it is possible to include feedback loops to the model in Fig. 6 to account for internal transport. Application of this model to a real example is necessary to demonstrate its full potential. The authors have indicated that they are pursuing that objective (see also Nowell et al., 1981).

3. Fluid Transport Models

The activities of benthos not only redistribute sediment particles but also influence the exchange of water and solutes across the sediment–water interface. Burrows act as channels for the direct communication between interstitial and overlying waters (irrigation) (Aller, 1977); fecal pellets deposited on the sediment surface increase the porosity of the uppermost sediment and therefore increase the diffusional exchange (Fisher et al., 1980); and the injection of surface water for burrowing and feeding modifies the porosity and chemical conditions in the sediment and increases the exchange of water and solutes across the sediment–water interface (McCall et al., 1979; Aller, 1978). Each process may be caused by different organisms, although some benthos may exhibit more than one exchange technique and may exchange fluids and solutes to different degrees and from different depths. The exchange may also occur on a variety of time scales. For example, diffusional exchange of solutes may be very small even in the presence of high organism densities, while exchange caused by the turbulent mixing of surficial sediment during initial burrowing may be many times larger (G. Matisoff and P. L. McCall, unpublished data). Understanding the exchange processes and the time scales on which they occur is essential to a quantitative description of chemical mass transfer between sediments and water.

Mathematical models of the exchange of pore fluids in the absence of benthos are well developed (Berner, 1980). These are most commonly one-dimensional diffusion models, where the ionic dispersion coefficient represents the molecular diffusivity in the bulk sediment. These models have been extrapolated for use in the presence of macroinfauna and are similar to the particle diffusion models of Section 2.1. In these models the apparent diffusion coefficient in the bioturbated zone is elevated over the value in the bulk sediment (Hammond et al., 1975; Schink et al., 1975; Goldhaber et al., 1977; Schink and Guinasso, 1977, 1978, 1982; Vanderborght et al., 1977; Aller, 1978; Wong and Grosch, 1978; Cochran and Krishanaswami, 1980; Filipek and Owen, 1980; McCall and Fisher, 1980). These models will be discussed in more detail in Section 3.1.1.

Permanent or semipermanent tube structures or significant directed exchange of water cannot necessarily be described by simple one-dimensional models or by quantitatively modifying a molecular diffusivity (Aller, 1978). These conditions require diffusional models other than one-dimensional vertical diffusion and/or other mathematical expressions to describe the exchange. "Biopumping" models (Grundmanis and Murray, 1977; Hammond and Fuller, 1979; McCaffrey et al., 1980) describe the exchange of solutes between well-mixed and distinct reservoirs of sediment and overlying water as an advection velocity. The signal processing model (Goreau, 1977) formulates the volume of pore water exchanged with overlying water owing to bioturbational mixing of sediment and faunal water pumping. These models characterize pore water exchange as flow and may be classified as "advection" models. They are discussed in more detail in Section 3.2. Recently, Aller (1980a,b) has presented a model that does not account for the effects of macrobenthos with empirically elevated diffusion coefficients or advection velocities. His model assumes that at any instant the sediment is a mosaic of microenvironments represented on the average by a central irrigated tube or burrow and its immediately surrounding sediment. Solute exchange is described by three-dimensional diffusion (cylindrical coordinates). This model is discussed in more detail in Section 3.1.2.

3.1. Diffusion–Reaction Models

Although the number of mathematical models of fluid transport is small, models of the diffusion–reaction type have been selected in almost all cases for applications involving benthos and fluid transport. The mathematics of one-dimensional vertical diffusion models are identical to those of the particle diffusion models discussed in Section 2, although the diffusion coefficient refers to the apparent diffusivity of the solute in the sediment. Aller (1978) notes that sedentary infauna can alter sediments by building tubes that are more accurately characterized by cylindrical diffusion geometry in the sediment immediately surrounding the burrows. He also demonstrates (Aller, 1977) that mobile infauna do not remain stationary long enough to allow establishment of fixed diffusion–reaction geometry. In this case, a one-dimensional vertical diffusion model with an apparent diffusion coefficient may be used. The choice of a one-dimensional vertical diffusion model or another diffusion model that more accurately describes the benthic processes is dependent upon the spatial and temporal scale of investigation.

3.1.1. One-Dimensional Vertical Diffusion Models

Particle reworking near the sediment–water interface can create a mixed layer in which the pore water solutes migrate more rapidly than can be accounted for by molecular diffusion. The solid particles would be distributed as in Fig. 1. The solute transport within the sediment may be described as a two-layer, one-dimensional system undergoing vertical diffusion, sedimentation, compaction, adsorption, chemical reaction, and radioactive decay. The upper, mixed layer has an apparent molecular diffusivity higher than the molecular diffusivity in the lower layer. Mathematically, this may be described as (modified from Berner, 1980)

$$\text{Zone 1: } \frac{\partial C_1}{\partial t} = \left[\frac{1}{\phi(1+K)}\right]\left[\frac{\partial(\phi D_1 \partial C_1/\partial z)}{\partial z}\right] - \omega\left(\frac{\partial C_1}{\partial z}\right) - \lambda C_1 + \left(\frac{1}{1+K}\right)\Sigma R$$

$$\text{Zone 2: } \frac{\partial C_2}{\partial t} = \left[\frac{1}{\phi(1+K)}\right]\left[\frac{\partial(\phi D_2 \partial C_1/\partial z)}{\partial z}\right] - \omega\left(\frac{\partial C_2}{\partial z}\right) - \lambda C_2 + \left(\frac{1}{1+K}\right)\Sigma R$$

(17)

where C is concentration, t is time, ϕ is porosity, z is the depth below the sediment–water interface, D is the apparent diffusion coefficient, K is the adsorption coefficient, ω is the sediment rate, λ is the radioactive decay constant, and ΣR represents all nonequilibrium slow reactions affecting C. Ignoring compaction, adsorption, chemical reaction, and radioactive decay, and assuming the sediment properties to be homogeneous within each zone, the equations simplify to

$$\text{Zone 1: } \frac{\partial C_1}{\partial t} = D_1\left(\frac{\partial^2 C_1}{\partial z^2}\right) - \omega\left(\frac{\partial C_1}{\partial z}\right)$$

$$\text{Zone 2: } \frac{\partial C_2}{\partial t} = D_2\left(\frac{\partial^2 C_2}{\partial z^2}\right) - \omega\left(\frac{\partial C_2}{\partial z}\right)$$

(18)

The two zones are coupled by requiring that the concentrations at the boundary between the zones and the fluxes across the boundary be continuous. Assuming sedimentations to be small over the course of an investigation, the flux boundary condition becomes

$$D_1\left(\frac{\partial C_1}{\partial z}\right) = D_2\left(\frac{\partial C_2}{\partial z}\right)$$

(19)

Thus, if $D_1 > D_2$, the concentration gradient in the upper zone is less than the gradient in the lower zone and a break in curvature occurs at the boundary (Fig. 7). Some of the models of this type are now discussed in more detail.

Hammond et al. (1975) were the first to utilize this approach. They discovered that they could not explain observed fluxes of radon from sediments with a semiinfinite (single-layer, $D_1 = D_2$) model, but that the

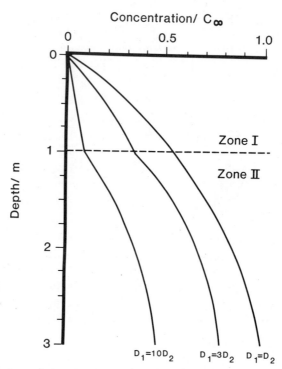

Figure 7. Comparison of the effect on steady-state pore water profiles of increasing the apparent diffusivity in the mixed layer (Zone I). No burrowing is described when $D_1 = D_2$. m is the depth of the mixed layer. C_∞ is the concentration observed at depth, when dC/dz approaches zero. The profile shape is maintained at steady state by a source of the substance in the sediment. After Aller (1978).

two-layer model, which incorporates an upper layer stirred by currents and organisms, could account for the higher flux. They estimated that the apparent diffusion coefficient in the mixed layer was 3–7 times the value in the unmixed lower layer. An examination of the interstitial water data alone would not have led them to this observation. Their results demonstrate the importance of obtaining flux data in addition to interstitial water data in order to identify the important processes occurring at the sediment–water interface.

Schink et al. (1975) presented a unique model in which the redistribution of sediment solids by particle biodiffusion is coupled with the pore water concentrations. In their model, silica is input to the sediment–water interface as particulate biogenic silica (diatom and radiolarian frustules). In the steady state, the interstitial water concentration is determined by a balance between the rate of dissolution of the particulate silica and the rate of diffusional flux of dissolved silica from the sediment.

They do not assign an enhanced value to the diffusion coefficient of the solute in the mixed zone (i.e., $D_1 = D_2$). The dissolution rate depends upon the degree of undersaturation ($C_s - C$), which decreases with depth in the sediment. Downward mixing will move particles from a zone of rapid dissolution and rapid flux from the sediment to a level of slower dissolution and less flux from the sediment. Subsequent models of theirs (Schink and Guinasso, 1977, 1978, 1982) have included terms for adsorption, sedimentation, water content, and sediment composition, and the models have been applied also to calcium carbonate in sediments. In order to illustrate the basics of their models, we will only examine their early, simpler work on the effects of mixing of the solid particles on the interstitial water concentrations (Schink et al., 1975).

Their coupling of interstitial water concentrations and solid particles was expressed as:

For dissolved silica: $$D\left(\frac{\partial^2 C}{\partial z^2}\right) + \left(\frac{K_B}{\phi}\right)\left(\frac{C_f - C}{C_f}\right)B = 0 \quad (20)$$

For reactive solid silica: $$D_B\left(\frac{\partial^2 B}{\partial z^2}\right) - K_B\left(\frac{C_f - C}{C_f}\right)B = 0 \quad (21)$$

where D is the aqueous molecular diffusion coefficient of silica modified for the effects of tortuosity, C is the concentration of dissolved silica, z is the depth below the sediment–water interface, K_B is the first-order dissolution rate constant of solid, reactive opaline silica, ϕ is porosity, C_f is the solubility concentration of solid, reactive opaline silica, B is the concentration of solid, reactive opaline silica, and D_B is the particle biodiffusion coefficient ($= 0$ below the mixed layer).

Figure 8, adapted from their work, illustrates some of their results. For a given input flux of particulate biogenic silica, the concentration at depth in the interstitial water (balance concentration, C_b) increases with increasing biological mixing, other parameters (porosity, dissolution rate, aqueous diffusion coefficient, concentration at the sediment–water interface) being fixed. The downward mixing moves biogenic siliceous sediment particles into a region where the supply of silica to the pore waters by dissolution exceeds the removal by diffusional flux to the water column, and so the interstitial water concentrations increase.

The results in Fig. 8 demonstrate the important effects of biological mixing on the calculated interstitial water concentrations. Schink and Guinasso (1977, 1982) have subsequently extended the model to permit the calculation of the fraction of reactive solid remaining as a function of the other parameters. They do not believe, however, that variations in bioturbation rates will significantly alter the $CaCO_3$ (and perhaps parti-

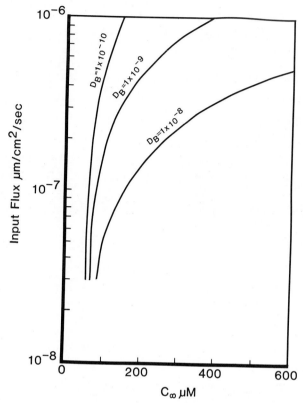

Figure 8. Effect of the biological particle mixing coefficient, D_B, on the interstitial water silica concentration at depth ($C_\infty >$ 10–100 cm). After Schink et al. (1975).

culate silica) contents of accumulating sediments, and therefore have fixed D_B as constant in their newer models.

Aller (1978) applied the two-layer approach (with a source term) to the pore water ammonium distribution in laboratory tanks with and without *Yoldia limatula*. He estimated the apparent diffusion coefficient to be 1×10^{-5} cm²/sec in the mixed zone, compared to 7×10^{-6} cm²/sec below the mixed zone. His calculated profiles did not agree with the shapes expected (Fig. 7), i.e., the interstitial water concentrations below the mixed zone should be lower in the presence of mixing by *Yoldia* because of enhanced diffusive loss in the upper layer. He concluded that microbial production rates of NH_4^+ are greater below the mixed zone in the presence of *Yoldia* than in its absence, and suggested that this result was caused by the provision of a nutrient source or a flushing from the sediment of an inhibitor.

Cochran and Krishnaswami (1980) examined concentration–depth profiles of radium, thorium, uranium isotopes, and ^{210}Pb in northern equatorial Pacific sediments and sediment pore waters. They divided the sediment column into two zones: an upper zone, which is biologically mixed, and deeper sediment. Throughout the sediment column ^{226}Ra is mobilized to the pore water following its production from radioactive decay of ^{230}Th. Radium in the pore water can be readsorbed onto sediment particles, diffuse, and decay. Mixing redistributes the ^{230}Th activity (hence ^{226}Ra production rate) such that a constant ^{230}Th activity is present in the mixed zone. They found good agreement between the deduced parametric constants for the "good fits" of their data and those expedited from independently available data.

McCall and Fisher (1980) adopted the two-layer approach in the examination of chloride transport across the sediment–water interface from NaCl-spiked pore waters. They found that the transport could be described in the absence of infauna by one-dimensional vertical diffusion in a semi-infinite medium, with $D_{Cl^-} = 5.5 \times 10^{-6}$ cm^2/sec. When tubificid oligochaetes were present, the transport could not be explained by assuming an apparent diffusion coefficient that was greater than the molecular diffusion coefficient over the entire tubificid life zone. They found that best agreement was obtained by applying the two-layer model of equations (18) and (19). The upper layer was 0.75 cm thick and constituted only the pelletal layer, even though the mixed layer was 6–8 cm thick. By setting D_{Cl^-} equal to 11×10^{-6} cm^2/sec at the sediment–water interface and decreasing it linearly to 5.5×10^{-6} cm^2/sec (the value in the absence of benthos) at the base of the pelletal layer, they obtained excellent agreement with the observed data.

Table III compares the ratios of the apparent diffusion coefficients in

Table III. Comparison of the Ratios of the Apparent Molecular Diffusivities in the Mixed and Unmixed Layers

Ratio D_1/D_2	m (cm)	Parameter	Organism(s)/process	Reference
3	5	Radon-222	Polychaetes	Hammond et al. (1975)
7	2	Radon-222	Polychaetes	Hammond et al. (1975)
7	8	Sulfate	*Nephtys incisa*	Goldhaber et al. (1977)
100	3.5	Silica	Waves and currents	Vanderborght et al. (1977)
1.4	3	Ammonium	*Yoldia limatula*	Aller (1978)
1.4–3.5[a]	7–12	Radium-226		Cochran and Krishnaswami (1980)
6	20.75	Sodium	Mixed assemblage	McCaffrey et al. (1980)
1–2	0	Chloride	*Tubifex tubifex*	McCall and Fisher (1980)

[a] Calculated as $(D + KD_B)/D$ (Schink and Guinasso, 1978).

the mixed layer with the values of the molecular diffusivity in the underlying sediments. As in Table II, these numbers should be normalized to the biomass density, but insufficient data do not permit this modification. The experimental studies of Aller (1978) and McCall and Fisher (1980) agree very well with $D_1/D_2 = 1.5$, but the results from the field studies are a factor of 1–67 higher than this value. This may be the result of additional stirring of the upper most sediments in natural conditions by processes other than one-dimensional vertical diffusion, such as radial diffusion, water pumping, and winds and waves. Vanderborght et al. (1977) showed that mixing by waves and currents in the upper 3.5 cm of North Sea sediments (20 m depth) requires a physical mixing coefficient in the upper layer of 10^{-4} cm^2/sec. This value is about 100 times greater than that of molecular diffusion and demonstrates the important effects of physical processes on the sediments in shallow-water environments.

3.1.2. Three-Dimensional Diffusion Models

Aller (1977) observed that permanent tube dwellings formed by sedentary benthos produced three-dimensional chemical gradients in the pore waters and solid phases of sediment and increased the flux of material across the sediment–water interface. He conducted laboratory and field studies of *Amphitrite ornata*, a sedentary, surface deposit-feeding polychaete, and *Clymenella torquata*, a sedentary, subsurface deposit-feeding polychaete. He found that chemical effects caused by the structure of the tube are the result of the distribution of bacteria around the tube, the way the burrow is maintained, and a change in the geometry of diffusion because of the presence of a hollow tube maintained at seawater concentrations.

For ammonium, Aller (1977, 1980a) considers the burrows to be vertical cylinders that are uniformly spaced and constantly irrigated with overlying oxygenated seawater (Fig. 9). He notes that the exact geometry of the burrows is three-dimensionally complex and time-dependent, but that the sediment may be approximated as a mosaic of microenvironments represented on the average by a central irrigated vertical burrow and its immediately surrounding sediment. The ammonium produced by a microbial activity diffuses toward areas of low ammonium concentration (burrows and overlying water). Mathematically, he expressed these processes at steady state in terms of cylindrical and vertical diffusion:

$$\left(\frac{D}{r}\right)\left[\frac{\partial(r\partial C/r)}{\partial r}\right] + D\left(\frac{\partial^2 C}{\partial z^2}\right) + R(z) = 0 \qquad (22)$$

where D is the aqueous molecular diffusion coefficient of ammonium in

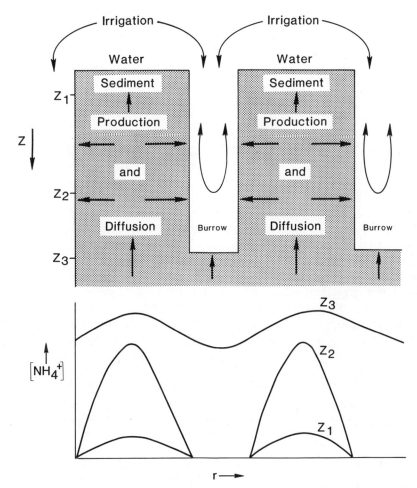

Figure 9. Schematic representation of the cylindrical burrow irrigation model of Aller (1977, 1980a). Note that diffusion occurs down the greatest concentration gradient, i.e., toward burrows and the sediment–water interface. Burrows are assumed to be continuously and instantaneously mixed with overlying water. After Berner (1980).

the sediment, r is the radial distance from the nearest burrow axis, and $R(z)$ is the rate of microbial production of ammonium.

Aller (1980a) solved equation (22), calculated the average ammonium concentrations for discrete depth intervals, and found good agreement between calculated and observed ammonium profiles. He concluded that his burrow irrigation model more closely simulates the processes affecting the solute distributions in the bioturbated zone than do other pore water models. His model is particularly significant because it represents a case in which both diffusion and faunal water pumping are considered. He treats the pumping passively by assuming instantaneous communication between the burrows and the overlying water, an assumption for which he has supporting observations.

3.2. Advection Models

In all the diffusion–reaction models presented in the previous section, mass transport has been examined only as a diffusional process. Not all of the processes are diffusional in nature. Some organisms pump and direct the flow of water as well as sediment. Overlying water that is drawn into burrows during irrigation may be mixed with surrounding pore waters and returned to the overlying water (Fig. 10). Models of this process require that solutes be transported by advection between one or two well-mixed sediment reservoirs (Grundmanis and Murray, 1977; Hammond and Fuller, 1979; McCaffrey et al., 1980) or as a function of depth (Goreau, 1977) and overlying water.

Hammond and Fuller (1979) observed radon deficiencies in surficial sediments of San Francisco Bay. They suggested that this might be caused by the pumping of overlying water into the sediments through worm

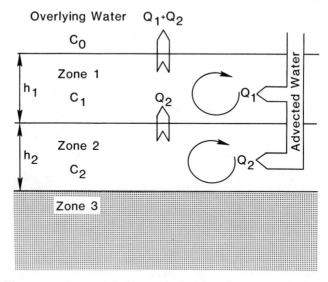

Figure 10. Worm pumping model of Hammond and Fuller (1979). Two sediment layers defined by the radon concentration profiles receive overlying water during irrigation. This water is mixed uniformly with the pore waters within each zone and fluxed to the overlying water.

tubes. In their model (Fig. 10) the sediments were divided into irrigation zones defined by the radon concentration profile. Each zone was assumed to be well mixed. The upper zone receives flow directly from the overlying water and from the underlying zone, whereas the lower layer receives advected water only from the overlying water. They characterized the radon balances as

$$\text{Zone 1: } Ph_1A + Q_2C_2 = (Q_1 + Q_2)C_1 + \lambda h_1 AC_1 + J_1A$$
$$\text{Zone 2: } Ph_2A = Q_2C_2 + \lambda h_2 AC_2 + J_2A$$
(23)

where $C_{1,2}$ are the radon concentrations in Zones 1 and 2, $h_{1,2}$ are the thicknesses of Zones 1 and 2, A is the area, P is the production rate of radon from radium, λ is the decay constant for radon, $J_{1,2}$ is the net loss by molecular diffusion from Zones 1 and 2 per unit area, and $Q_{1,2}$ are the net rates of inflow of water to Zones 1 and 2. They calculated an irrigation velocity from the observed radon concentration profiles and estimated fluxes of nutrients owing to irrigation. The irrigation velocities were $\simeq 3$ cm/day, and the calculated fluxes about an order of magnitude greater than can be attributed to molecular diffusion.

McCaffrey et al. (1980) examined Narrangansett Bay sediments that were homogenized to a depth of 20 cm. They determined that infaunal bivalves, gastropods, amphipods, and some polychaetes rapidly mix the upper 5–6 cm of sediment, and that sediment from 15–20 cm was recycled to the sediment surface by the "conveyor belt" (Rhoads, 1974) species *Maldanopsis elongata*, a holothurian. The rate of exchange between overlying and pore waters was determined by measuring the rate at which ^{22}Na introduced into the overlying water was removed by transport into pore waters and the vertical distribution of ^{22}Na in the sediment. They found that the transfer of tagged overlying water was far greater than would be predicted from vertical molecular diffusion alone. They modeled the transport of dissolved species in two ways. In both models they assumed that species released below 25 cm are transported into the zone of bioturbation by simple diffusion. In the first model, they assumed that the species are transported across the sediment–water interface by enhanced diffusion. This is identical to the approach of Hammond et al. (1975), Aller (1978), and McCall and Fisher (1980), discussed earlier. The second model assumes that dissolved substances are transported across the sediment–water interface by both simple diffusion and advection.

In the advection model for ^{22}Na, McCaffrey et al. (1980) assumed that organisms pumped water at a constant rate independent of depth in the sediment between the sediment and the overlying water. They treated the process as an exchange between two well-mixed reservoirs: pore water and overlying water. The decrease of ^{22}Na activity in the overlying water was then described by

$$\frac{dC_s(t)}{dt} = -KC_s(t) + K\left(\frac{V_s}{V_p}\right)C_s(0) - K\left(\frac{V_s}{V_p}\right)C_s(t) \quad (24)$$

where $C_s(0)$ is the normalized ^{22}Na concentration in the overlying water at time $t = 0$, $C_s(t)$ is the normalized ^{22}Na concentration in the overlying water at any time t, K is the exchange constant, V_s is the volume of the overlying water, and V_p is the volume of pore water. Similarly, $C_p(t)$, the normalized ^{22}Na concentration in the pore water at any time t, may be calculated:

$$C_p(t) = -\left(\frac{V_s}{V_p}\right)\left[\frac{dC_s(t)}{dt}\right] \quad (25)$$

K was determined as the fraction of overlying water assumed to be pumped across the sediment–water interface that is necessary to account for the observed decrease in C_s with time. The advection rate is equal to $K(V_s/A)$, where A is the cross-sectional area of the sediment, and was found to be 0.7 ml/cm^2 per day. The advective flux is equal to the product of the average increase in the concentration of a species in the pore water relative to bottom water times the pumping rate. A comparison of the mass transport calculated from the diffusion and advection models is given in Table IV. Note that the pumping model does not include transport across the sediment–water interface by molecular diffusion, whereas the Hammond and Fuller (1979) model does. These calculations indicate that transport by diffusion and by advection are of the same order (*if the assumption that the irrigated water mixes with pore waters is valid*). Smethie et al. (1981) reached a similar conclusion by examining ^{222}Rn in Washington continental shelf sediments.

Table IV. Comparison of Fluxes across the Sediment–Water Interface Calculated from the Diffusion and Advection Models of McCaffrey et al. (1980) and the Advection Model of Hammond and Fuller (1979)

Species	Diffusional flux[a] (μmol/cm^2 per day)	Advective flux[a] (μmol/cm^2 per day)	Advective flux[b] (μmol/cm^2 per day)
H_4SiO_4	0.3	0.3	0.6
NH_4^+	0.19	0.07	0.4
PO_4^{3-}	0.018	0.02	0.003
ΣCO_2	0.09	0.6	4.0
Mn^{2+}	0.02	0.01	

[a] McCaffrey et al. (1980) (Narragansett Bay).
[b] Hammond and Fuller (1979) (San Francisco Bay).

Goreau (1977) suggests that a signal theory approach to the study of sediments may be used to extract this information in the presence of biological mixing. He shows that pore waters may be analyzed in a manner similar to the approach he used for the bulk sediment. He defines a depth-dependent mixing function for the pore water, $F_w(x)$, as the volume of pore water exchanged with overlying water per unit area per unit time (hence his model has been loosely classified as an advection model). Goreau shows that the flux of a dissolved species across the sediment–water interface is given by

$$\text{Flux [g/cm}^2 \text{ per sec]} = \int_0^{x\ max} F_x(x)[C(x) - C_w]dz \qquad (26)$$

where $C(x)$ is the concentration of the dissolved species at depth x and C_w is the concentration of dissolved species in the overlying water. When $[C(x) - C_w]$ is positive the flux will be from the sediment to the water column, and when $[C(x) - C_w]$ is negative the net flux will be into the sediment.

McCaffrey et al. (1980) essentially employed equation (26) but treated $F_w(x)$ as the advection rate and a constant in depth, and $C(x)$ as the mean concentration in the mixed zone. Carefully controlled laboratory experiments are necessary to obtain $F_w(x)$ and permit the use of equation (26).

The mass transfer of dissolved species across the sediment–water interface can be a factor of two or more higher than that predicted by one-dimensional vertical diffusion alone. The results of Hammond and Fuller (1979) and McCaffrey et al. (1980) demonstrate that the additional flux may be the result of the direct exchange of pore water and overlying water during mixing. Goreau (1977) has presented a general formula for this exchange. Aller (1977, 1980a) has indicated that the nature of the exchange may be diffusion in three dimensions into burrows followed by mixing of overlying and burrow waters by irrigation (pumping). Clearly, the nature of the exchange is dependent upon the activities of the macrobenthos. As McCall and Fisher (1980) show, *Tubifex tubifex* does not pump water, so that, in a sediment inhabited only by them, it does not make sense to employ a pumping model. Indeed, they were able to account for chemical fluxing across the sediment–water interface by one-dimensional vertical diffusion alone.

4. Conclusions

The purpose of constructing mathematical models of bioturbation has been, so far, to describe the redistribution of sediment particles caused by biological mixing and the enhancement of the chemical mass transfer between sediment pore fluids and overlying waters. Particle mixing models are applicable to stratigraphic problems. They may provide an absolute time framework for the accumulation of the sediments and may be used to determine the sedimentologic and postdepositional history of the deposit. Pore water models are usually applied to chemical cycling problems. They permit estimates of chemical fluxing across the sediment–water interface and/or rates of mineralogical and biogeochemical reactions occurring in the sediment.

As we acquire a more detailed knowledge of processes that occur near the sediment–water interface we learn that the animal–sediment–water system is very complex. There are many ways of explaining experimental and field observations. We may need more complicated models, but the utility of the models may actually increase, because they can be used to rationalize anomalous observations or limit the number of possible processes producing the observed pattern. McCaffrey et al. (1980) attempted this (Table IV), but they did not employ a very sophisticated diffusion model. If they had developed the diffusion model in the same manner as Aller (1980a), they might have found that diffusion with irrigation could also account for the observed net flux. Nevertheless, a detailed comparison of different models on the same data should permit a more accurate evaluation of the actual processes that are occurring.

The differences between the models are more pronounced in the particle transport problem, and such a comparison is even more important in these situations. Failure to recognize biological mixing in sediments has been common in the historical geochemical literature because some of the parameters analyzed do not readily indicate a mixed layer. For example, ^{210}Pb is commonly uniformly distributed in the mixed layer and then decreases exponentially below, while organic carbon and nitrogen usually decrease continuously from the surface. This apparent discrepancy is of fundamental importance when attempting to interpret the sediment record. Clearly, the relative rates of sedimentation, mixing, decomposition/decay, and depositional flux, as well as selective feeding, must result in a homogeneous profile of some parameters in the mixed layer and a profile that appears unmixed for other parameters. This question must be quantitatively resolved.

ACKNOWLEDGMENTS. Juan C. Heinrich (Department of Aerospace and Mechanical Engineering, University of Arizona, Tucson, Arizona 85721) assisted in the discussions and early writing of this chapter. This work was supported in part by NSF Grant OCE-8005103 and NOAA Grant NA80RAD00036. This is Contribution No. 140 from the Department of Geological Sciences, Case Western Reserve University.

References

Aller, R. C., 1977, The influence of macrobenthos on chemical diagenesis of marine sediments, Ph.D. dissertation, Yale University, New Haven, Connecticut, 600 pp.
Aller, R. C., 1978, The effects of animal–sediment interactions on geochemical processes near the sediment–water interface, in: *Estuarine Interactions* (M. L. Wiley, ed.), pp. 157–172, Academic Press, New York.
Aller, R. C., 1980a, Quantifying solute distributions in the bioturbated zone of marine sediments by defining an average microenvironment, *Geochim. Cosmochim. Acta* **44**:1955–1965.
Aller, R. C., 1980b, Diagenetic processes near the sediment–water interface of Long Island Sound. I. Decomposition and nutrient element geochemistry (S, N, P), in: *Estuarine Physics and Chemistry: Studies in Long Island Sound* (B. Saltzman, ed.), pp. 237–350, Advances in Geophysics, Volume 22, Academic Press, New York.
Aller, R. C., and Cochran, J. K., 1976, ^{234}Th/^{238}U disequilibrium in near-shore sediment: Particle reworking and diagenetic time scales, *Earth Planet. Sci. Lett.* **29**:37–50.
Aller, R. C., and Dodge, R. E., 1974, Animal–sediment relations in a tropical lagoon, Discovery Bay, Jamaica, *J. Mar. Res.* **32**:209–232.
Aller, R. C., Benninger, L. K., and Cochran, J. K., 1980, Tracking particle associated processes in nearshore environments by use of ^{234}Th/^{238}U disequilibrium, *Earth Planet. Sci. Lett.* **47**:161–175.
Amiard-Triquet, C., 1974, Etude experimentale de la contamination par le cérium 144 et le fer 59 d'un sédiment à *Arenicola marina* L. (Annelide Polychete), *Cah. Biol. Mar.* **15**:483–494.
Benninger, L. K., Aller, R. C., Cochran, J. K., and Turekian, K. K., 1979, Effects of biological sediment mixing on the ^{210}Pb chronology and trace metal distribution in a Long Island Sound sediment core, *Earth Planet. Sci. Lett.* **43**:241–259.
Benoit, G. J., Turekian, K. K., and Benninger, L. K., 1979, Radiocarbon dating of a core from Long Island Sound, *Estuarine Coastal Mar. Sci.* **9**:171–180.
Berger, W. H., and Heath, G. R., 1968, Vertical mixing in pelagic sediments, *J. Mar. Res.* **26**:134–143.
Berner, R. A., 1980, *Early Diagenesis: A Theoretical Approach*, Princeton University Press, Princeton, New Jersey, 241 pp.
Blanc-Lapierre, A., and Fortet, R., 1965, *Theory of Random Functions* (J. Gani, transl.), Volume 1, Gordon and Breach, New York, 443 pp.
Bruland, K. W., 1974, Pb-210 geochronology in the coastal marine environment, Ph.D. thesis, University of California, San Diego.
Christensen, E. R., 1982, A model for radionuclides in sediments influenced by mixing and compaction, *J. Geophys. Res.* **87**:566–572.
Cochran, J. K., 1980, The flux of Ra-226 from deep-sea sediments, *Earth Planet. Sci. Lett.* **49**:381–392.
Cochran, J. K., and Aller, R. C., 1980, Particle reworking in sediments from the New York Bight apex—Evidence from the Th-234–U-238 disequilibrium, *Estuarine Coastal Mar. Sci.* **9**:739.
Cochran, J. K., and Krishnaswami, S., 1980, Radium, thorium, uranium, and ^{210}Pb in deep-sea sediments and sediment pore waters from the north equatorial Pacific, *Am. J. Sci.* **280**:849–889.
Davis, R. B., 1974, Stratigraphic effects of tubificids in profundal lake sediments, *Limnol. Oceanogr.* **19**:466–488.
Davison, C., 1891, On the amount of sand brought up by Lobworms to the surface, *Geol. Mag.* **8**:489–493.

Demaster, D. J., Nittrouer, C. A., Cutshall, N. H., Larsen, I. L., and Dion, E. P., 1980, Short lived radionuclide profiles and inventories from Amazon continental shelf sediments, Eos **61**:1004.

Duursma, E. K., and Gross, M. G., 1971, Marine sediments and radioactivity, in: *Radioactivity in the Marine Environment* (Committee on Oceanography, National Research Council), pp. 147–160, National Academy of Sciences, Washington, D.C.

Filipek, L. H., and Owen, R. M., 1980, Early diagenesis of organic carbon and sulfur in outer shelf sediments from the Gulf of Mexico, *Am. J. Sci.* **280**:1097–1112.

Fisher, J. B., 1978, Effects of tubificid oligochaetes on sediment movement and the movement of materials across the sediment–water interface, Ph.D. dissertation, Case Western Reserve University, Cleveland, Ohio, 132 pp.

Fisher, J. B., Lick, W. J., McCall, P. L., and Robbins, J. A., 1980, Vertical mixing of lake sediments by tubificid oligochaetes, *J. Geophys. Res.* **85**:3997–4006.

Fox, D. L., Crane, S. C., and McConnaughey, B. H., 1948, A biochemical study of the marine annelid worm, *Thoracophelia mucronata*, *J. Mar. Res.* **7**:567–585.

Goldberg, E. D., and Koide, M., 1962, Geochronological studies of deep sea sediments by the ionium/thorium method, *Geochim. Cosmochim. Acta* **26**:417–450.

Goldhaber, M. B., Aller, R. C., Cochran, J. K., Rosenfeld, J. K., Martens, C. S., and Berner, R. A., 1977, Sulfate reduction, diffusion and bioturbation in Long Island Sound sediments: Report of the FOAM Group, *Am. J. Sci.* **277**:193–237.

Gordon, D. C., Jr., 1966, The effects of the deposit feeding polychaete *Pectinaria gouldii* on the intertidal sediments of Barnstable Harbor, *Limnol. Oceanogr.* **11**:327–332.

Goreau, T. J., 1977, Quantitive effects of sediment mixing on stratigraphy and biogeochemistry: A signal theory approach, *Nature* **265**:525–526.

Guinasso, N. L., and Schink, D. R., 1975, Quantitative estimates of biological mixing rates in abyssal sediments, *J. Geophys. Res.* **80**:3032–3043.

Grundmanis, V., and Murray, J. W., 1977, Nitrification and denitrification in marine sediments from Puget Sound, *Limnol. Oceanogr.* **22**:804–813.

Håkanson, L., and Källström, A., 1978, An equation of state for biologically active lake sediments and its implications for interpretations of sediment data, *Sedimentology* **25**:205–226.

Hammond, D. E., and Fuller, C., 1979, The use of radon-222 as a tracer in San Francisco Bay, in: *San Francisco Bay: The Urbanized Estuary* (T. J. Comomos, ed.), pp. 213–230, American Association for the Advancement of Science, San Francisco.

Hammond, D. E., Simpson, H. J., and Mathieu, G., 1975, Methane and radon-222 as tracers for mechanisms of exchange across the sediment–water interface in the Hudson River Estuary, in: *Marine Chemistry in the Coastal Environment* (T. M. Church, ed.), pp. 119–132, American Chemical Society Symposium Series 18, American Chemical Society, Washington, D.C.

Hanor, J. S., and Marshall, N. F., 1971, Mixing of sediment by organisms, in: *Trace Fossils: A Field Guide to Selected Localities in Pennsylvanian, Permian, Cretaceous, and Tertiary Rocks* (B. F. Perkins, ed.), pp. 127–135, School of Geoscience Miscellaneous Publication 71-1, Louisiana State University, Baton Rouge.

Imboden, D. M., and Stiller, M., 1982, The influence of radon diffusion on the Pb-210 distribution in sediments, *J. Geophys. Res.* **87**:557–565.

Jumars, P. A., Nowell, A. R. M., and Self, R. F. L., 1981, A simple model of flow–sediment–organism interaction, in: *Sedimentary Dynamics of Continental Shelves* (C. A. Nittrouer, ed.), *Mar. Geol.* **42**:155–172.

Kadko, D., 1980a, ^{230}Th, ^{226}Ra, and ^{222}Rn in abyssal sediments, *Earth Planet. Sci. Lett.* **49**:360–380.

Kadko, D., 1980b, A detailed study of some uranium series nuclides at an abyssal hill area near the East Pacific Rise at 8°45'N, *Earth Planet. Sci. Lett.* **51**:115–131.

Kemeny, L. G., and Snell, J. L., 1960, *Finite Markov Chains*, D. Van Nostrand, Princeton, New Jersey, 210 pp.

Kowal, N. E., 1971, A rationale for modeling dynamic ecological systems, in: *Systems Analysis and Simulation in Ecology* (B. C. Patten, ed.), pp. 123–194, Academic Press, New York.

Krishnaswami, S., Benninger, L. K., Aller, R. C., and Van Damm, K. L., 1980, Atmospherically-derived radionuclides as tracers of sediment mixing and accumulation in nearshore marine and lake sediments: Evidence from ^{7}Be, ^{210}Pb, and ^{239,240}Pu, *Earth Planet. Sci. Lett.* **47**:307–318.

Levinton, J. S., and Lopez, G. R., 1977, A model of renewable resources and limitation of deposit-feeding benthic populations, *Oecologia (Berlin)* **31**:177–190.

McCaffrey, R. J., Meyers, A. C. Davey, E., Morrison, G., Bender, M., Luedtke, N., Cullen, D., Froelich, P., and Klinkhammer, G., 1980, The relation between pore water chemistry and benthic fluxes of nutrients and manganese in Narragansett Bay, Rhode Island, *Limnol. Oceanogr.* **25**:31–44.

McCall, P. L., 1977, Community patterns and adaptive strategies of the infaunal benthos of Long Island Sound, *J. Mar. Res.* **38**:221–266.

McCall, P. L., and Fisher, J. B., 1980, Effects of tubificid oligochaetes on physical and chemical properties of Lake Erie sediments, in: *Aquatic Oligochaete Biology* (R. O. Brinkhurst and D. G. Cook, eds.), pp. 253–317, Plenum Press, New York.

McCall, P. L., Tevesz, M. J. S., and Schwelgien, S. F., 1979, Sediment mixing by *Lampsilis radiata siliquoidea* (Mollusca) from Western Lake Erie, *J. Great Lakes Res.* **5**:105–111.

Marzolf, G. R., 1965, Substrate relations of the burrowing amphipod *Pontoporeia affinis* in Lake Michigan, *Ecology* **46**:579.

Nowell, A. R. M., Jumars, P. A., and Eckman, J. E., 1981, Effects of biological activity on the entrainment of marine sediments, in: *Sedimentary Dynamics of Continental Shelves* (C. A. Nittrouer, ed.), *Mar. Geol.* **42**:133–153.

Nozaki, Y., Cochran, J. K., Turekian, K. K., and Keller, G., 1977, Radiocarbon and ^{210}Pb distribution in submersible-taken deep sea cores from Project Famous, *Earth Planet. Sci. Lett.* **34**:167–173.

Olsen, C. R., Simpson, H. J., Peng, T.-H., Bopp, R. F., and Trier, R. M., 1981, Sediment mixing and accumulation rate effects on radionuclide depth profiles in Hudson estuary sediments, *J. Geophys. Res.* **86**:11020–11028.

Peng, T.-H., Broecker, W. S., Kipphut, G., and Shackleton, N., 1977, Benthic mixing in deep sea cores as determined by ^{14}C dating and its implications regarding climate stratigraphy and the fate of fossil fuel CO_2, in: *The Fate of Fossil Fuel CO_2 in the Oceans* (N. R. Andersen and A. Malahoff, eds.), pp. 355–373, Plenum Press, New York.

Peng, T.-H., Broecker, W. S., and Berger, W. H., 1979, Rates of benthic mixing in deep-sea sediment as determined by radioactive tracers, *Quat. Res.* **11**:141–149.

Piper, D. J. W., and Marshall, N. F., 1969, Bioturbation of Holocene sediments on La Jolla Deep Sea Fan, California, *J. Sediment. Petrol.* **39**:601–606.

Rhoads, D. C., 1963, Rates of sediment reworking by *Yoldia limatula* in Buzzards Bay, Massachusetts and Long Island Sound, *J. Sediment. Petrol.* **33**:723–727.

Rhoads, D. C., 1974, Organism–sediment relations on the muddy sea floor, *Oceanogr. Mar. Biol. Annu. Rev.* **12**:263–300.

Robbins, J. A., 1978, Geochemical and geophysical applications of radioactive lead, in: *Biogeochemistry of Lead in the Environment* (J. O. Nriagu, ed.), pp. 285–393, Elsevier, Amsterdam.

Robbins, J. A., Krezoski, J. R., and Mozley, S. C., 1977, Radioactivity in sediments of the Great Lakes: Post-depositional redistribution by deposit-feeding organisms, *Earth Planet. Sci. Lett.* **36**:325–333.

Robbins, J. A., McCall, P. L., Fisher, J. B., and Krezoski, J. R., 1979, Effect of deposit feeders on migration of ^{137}Cs in lake sediments, *Earth Planet. Sci. Lett.* **42**:277–287.

Rowe, G. T., Palloni, P. T., and Horner, S. G., 1974, Benthic biomass estimates from the northwestern Atlantic Ocean and the Northern Gulf of Mexico, *Deep Sea Res.* **21**:641–650.

Ruddiman, W. F., and Glover, L. K., 1972, Vertical mixing of ice-rafted volcanic ash in North Atlantic sediments, *Bull. Geol. Soc. Am.* **83**:2817–2836.

Santchi, P. H., Li, Y.-H., Bell, J. J., Trier, R. M., and Kawtaluk, K., 1980, Pu in coastal marine environments, *Earth Planet. Sci. Lett.* **51**:248–265.

Schink, D. R., and Guinasso, N. L., Jr., 1977, Modelling the influence of bioturbation and other processes on calcium carbonate dissolution at the sea floor, in: *The Fate of Fossil Fuel CO_2 in the Oceans* (N. R. Andersen and A. Malahoff, eds.), pp. 375–398, Plenum Press, New York.

Schink, D. R., and Guinasso, N. L., 1978, Redistribution of dissolved and adsorbed materials in abyssal marine sediments undergoing biological stirring, *Am. J. Sci.* **278**:687–702.

Schink, D. R., and Guinasso, N. L., 1982, Processes affecting silica at the abyssal sediment–water interface, in: *Actes des Colloques du C.N.R.S.: Biogeochimie de la Matière Organique à l'Interface Eau–Sediment Marin* (R. Daumas, ed.) (in press).

Schink, D. R., Guinasso, N. L., Jr., and Fanning, K. A., 1975, Processes affecting the concentration of silica at the sediment–water interface of the Atlantic Ocean, *J. Geophys. Res.* **80**:3013–3031.

Shokes, R. F., 1976, Rate-dependent distributions of lead-210 and interstitial sulfate in sediments of the Mississippi River delta, Ph.D. thesis, Texas A&M University, College Station, Texas.

Smethie, W. M., Jr., Nittrouer, C. A., and Self, R. F. L., 1981, The use of radon-222 as a tracer of sediment irrigation and mixing on the Washington continental shelf, in: *Sedimentary Dynamics of Continental Shelves* (C. A. Nittrouer, ed.), *Mar. Geol.* **42**:173–200.

Spencer, D. W., 1975, Distribution of lead-210 and polonium-210 between soluble and particulate phases in seawater, USAEC Report, COD-3566-11, Woods Hole Oceanographic Institute, Woods Holes, Massachusetts, 105 pp.

Sundquist, E., Richardson, D. K., Broecker, W. S., and Peng, T.-H., 1977, Sediment mixing and carbonate dissolution in the southeast Pacific Ocean, in: *The Fate of Fossil Fuel CO_2 in the Oceans* (N. R. Andersen and A. Malahoff, eds.), pp. 429–454, Plenum Press, New York.

Turekian, K. K., Cochran, J. K., and DeMaster, D. J., 1978, Bioturbation in deep sea deposits: Rates and consequences, *Oceanus* **21**:34–41.

United States Army Corps of Engineers, 1975, Lake Erie wastewater management study: Preliminary feasibility report, Corps of Engineers, Buffalo District, Buffalo, New York.

Vanderborght, J. P., Wollast, R., and Billen, G., 1977, Kinetic models of diagenesis in disturbed sediments. Part 1. Mass transfer properties and silica diagenesis, *Limnol. Oceanogr.* **22**:787–793.

Wong, G. T. F., and Grosch, C. E., 1978, A mathematical model for the distribution of dissolved silicon in interstitial waters—An analytical approach, *J. Mar. Res.* **36**:735–750.

Index

Amphipods, 116–118
Amphitrite burrow walls, 199
Amphitrite ornata, 70–71, 90, 320
Aquatic nonmarine trace fossils, geological significance, 257–285
   conclusions, 280–281
   depositional environment, 271–273
   paleoenvironmental information, usefulness, 278–280
   paleoenvironmental reconstruction, 271–276
   systematic information, usefulness, 277–278
   taxonomic, lithologic, and paleoenvironmental diversity, 260–265
   traces, identification of, 259–270
   trophic level reconstruction, 270–271
   vectorial features, 273–276
Archanodon, 275
Azoic values, 36

Bacterial colonization of surfaces, 27
Bacterially mediated diagenetic reactions, free energy change and schematic chemistry, 182
Bed roughness, 19–24
   bioturbational activity, 19–20
   blocks, 21
   fluids literature, 20–21
   natural tube problem, 21
   projecting tubes, 20
   shell material, 23–24
Belt Supergroup, 250
Benthic invertebrates, classification and terminology of feeding, 179

Benthos effects on physical properties of freshwater sediments, 105–176
   ecological interactions and sediment properties, 161–166
   freshwater sediments and macrobenthos, 106–113
   future work, 166–167
   macrobenthos life-styles, 113–124
   sediment mixing, 124–150
   sediment transport, biogenic modification, 150–161
Biological benthic boundary layer, 4
Bioturbation, 4
Bioturbation, mathematical models, 289–330
   advection models, 322–325
   box models, 292
   conclusions, 325–326
   diffusion models, 292, 293
   diffusion–reaction models, 314
   equations, 295, 296, 298, 299, 306, 307, 309
   fluid transport models, 313–325
   kinds of models and objectives, 291–293
   Markov models, 310–312
   one-dimensional vertical diffusion models, 314–320
   particle transport models, 293–312
   processes to be modeled, 289–291
   signal-theory-based model, 292, 309–310
   three-dimensional diffusion models, 320–322
Bivalves, 120–124
   Pisidiidae, 123–124
   Unionacea, 120–123

Black Sea sediments, 223
Bottom communities, 5–7
Box models, 292, 305–309
  equations, 306, 307
Branchyura sowerbi, 164
Buchholzbrunnichnus, 249
Bunyerichnus, 249
Burrow habitat, chemistry of, 89–92
  animal adaptations, 91–92
  metabolic activity, 90
  model equations, 89, 91
Burrowing in muds, Buzzards Bay, 30–31

Celliforma spirifer, 266
Chironomids, 113–116
  sediment mixing, depth and rate of, 135–137
Chironomus americanus, 31
Chironomus anthracinus, 163
Chironomus plumosus, 136
Chironomus riparius, 206
Chironomus tentans, 206
Chitinous tubes, 228
Chondrities, 229
Clymenella torquata, 70–71, 320
Cohesive sediment erosion, 32
Coquinas, 226
Corbicula fluminea, 144
Critical excitation velocity, 28
Cruziana facies, 236

Diagenetic reactions, 54–56
  biogeochemical successions, 55
  macrobenthos influence, 56
Decomposition reactions, idealized, 54
Detritus-feeding invertebrates, potential food, 24–25
Diffusion models, 292, 293–305
  equations, 295, 296, 298, 299
Diplocraterion, 236
Dominant profundal macrobenthos, trophic type and life habitat, 180
Dreissena clam, 163

Eau Claire Formation, 245
Ecological interactions and sediment properties, 161–166
  fauna, spatial–temporal variations, 161–162
  macrofauna interactions, 163–166

Ecological succession and geotechnical properties, 10–19
  conclusions, 17–19
  experimental design, 11
  FOAM sediments, 15–16
  FOAM study sites, 10
  methods, 13–15
  NWC sediments, 16–17
  NWC station, 11
  results, 15–17
  water content profiles, 19
Ecological succession and organism, 5–10
  bottom communities, 5–7
  equilibrium stages, 9–10
  pioneering stages, 7–9
Ediacaran fauna, 244
Equilibrium stages, 9–10
  coexisting species, 10
  late successional stage, 9
  sedimentary effects, 9–10
Erosion resistance and sediment water content, 34

"Far-field" phenomena, 18, 149
Filter-feeding clams, 143–144
Fluid bioturbation, 37
Fluid transport models, 313–325
  advection models, 322–325
  diffusion–reaction models, 314
  equations, 315, 317, 320, 323, 324, 325
  one-dimensional vertical diffusion models, 314–320
  three-dimensional diffusion models, 320–322
FOAM study, *see* Ecological succession and geotechnical properties
Freshwater sediment chemical diagenesis, macrobenthos effects on, 177–218
  chemical diagenesis of sediments, 182–186
  conclusions, 209–211
  diagenetic rates, 204–205
  fluid transport, 189–194
  freshwater sediment system, 178–186
  macrobenthos, 178–182
  materials flux, 205–209
  mechanism affecting chemical diagenesis, 186–199
  model equations, 187, 189, 190, 194
  observed effects on chemical diagenesis, 200–209

Freshwater sediment chemical diagenesis,
    macrobenthos effects on (cont.)
  particle transport, 186–189
  reactive materials addition, 199
  sediment fabric alteration, 194–199
  sedimentary chemical milieu, 200–204
Freshwater sediments and macrobenthos,
    106–113
  lacustrine and marine environments, 106
  salinity fluctuations and tidal motions,
    106
  soft-bottom communities, 108
Freshwater sediment physical properties,
    benthos effects on, 105–176
  ecological interactions and sediment
    properties, 161–166
  freshwater sediments and macrobenthos,
    106–113
  future work, 166–167
  macrobenthos life-styles, 113–124
  sediment mixing, 124–150
  sediment transport, biogenic
    modification, 150–161

Gammarus, 117
Geotechnical mass properties, 29–36
Glycoproteinaceous films, 27
Great Slave Lake, Canada, 111–112

"Head-down" conveyor belt feeders, 9,
    125, 188, 297
Helminthoidichnites, 250
Heteromastus filiformis, 21
Hexagenia mayfly nymph, 140
High surface-area-to-volume ratios, 25
"Hummocky" relief on surface, 24
Hyalella azteca, 116–117
Hydrobia sp. pellets, 28

Ichnogenera, 246
Ichnotaxa, 268
  diverse, 277
Infaunal deposit feeders, 31
Isopodichnus, 268

Lacustrine and marine environments, 106
Lake Erie, Western Basin, 141–143
  box cores, 152, 155
Lake Esrom, Denmark, 108
Lake Mälaren, Sweden, 111–112
Lake Vôrtsjärv, Estonia, 110, 111

Lampsilis radiata siliquoidea, 199
Lepidurus, 273
Lepidurus arcticus, 266
Limnodrilus cervix, 207
Limnodrilus hoffmeisteri, 163, 164, 207
  Green Bay, Michigan, 130
  Toronto Harbor, 163
Littoral community, 108
Lumbrineris–Alpheus–Diolodonta
    community, Kingston Harbor, Jamaica,
    9

Macoma baltica, 29
Macrobenthos effects on chemical
    diagenesis in freshwater sediments,
    177–218
  chemical diagenesis of sediments,
    182–186
  conclusions, 209–211
  diagenetic rates, 204–205
  fluid transport, 189–194
  freshwater sediment system, 178–186
  macrobenthos, 178–182
  materials flux, 205–209
  mechanisms affecting chemical
    diagenesis, 186–199
  model equations, 187, 189, 190, 194
  observed effects on chemical diagenesis,
    200–209
  particle transport, 186–189
  reactive materials addition, 199
  sediment fabric alteration, 194–199
  sedimentary chemical milieu, 200–204
Macrobenthos effects on chemical
    properties of marine sediments and
    overlying water, 53–101
  burrow habitat, chemistry of, 89–92
  diagenetic reactions, 54–56
  macrofaunal influence on
    sediment–water exchange rates, 83–87
  model equations, solutions, 94–95
  reaction rates, 87–89
  reactive particle redistribution, 56–71
  sediment chemistry, spatial and
    temporal patterns, 92–93
  solute transport, 71–83
Macrobenthos life-styles, 113–124
  amphipods, 116–118
  bivalves, 120–124
  chironomids, 113–116
  oligochaetes, 118–120

Macrobenthos life-styles (cont.)
    Pisidiidae, 123–124
    Unionacea, 120–123
Maldane–Amphiura community, Ria de Muros, Spain, 9
Maldanid–Nucula–Syndosmya community, Clyde Sea, 9
Maldanidae polychaete family, 9
Margaritifera margaritifera, 275
Marine and lacustrine environments, 106
Marine benthos effects on sediment physical properties, 3–52
    bioturbation, 4
    equilibrium stages, 9–10
    future work recommendations, 40–43
    invertebrate benthos, 4
    laboratory experiments, 10–19
    ultrasound imaging, 41
Marine biogenic sedimentary structures, geological significance, 221–256
    paleobathymetry, 233–243
    Precambrian traces, 243–252
    traces and sedimentology, 222–233
Marine sediment and overlying water, chemical properties, macrobenthos effects on, 53–101
    burrow habitat, chemistry of, 89–92
    diagenetic reactions, 54–56
    macrofaunal influence on sediment–water exchange rates, 83–87
    model equations, solutions, 94–95
    reaction rates, 87–89
    reactive particle redistribution, 56–71
    sediment chemistry, spatial and temporal patterns, 92–93
    solute transport, 71–83
Markov models, 310–312
Microbially mediated decomposition reaction rates, 87–89
Midges, 113
Mississippi Delta sedimentation, 223
Molluscs, Movice Lake, 163
Molpadia–Euchone community, Cape Cod Bay, 9
Molpadia oolitica, 20
Mt. Simon foundation, 245
Mucopolysaccharides, 28–29
Mucus production and degradation, seasonal changes, 28
Mud facies, Buzzards Bay, 20
Mulinia lateralis, 11

Nearshore marine muds, geotechnical and transport properties, 38
Nephtys incisa, 11
Nucula annulata, 11
Nucula–nephtys community, Buzzards Bay, 9

Oceanic water chemistry, 184
Oligochaetes, 118–120
    depth and rate of sediment mixing, 127–135
    feeding rate, 131
Ophiomorpha, 228
Organic–mineral aggregates, 27
Organism–sediment–fluid interactions, 37
Owenia filiformis, 21
Owenia flume experiment, 22

Paleobathymetry, 233–243
    graphoglyptid controversy, 241–243
    ocean evidence, 238–241
    Seilacher's model, 234
        problems with, 234–238
Paleodictyoa, 241
Pellet textures, 24–29
Peloscolex multisetous, 163
Physiochemical properties, interface, 32–34
Pioneering stages, 7–9
    adaptive effects, 8–9
    pore water chemistry, 8
    sedimentary effects, 8–9
Pisidiidae, 123–124
    depth and rate of sediment mixing, 137
Planolites, 248, 250
Planolites rugulosus, 268
Pomatothrix, 163
Pontoporeia, 117
Pontoporeia hoyi, 137, 197
Pore water irrigation, 31
Precambrian traces, 243–252
    Alpert's proposal, 246–248
    chronology, 248
    oldest trace fossil, 248–252
    Precambrian–Cambrian boundary, 243–248
    trace fossil zonation, 247
Profundal community, 109

Radionuclide distribution, 295

Index   **335**

Reactive particle redistribution, 56–71
  biogenic reworking and decay rate, 64
  homogeneous reworking, 56–67
  model equations, 57, 59, 60, 61
  organic matter subduction, 66
  particle eddies, 59
  particle motion, 58
  particle transport, isotopes, 60
  representative particle reworking coefficients, 62–63
  resuspension and current transport, 64
  selective reworking, 67–71
  vertical properties distribution, 61
Redox potential discontinuity (RPD), 8–10
Rhizocorallium, 236
Rhoads–Cande Profile Camera, 41
Rugoinfractus, 250
Rusophycus, 246

Scoyenia association, 271
Seaflume, *in situ*, 36
Sediment chemistry, spatial and temporal patterns, 92–93
Sediment mixing, 124–150
  depth and rate, 127–137
  feeding rate, 131
  kinds of and factors affecting, 124–127
  particle size distribution effects, 139–146
  permeability, 149–150
  sediment stratigraphy, 124–139
  sediments, mass property effects, 146–150
  stratigraphic effects, 137–139
  tubificid-induced subduction velocities, 132–133
  water content, 146–149
Sediment physical properties, effect of marine benthos on, 3–52
  bed roughness, 19–24
  biological benthic boundary layer, 4
  bioturbation, 4, 34
  cohesive sediment erosion, 32
  equilibrium stages, 9–10
  fluid shear stress, 4–5
  future work recommendations, 40–43
  geotechnical mass properties, 29–36, 38–39
  laboratory experiments, 10–19, 157–160
  macrofaunal effects, 5, 163–166

Sediment physical properties, effect of marine benthos on (*cont.*)
  organism–sediment–fluid interactions, 37
  pelletal textures, 24–29
  physical properties, 4, 151
  pioneering stages, 7–9, 36
  qualitative predictive model, 36–40
  sediment relations, 5–10
  sediment transport and biogenic activity, 19–36
  substratum stability, 20–23, 161
  ultrasound imaging, 41
  water content and erosion resistance, 34, 153
Sediment transport and biogenic activity, 19–36
  bed roughness, 19–24
  cohesive sediment erosion, 32
  fluids literature, 20–21
  geotechnical mass properties, 29–36, 38–39
  pellet textures, 24–29
  substratum stability, 20–23, 161
  tube fields, 20, 23
Sediment transport, biogenic modification of, 150–167
  anoxia effects, 159
  biological activities, importance in, 150
  comparisons, 160–161
  field observations, 150–156
  future work, 166–167
  laboratory experiments, 26–27, 157–160
  macrofauna, spatial–temporal variations, 161–167
  substratum properties, 161
  summer–fall contrast, 156
Sediment water content, and erosion resistance, 34, 153
  oxygen content of water, 156
Shrimp mounds, Discovery Bay, Jamaica, 20
Signal processing models, *se* Signal-theory-based model
Signal-theory-based model, 292, 309–310
  equations, 309
Skolithos–Planolites couplet, 236
Solute transport, 71–83
  apparent diffusion, 71–75
  apparent transport coefficients, 75
  average diffusion geometry, 77–83

Solute transport (cont.)
  biogenic advection, 75–77
  conclusions, 74
  model equations, 72, 78
  muddy sediments, 76
Species, identifying characterizing, 6
Spriggina, 249
Stylodrilus heringianus, 195, 197
Subtidal mud erodibility, Long Island Sound, 35
Swedish entrophic lakes, 206

Taenidium, 270
Taxonomic resolution, 277
Teichichnus, 236
Temporal mosaics, taxonomic and functional structure, 7
Terebellina, 228
Terrigenous sediments, Mud Bay, South Carolina, 73
Traces and sedimentology, 222–233
  bioturbation in recent muds, 223–225
  diagenesis, 230–233
  sedimentation, rate and continuity, 225–227
  substrate consistency, 227–230
Transenella tantilla, 24

Tubifex tubifex, 196
  Toronto Harbor, 163
Tubificid life histories, 119–120
Tubificid oligochaetes
  freshwater environment, 188
  pelletizers, 141

Upper Cambrian trilobites, 245
Upper Devonian shelf, New York, 226
Unionacea, 120–123
  depth and rate of sediment mixing, 137

Water content of sediment, and erosion reistance, 34, 153
  oxygen content of water, 156
Water exchange rates, macrofaunal influence on sediment, 83–87
  model equations, 84, 86
  solute flux ratio, 85
Western Basin, Lake Erie, 116

Xenohelix, 268

Yoldia limatula, 11

Zoophycos facies, 236